Alpine Space Programme: Approaches and Environmental Risks

Alpine Space Programme: Approaches and Environmental Risks

Edited by **Lee Zieger**

R CALLISTO REFERENCE

New York

Published by Callisto Reference,
106 Park Avenue, Suite 200,
New York, NY 10016, USA
www.callistoreference.com

Alpine Space Programme: Approaches and Environmental Risks
Edited by Lee Zieger

International Standard Book Number: 978-1-63239-066-0 (Hardback)

Printed in the United States of America.

Contents

Preface

This book presents an overview of the approaches and challenges associated with climate change risks in the Alpine region. Climate variations indicate that current forest resources will face drastically different temperature and precipitation conditions in the future. Developing prospective strategies for forest management in spite of uncertain and highly variable forecasts of future site conditions is a great challenge. Transnational case studies dealing with different manifestations of climate change effects have been analyzed in this book. It intends to encourage discussions on management strategies to adapt forests in the Alps to climate variation risks. The results have been derived from the INTERREG project "Management Strategies to Adapt Alpine Space Forests to Climate Change Risks" that was enforced within the framework of the European Territorial Cooperation "Alpine Space Programme 2007-2013".

After months of intensive research and writing, this book is the end result of all who devoted their time and efforts in the initiation and progress of this book. It will surely be a source of reference in enhancing the required knowledge of the new developments in the area. During the course of developing this book, certain measures such as accuracy, authenticity and research focused analytical studies were given preference in order to produce a comprehensive book in the area of study.

This book would not have been possible without the efforts of the authors and the publisher. I extend my sincere thanks to them. Secondly, I express my gratitude to my family and well-wishers. And most importantly, I thank my students for constantly expressing their willingness and curiosity in enhancing their knowledge in the field, which encourages me to take up further research projects for the advancement of the area.

Editor

Management Strategies to Adapt Alpine Space Forests to Climate Change Risks – An Introduction to the Manfred Project

Robert Jandl, Gillian Cerbu, Marc Hanewinkel,
Fred Berger, Giacomo Gerosa and Silvio Schüler

Additional information is available at the end of the chapter

1. Introduction

The challenges posed by climate change for forests in the Alpine Space are considerable. Climatologists have identified mountains as types of ecosystems that are facing an above-average warming trend. However, it is not entirely clear why mountains are warming up quicker than other landscapes, and whether the the past trend will continue in the future [1]. The currently expected warming is at least 3 to 4°C in the next century and thus far above the warming that is considered to be controlable by mankind [2, 3]. Despite the uncertainty of the future development several consequences of climate change on mountains are well understood. The duration of snow cover is expected to decrease, the water discharge regime will be altered, and the frequency of rock falls may increase, the distribution of plant species will change, and invasive species, such as pests and pathogens, will change the environmental setting of forestry. Trees are especially vulnerable when they are growing at the margins or even outside of their natural range [4, 5].

The human demands on the utilisation of mountain forests are manifold. A harmonized approach is required for the settlement of conflicting demands on land management [6, 7]. When maintaining the concept of a multi-purpose forestry the provision of ecosystem services need to be harmonized solutions from both demand and supplier side. Concepts for forest management may be insufficient when they are focused solely on timber production. An increasingly demanding society asks for biodiversity and scenic beauty, and is heavily dependent on the provision of drinking water and the protective function of mountain forests, even in regions far away from mountains [8, 9].

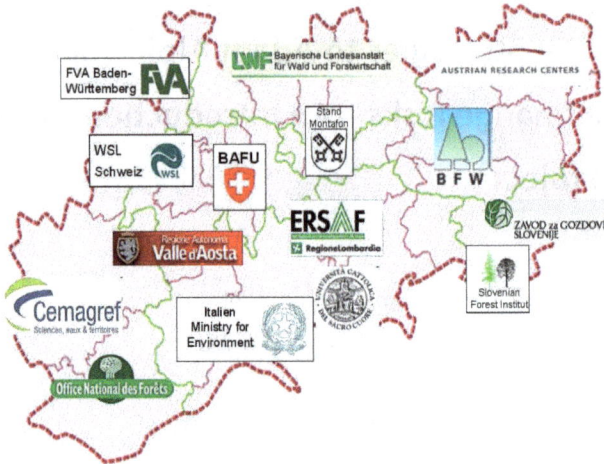

Figure 1. Geographical distribution of the partner institutions of the Interreg project "MANFRED - Management strategies to adapt Alpine Space forests to climate change risks".

The forests have a dual function within climate change. They are contributing to the *mitigation* of climate change, and their *adaptation* to climate change effects ensures their sustainable development under future site conditions. When the conservation of the huge carbon stock of forests in the Alpine Space is on the agenda, it is necessary to enhance the resilience of forests. Many forests, particularly in mountains, are going to benefit from climate change effects in the near future. Their productivity is currently limited by the duration of the growing season, which is expected to be elongated in the future. However, climate change is expected to manifest itself in many ways and the increase in the temperature is only one aspect. The rising CO_2 level affects the growth rate as well and altered temporal patterns of drought, mortality, and disturbances are to be taken into account. Moreover, the nitrogen supply has been increasing in the last decades and may in some regions exceed the demand of forests [10, 11]. Superimposed on these effects are potentially adverse effects such as extreme meteorological events, ecosystem disturbances, air pollution, and an increasing pressure from pests and pathogens on forests.

The Interreg project "MANFRED - Management strategies to adapt Alpine Space forests to climate change risks" is a collaboration between numerous institutions, comprising researchers, stakeholders, and forest practitioners who are engaged in the Alps (Figure 1). The intention of MANFRED is creating a mutual understanding of the recognized challenges due to global change, and elaborating guidelines for forest management in a joint effort.

Efforts of responding to climate change are embedded in a particular political setting. There are many processes, conventions on different geographical scales, from regional to international policies. The political framework is described in one module of the project [12]

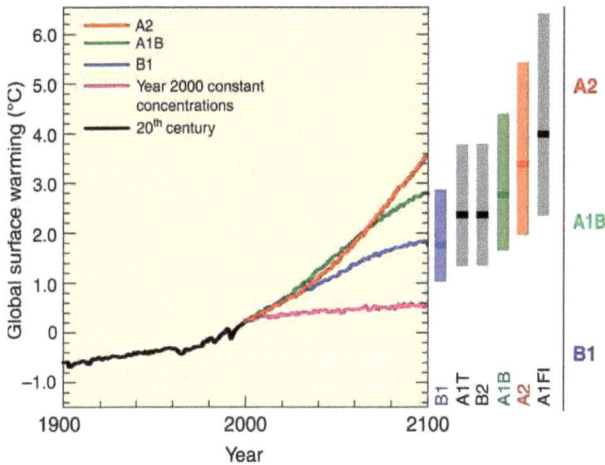

Figure 2. Scenarios of the global surface warming according to different emission rates of greenhouse gases; Source: IPCC Climate Change 2007: Synthesis Report Fig. 3-2.

2. Biotic and abiotic drivers call for an adaptation of forest management

The Fourth Assessment Report of the IPCC has provided global scenarios for the warming trend under different emission scenarios (Figure 2). In the Fifth Assessment Report even more comprehensive scenarios are developed that do not only include changes in radiative forcings, but also the implications of various responses of mankind in terms of technology, mitigation, and adaptation [13].The IPCC uses several climate models. They are yielding widely different results, and come in a rough spatial resolution. In order to include the differences between the models, often ensembles of simulation results are used. For a regionally valid estimate on the expected climate change, the large-scale scenarios need to be regionalized.

The climate scenarios give allow to interpretations, which tree species can best cope with future conditions. Trees will either go extinct, or migrate to sites where their requirements are fulfilled, or adapt to climate change. The extent of climate change and the expected future distribution of trees has been assessed in two comprehensive modules of the Manfred project [14, 15]. The analysis included the prediction of the potential occurrence of a large number of tree species by different statistical models and several climate scenarios. For the unique effort field data from national forest inventories of countries within the Alpine Space were made available by the participating institutions (Figure 1).

The successful regeneration of trees depends to a large extent on the average climatic conditions. However, extreme events can reset the success of tree seedlings. Extreme events come in many manifestations. Despite the uncertainty of the future climate there is sufficient evidence that they will be more common in the future [16]. The IPCC has even published a special report on extreme events [17]. Particularly harmful situations for tree seedlings are created by frosts in late spring and early summer, and by elongated drought events. Other types of extreme events are forest fires, that are triggered by increasingly dry conditions. Due to the history of active fire suppression the fire risk does not receive the full attention in the Alps. Nevertheless, future climate conditions may well call for an improved alerting system for the prevention of devastating forest fires [18]. These topics are covered in several chapters in this book [19, 20]. The knowledge of the past is instrumental when learning lessons for the future. One of the outcomes of the Manfred project is an online-database of documented effects of extreme events in the Alpine Space [21]. It is our true hope that the database can be kept alive beyond the lifetime of the research project.

A particularly difficult issue is the topic whether storm damages are becoming more problematic in the Alpine Space. There is sufficient evidence that European forests have been hit by a series of major storms in only a few years [22–25]. However, it is scientifically not established whether this trend of damages continues or even attenuates in the future, or whether the devasting storms where only a coincidence. Judging extreme events only by the local effects is clearly insufficient. It requires the geophysical expertise to evaluate whether a particular event is caused by global changes of the disturbance regime or whether it is caused by severe local conditions with a low probability. To further complicate the case it has been shown that the recorded damages due to storms are not only driven by climate change. Instead, the practised form of forest management such as the maturing of increasingly dense forests has created types of forest stands that are quite vulnerable to storm damages [26]. In Germany an exceptional dataset on a storm event has been collected. The findings could not be extended to the entire Alpine Space due to the relevance of regional factors for the vulnerability of storms. The regionally valid results are elaborated for a particular region [27, 28].

Biotic threats to forests are also triggered by global change. Climatic change clearly modifies the potential habitat of insects and pathogens. Warmer winters and longer growing seasons enable insects extending their natural occurrence into regions where they have not been encountered before, and to develop more generations in regions where they are already present. Infamous examples are the gradation of mountain pine beetles in the Canadian forests with unfavorable effects on the Canadian greenhouse-gas emission budget and the increased virulence of fungi [29, 30]. The second aspect of global change is the increase in international, even global, commerce. Unintentionally, pests and pathogens are distributed globally. Sometimes, these invaders encounter at least temporarily favorable conditions and can cause havoc on resident tree species [31–33]. – Within the Manfred project the propagation of potentially problematic pests and pathogens was analyzed [34]. In addition, a transnational web-based alerting system was created. The intention is providing sufficient information that regionally arising entomological challenges are identified early on so that sufficient time for appropriate counter measures is available. In recognition of the limited experience of local foresters with emerging pests and pathogens the online information is supported with a handbook specifically designed for field use in order to offer instantenously available information on indicators of pest and pathogen propagations.

Air pollution research has a long tradition in Central Europe. Although the attention has shifted towards other topics, ozone damages to forests are still problematic [35–37]. The emission of oxidating precursors favors the formation of ozone and high rates of incoming solar radiation are potentially a problem in mountain forests, especially in the southern part of the Alps the risk of increasing ozone damages is increasing. The topic is covered within the Manfred project in focused research effort [38–40].

3. Forest engineering approaches – Adaptive forest management

Eco-engineering methods are invoked when the protective function of forests is jeopardized. Climate change can affect forests in many ways so that they are more vulnerable to rockfall, avalanches, and landslides. This is a serious threat, given that the protection function is by far the most important ecosystem service of mountain forests [41]. For the development of appropriate counter measures the potential threat to these ecosystem services needs to be mapped and evaluated. Efficient methods are described and applied [42, 43]. The developed methods meet the challenge of being easily applicable in order to be useful even when extensive data are not available.

Another engineering approach is the utilization of the genetic adaptive capacity of trees. Many species are successfully growing on a wide geographical range and it is left to the expertise of forest geneticists to identify the most useful provenances. With the knowledge of the anticipated future site conditions it is possible to establish new forests. The full use of existing long-term provenance trials that have been established globally as an adaptive measure is pursued since a long time and promising populations have been identified for several tree species [44, 45]. Within the Manfred project such provenance trials have been evaluated for *Picea abies* [46, 47]. It has been shown that the potential productivity of forests can be greatly increased when chosing the appropriate provenance. Including these informations on the genetic amplitude of tree species can help to mitigate climate-induced productivity losses and can reduce the risk for the chosen tree species in their lifetime of presumably accentuated changes in the local climatic conditions.

A straightforward engineering tool is the use of simulation models for forest growth. These models are popular because forest management decisions can be instantenously tested. In traditional forest growth models the focus is usually on the temporal trend of the stem volume as dependent on different silvicultural practices. For a more comprehensive evaluation of different forest treatments a growth model has been combined with a soil carbon model. Hence, it is possible to extend the evaluation from timber production to the carbon sequestration potential of forests [48].

In four transnational case study regions, comprising 6 countries and 15 partners of the consortium, an analysis of land use, land-use change, and the demand and supply of ecosystem services was assessed. Local authorities and forest owners played an important role in these projects in recognition that their experience harbors a wealth of information, and yields valuable observations on early indicators of climate change effects. The study sites were chosen such that diverse situations for practical forestry are covered, such as regions with a peculiar threat of biotic and abiotic damages of forests, regions were the protective function of forests may suffer from climate change effects, and regions with a high risk of adverse effects of air pollution [40, 43, 49–51]. The different aspects should give a broad overview on the challenges for forestry under future conditions and the proposed solutions are intended to provide a range of strategies serving as reference for other regions as well.

The final chapter develops strategies for adaptive forest management, based on the experiences derived from the case study areas and based on previous concepts [52–58]. The concepts are intended to build bridges between research and practical forest management. The trans-national collaboration ensures that regional experiences are brought to the attention of a wider community. The suggested strategies of adaptive forest management are conceptually not necessarily new. Like in earlier periods were forests were under pressure due to acidic precipitation and elevated rates of nitrogen deposition, the concepts aim at increasing the stability of the forests in order to distribute potential risks within a forest stand among different tree species and cohorts of age groups. The suggested measures are rather conservative with hindsight to the high uncertainty of the future climatic conditions. Nevertheless, we want to emphasize that the suggested concepts only can be successfully implemented when damages of forests due to high population densities of ungulates are kept at an acceptable level. Presently, damages caused by deer are inhibiting the establishment of species-rich forest ecosystems.

In recognition of the limited distribution of textbooks among forest practitioners, additional avenues for education and dissemination were followed. The generated experience has been distributed in guidelines and handbooks, and was brought forward in roundtables with different stakeholder groups. It was recognized that forests in the Alpine Space always have to be seen in the context of the regional land-use system, embedded in agriculture and recreational demands for the landscape. There is no generally valid concept of forest management. Driven by land-management policies, forestry is an integral part of the green economy where land owners have both a lot freedom and responsibility for the general public in their management decisions. The increasing demand for renewable energy will have an impact on the form of silviculture. Nevertheless, forests are satisfying many needs of humans for goods and services in a sustainable way, and are also safeguarding the conservation of habitats for plants and animals.

Acknowledgements

This chapter is an outcome of the Interreg project Manfred, conducted within the Alpine Space programme.

Author details

Robert Jandl[1], Gillian Cerbu[2],
Marc Hanewinkel[3], Fred Berger[4],
Giacomo Gerosa[5] and Silvio Schüler[1]

1 Forest Research Center (BFW), Vienna, Austria

2 Forstliche Versuchs- und Forschungsanstalt Baden Württemberg (FVA), Freiburg i. Breisgau, Germany

3 Eidgenössische Forschungsanstalt für Wald, Schnee und Landschaft (WSL), Birmensdorf, Switzerland

4 National Research Institute of Science and Technology for Environment and Agriculture (IRSTEA), Grenoble, France

5 DMF, Università Cattolica del Sacro Cuore, Brescia, Italy

References

[1] J. Arblaster, G. Brasseur, J.H. Christensen, K.L. Denman, D.W. Fahey, P. Forster, E. Jansen, P.D. Jones, R. Knutti, H. Le Treut, P. Lemke, G. Meehl, P. Mote, D. Randall, D.A. Stone, K.E. Trenberth, J. Willebrand, and F. Zwiers. *Summary for Policymakers of the Synthesis Report of the IPCC Fourth Assessment Report – Contribution of Working Group I to the Fourth Assessment Report of the Intergovernmental Panel on Climate Change.* Cambridge University Press, Cambridge, UK, 2007.

[2] Malte Meinshausen, Nicolai Meinshausen, William Hare, Sarah C. B. Raper, Katja Frieler, Reto Knutti, David J. Frame, and Myles R. Allen. Greenhouse-gas emission targets for limiting global warming to 2°C. *Nature*, 458:1158–1162, 2009.

[3] Joeri Rogelj, Julia Nabel, Claudine Chen, William Hare, Kathleen Markmann, Malte Meinshausen, Michiel Schaeffer, Kirsten Macey, and Niklas Höhne. Copenhagen accord pledges are paltry. *Nature*, 464:1126–1128, 2010.

[4] Joseph Alcamo, José M. Moreno, Bela Nováky, Marco Bindi, Roman Corobov, Robert J.N. Devoy, Christos Giannakopoulos, Eric Martin, Jørgen E. Olesen, and Anatoly Shvidenko. *Europe. Climate Change 2007: Impacts, Adaptation and Vulnerability. Contribution of Working Group II to the Fourth Assessment Report of the Intergovernmental Panel on Climate Change,* chapter 12, pages 541–580. Cambridge University Press, Cambridge, UK, 2007.

[5] Lorenzo Marini, Matthew P. Ayres, Andrea Battisti, and Massimo Faccoli. Climate affects severity and altitudinal distribution of outbreaks in an eruptive bark beetle. *Climatic Change*, 115:327–341, 2012.

[6] Dagmar Schröter, Wolfgang Cramer, Rik Leemans, I. Colin Prentice, Miguel B. Araújo, Nigel W. Arnell, Alberte Bondeau, Harald Bugmann, Timothy R. Carter, Carlos A. Gracia, Anne C. de la Vega-Leinert, Markus Erhard, Frank Ewert, Margaret Glendining, Joanna I. House, Susanna Kankaanpää, Richard J. T. Klein, Sandra Lavorel, Marcus Lindner, Marc J. Metzger, Jeannette Meyer, Timothy D. Mitchell, Isabelle Reginster, Mark Rounsevell, Santi Sabaté, Stephen Sitch, Ben Smith, Jo Smith, Pete Smith, Martin T. Sykes, Kirsten Thonicke, Wilfried Thuiller, Gill Tuck, Sönke Zaehle, and Bärbel Zierl. Ecosystem service supply and vulnerability to global change in europe. *Science*, 310:1333–1337, 2005.

[7] Erik Nelson, Guillermo Mendoza, James Regetz, Stephen Polasky, Heather Tallis, D Richard Cameron, Kai MA Chan, Gretchen C Daily, Joshua Goldstein, Peter M Kareiva, Eric Lonsdorf, Robin Naidoo, Taylor H Ricketts, and M Rebecca Shaw. Modeling multiple ecosystem services, biodiversity conservation, commodity production, and tradeoffs at landscape scales. *Frontiers in Ecology and the Environment*, 7(1):4–11, 2009.

[8] Christian Körner and Masahiko Ohsawa. Mountain systems. In *Ecosystems and Human Well-being: Current State and Trends*, chapter 24, pages 681–716. Island Press, Washington DC, 2007.

[9] Anatoly Shvidenko, Charles Victor Barber, and Reidar Persson. *Forest and Woodland Systems*, chapter 21, pages 585–621. Island Press, Washington DC, 2007.

[10] Jan Willem Erisman. The european nitrogen problem in a global perspective. In *The European Nitrogen Assessment*, chapter 2, pages 9–31. Cambridge Univ Press, 2011.

[11] Robert Jandl, Stefan Smidt, Franz Mutsch, Alfred Fürst, Harald Zechmeister, Heidi Bauer, and Thomas Dirnböck. Acidification and nitrogen eutrophication of Austrian forest soils. *Applied Environmental Soil Science*, 2012.

[12] Luca Cetara and Federico Mannoni. *Developing a backgroung for Forest Adaptation Strategies in the Alps: a perspective for policy building*, chapter 2, In: Management Strategies to Adapt Alpine Space Forests to Climate Change Risks. InTech, 2013.

[13] Richard H. Moss, Jae A. Edmonds, Kathy A. Hibbard, Martin R. Manning, Steven K. Rose, Detlef P. van Vuuren, Timothy R. Carter, Seita Emori, Mikiko Kainuma, Tom Kram, Gerald A. Meehl, John F. B. Mitchell, Nebojsa Nakicenovic, Keywan Riahi, Steven J. Smith, Ronald J. Stouffer, Allison M. Thomson, John P. Weyant, and Thomas J. Wilbanks. The next generation of scenarios for climate change research and assessment. *Nature*, 463:747–756, 2010.

[14] Niklaus E. Zimmermann, Ernst Gebetsroither, Johannes Züger, Dirk Schmatz, and Achilleas Psomas. *Future Climate of the European Alps*, chapter 3, In: Management Strategies to Adapt Alpine Space Forests to Climate Change Risks. InTech, 2013.

[15] Niklaus E. Zimmermann, Robert Jandl, Mark Hanewinkel, Georges Kunstler, Christian Kölling, Patrizia Gasparini, Andrej Breznikar, Eliane S. Meier, Signe Normand, Ulrich Ulmer, Thomas Gschwandtner, Holger Veit, Maria Naumann, Wolfgang Falk, Karl Mellert, Maria Rizzo, Mitja Skudnik, and Achilleas Psomas. *Potential future ranges of tree species in the Alps*, chapter 4, In: Management Strategies to Adapt Alpine Space Forests to Climate Change Risks. InTech, 2013.

[16] Stefan Rahmstorf and Dim Coumou. Increase of extreme events in a warming world. *PNAS*, 108(44):17905–17909, 2011.

[17] IPCC. *Managing the Risks of Extreme Events and Disasters to Advance Climate Change Adaptation. A Special Report of Working Groups I and II of the Intergovernmental Panel on Climate Change*. Cambridge Univ Press, United Kingdom, 2011.

[18] Sara Marañón Jiménez. *Efecto del Manejo de la Madera Quemada Después de un Incendio sobre el Ciclo del Carbono y Nutrientes en un Ecosistema de Montaña Mediterránea*. PhD thesis, Universidad de Granada, 2011.

[19] Bruna Comini, Elena Gagliazzi, and Giampaolo Cocca. *Abiotic Stressors: Fire Hazard and Risk*, chapter 5, In: Management Strategies to Adapt Alpine Space Forests to Climate Change Risks. InTech, 2013.

[20] Ernst Gebetsroither and Johann Züger. *Drought hazard estimations according to climate change in the Alpine area*, chapter 11, In: Management Strategies to Adapt Alpine Space Forests to Climate Change Risks. Intech, 2013.

[21] Stefano Oliveri, Marco Pregnolato, and Giacomo Gerosa. *A new webGIS platform dedicated to forest extreme events in the Alps: aims and functionalities*, chapter 10, In: Management Strategies to Adapt Alpine Space Forests to Climate Change Risks. InTech, 2013.

[22] M. G. Donat, D. Renggli, S. Wild, L. V. Alexander, G. C. Leckebusch, and U. Ulbrich. Reanalysis suggests long-term upward trends in European storminess since 1871. *Geophysical Research Letters*, 38:L14703, 2011.

[23] Barry Gardiner, Kristina Blennow, Jean-Michel Carnus, Peter Fleischer, Frederik Ingemarson, Guy Landmann, Marcus Lindner, Mariella Marzano, Bruce Nicoll, Christophe Orazio, Jean-Luc Peyron, Marie-Pierre Reviron, Mart-Jan Schelhaas, Andreas Schuck, Michaela Spielmann, and Tilo Usbeck. *Destructive Storms in European Forests: Past and Forthcoming Impacts*. EFI, Joensuu, 2010.

[24] U. Ulbrich, G. C. Leckebusch, and J. G. Pinto. Extra-tropical cyclones in the present and future climate: a review. *Theoretical and Applied Climatology*, 96:117–131, 2009.

[25] Helga Van Miegroet and Mats Olsson. *Ecosystem disturbance and soil organic carbon - a review*, chapter 5, pages 85–117. Wiley-Blackwell, 2011.

[26] Rupert Seidl, Mart-Jan Schelhaas, and Manfred J Lexer. Unraveling the drivers of intensifying forest disturbance regimes in Europe. *Global Change Biology*, 17(9):2842–2852, 2011.

[27] Matthias Schmidt, Marc Hanewinkel, Gerald Kändler, Edgar Kublin, and Ulrich Kohnle. An inventory-based approach for modeling single-tree storm damage – experiences with the winter storm of 1999 in southwestern Germany. *Canadian Journal of Forest Research*, 40:1636–1652, 2010.

[28] Bin You and Mitja Skudnik. *Abiotic stressor: storms*, chapter 7, In: Management Strategies to Adapt Alpine Space Forests to Climate Change Risks. InTech, 2013.

[29] W. A. Kurz, C. C. Dymond, G. Stinson, G. J. Rampley, E. T. Neilson, A. L. Carroll, T. Ebata, and L. Safranyik. Mountain pine beetle and forest carbon feedback to climate change. *Nature*, 452:987–990, 2008.

[30] Alan C. Gange, Edward G. Gange, Aqilah B. Mohammad, and Lynne Boddy. Host shifts in fungi caused by climate change? *Fungal Ecology*, 4(2):184–190, 2011.

[31] Dana Blumenthal, Charles E. Mitchell, Petr Pyšek, and Voytěk Jarošík. Synergy between pathogen release and resource availability in plant invasion. *Proceedings of the National Academy of Sciences*, 106(19):7899–7904, 2009.

[32] Bethany A. Bradley, Dana M. Blumenthal, David S. Wilcove, and Lewis H. Ziska. Predicting plant invasions in an era of global change. *Trends in Ecology & Evolution*, 25(5):310 – 318, 2010.

[33] Christian Tomiczek and Ute Hoyer-Tomiczek. Der Asiatische Laubholzbockkäfer (*Anoplophora glabripennis*) und der Citrusbockkäfer (*Anoplophora chinensis*) in Europa - ein Situationsbericht. *Forstschutz Aktuell*, 38:2–5, 2007.

[34] Holger Griess, Holger Veit, and Ralf Petercord. *Risk assessment for biotic pests under prospective climate conditions*, chapter 5, In: Management Strategies to Adapt Alpine Space Forests to Climate Change Risks. InTech, 2013.

[35] R Fischer and M Lortenz. Forest condition in Europe. Work Report of the Institute for World Forestry 1, ICP Forests and FutMon, Hamburg, 2011.

[36] S.B. McLaughlin, P.A. Layton, M.B. Adams, N.T. Edwards, P.J. Hanson, E.G. O'Neill, and W.K. Roy. Growth responses of 53 open-pollinated loblolly pine families to ozone and acid rain. *Journal of Environmental Quality*, 23:247–257, 1994.

[37] Gerhard Wieser and Michael Tausz. *Trees at their upper limit - Treelife limitation at the Alpine Timberline*, volume 5 of *Plant Ecophysiology*. Springer, 2007.

[38] Angelo Finco, Stefano Oliveri, Giacomo Gerosa, Wilfried Winiwarter, Johann Züger, and Ernst Gebetsroither. *Assessing present and future ozone hazards to natural forests in the Alpine area. Comparison of a wide scale mapping technique with local passive sampler measurements*, chapter 9, In: Management Strategies to Adapt Alpine Space Forests to Climate Change Risks. InTech, 2013.

[39] Giacomo Gerosa, Angelo Finco, Antonio Negri, Stefano Oliveri, and Marco Pregnolato. *Ozone fluxes to a larch forest ecosystem at the timberline in the Italian Alps*, chapter 8, In: Management Strategies to Adapt Alpine Space Forests to Climate Change Risks. InTech, 2013.

[40] Giacomo Gerosa, Angelo Finco, Stefano Oliveri, Riccardo Marzuoli, Alessandro Ducoli, Giambattista Sangalli, Bruna Comini, Paolo Nastasio, Giampaolo Cocca, and Elena Gagliazzi. *Case study Valle Camonica and the Adamello Park*, chapter 18, In: Management Strategies to Adapt Alpine Space Forests to Climate Change Risks. InTech, 2013.

[41] Martin F Price, Georg Gratzer, Lalisa Alemayehu Duguma, Thomas Kohler, Daniel Maselli, and Rosalaura Romeo. Mountain forests in a changing world - realizing values, addressing challenges. Technical report, FAO/MPS and SDC, Rome, 2011.

[42] Fred Berger, Franck Bourrier, Luuk Dorren, Charly Kleemayr, Bernhard Maier, Spela Planinsek, Christophe Bigot, Franck Bourrier, Oliver Jancke, David Toe, and Gillian Cerbu. *Eco-engineering and protection forests against rockfalls and snow avalanches*, chapter 12, In: Management Strategies to Adapt Alpine Space Forests to Climate Change Risks. InTech, 2013.

[43] Laurent Borgniet, David Toe, Frédéric Berger, Marta Galvagno, Cinzia Panigada, Roberto Colombo, Umberto Morra di Cella, Simone Gottardelli, Ivan Rollet, Mario Negro, Flavio Vertui, and Cédric Fermont. *Monitoring climatic change impacts on protection forests in Aosta Valley (Italy) and in Drôme (France) using medium and high resolution remote sensing and mateloscopes plots*, chapter 16, In: Management Strategies to Adapt Alpine Space Forests to Climate Change Risks. InTech, 2013.

[44] Csaba Matyas. Climatic adaptation of trees: rediscovering provenance tests. *Euphytica*, 92:45–54, 1996.

[45] T. Wang, A Hamann, Y Yanchuk, GA O'Neill, and SN Aiken. Use of response functions in selecting lodgepole pine populations for future climates. *Global Change Biology*, 12:2404–2416, 2006.

[46] Stefan Kapeller, Manfred J. Lexer, Thomas Geburek, Johann Hiebl, and Silvio Schueler. Intraspecific variation in climate response of norway spruce in the eastern alpine range: Selecting appropriate provenances for future climate. *Forest Ecology and Management*, 271:46–57, 2012.

[47] Stefan Kapeller, Silvio Schueler, Gerhard Huber, Gregor Božič, Thomas Wohlgemuth, and Raphael Klumpp. *Provenance trials in Alpine range - review and perspectives for applications in climate change*, chapter 14, In: Management Strategies to Adapt Alpine Space Forests to Climate Change Risks. InTech, 2013.

[48] Klaus Dolschak, Robert Jandl, and Thomas Ledermann. *Coupling a forest growth model with a soil carbon simulator*, chapter 13, In: Management Strategies to Adapt Alpine Space Forests to Climate Change Risks. InTech, 2013.

[49] Holger Veit, Bernhard Maier, Holger Griess, and Bin You. *Case study Oberschwaben / Allgäu / Vorarlberg – Risk Assessment of Abiotic and Biotic Hazards*, chapter 19, In: Management Strategies to Adapt Alpine Space Forests to Climate Change Risks. InTech, 2013.

[50] Robert Jandl, Gillian Cerbu, Marc Hanewinkel, Fred Berger, Giacomo Gerosa, and Silvio Schüler. *Management strategies to adapt Alpine Space forests to climate change risks - an introduction to the Manfred project*, chapter 1, In: Management Strategies to Adapt Alpine Space Forests to Climate Change Risks. InTech, 2013.

[51] Robert Jandl, Andrej Breznikar, Marko Lekše, Christian Tomiczek, Silvio Schüler, Klaus Dolschak, and Hans Zöscher. *Case Study Carinthia / Slovenia – Productive Forests Affected by Climate Change*, chapter 17, In: Management Strategies to Adapt Alpine Space Forests to Climate Change Risks. InTech, 2013.

[52] Peter Brang, Harald Bugmann, Anton Bürgi, Urs Mühlethaler, Andreas Rigling, and Raphael Schwitter. Klimawandel als waldbauliche Herausforderung. *Schweizerische Zeitschrift für das Forstwesen*, 159:362–373, 2008.

[53] Peter Brang, Marc Hanewinkel, Robert Jandl, Andrej Breznikar, and Bernhard Maier. *Managing Alpine Forests in a Changing Climate*, chapter 20, In: Management Strategies to Adapt Alpine Space Forests to Climate Change Risks. InTech, 2013.

[54] M Frehner, B Wasser, and R Schwitter. *Nachhaltigkeit und Erfolgskontrolle im Schutzwald. Wegleitung für Pflegemassnahmen in Wäldern mit Schutzfunktion, Vollzug Umwelt.* Bundesamt für Umwelt, Wald und Landschaft, Bern, Schweiz, 2005.

[55] Robert Jandl, Lars Vesterdal, Mats Olsson, Oliver Bens, Franz Badeck, and Joachim Rock. Carbon sequestration and forest management. *CAB Reviews: Perspectives in Agriculture, Veterinary Science, Nutrition and Natural Resources*, 2(017):1–16, 2007.

[56] Robert Jandl. Adapting mountain forest management to climate change. In Axel Borsdorf, J Stötter, and E Veulliet, editors, *Managing Alpine Future II*, volume 4 of *IGF-Forschungsberichte*, pages 193–202. Verlag der Österreichischen Akademie der Wissenschaften, 2011.

[57] M. Lindner, Tim Green, C.W. Woodall, C.H. Perry, G.-J. Nabuurs, and M.J. Sanz. Impacts of forest ecosystem management on greenhouse gas budgets. *Forest Ecology and Management*, 256(3):191 – 193, 2008. Impacts of forest ecosystem management on greenhouse gas budgets.

[58] Andreas Rigling, Peter Brang, Harald Bugmann, Norbert Kräuchi, Thomas Wohlgemuth, and Niklaus Zimmermann. Klimawandel als Prüfstein für die Waldbewirtschaftung. *Schweizerische Zeitschrift für das Forstwesen*, 159:316–325, 2008.

Chapter 2

Future Climate of the European Alps

Niklaus E. Zimmermann, Ernst Gebetsroither,
Johann Züger, Dirk Schmatz and Achilleas Psomas

Additional information is available at the end of the chapter

1. Introduction

The global climate is currently warming and this trend is expected to continue towards an even warmer world, associated partly with drastic shifts in precipitation regimes (IPCC 2007). While the global temperature has roughly been warming by 0.6°C (±0.2°C) during the 20th century (IPCC 2001), the land masses have had a higher temperature increase during the same period, and some areas such as the Alps showed an exceptionally high warming trend, with increases reaching 1.7°C in some regions (Rebetez 2006, Rebetez & Reinhard 2008). Here, we report on the current state of the art in climate model projections for the Alps, with an outlook to the soon available 5th IPCC assessment report.

It is a challenging task to project how the climate might look like in 50-100 years, a duration that is relevant for forest management. Climatologists use a range of models that generate possible climate futures. Each model and each simulation run is considered a representation of how the climate development during the 21st century might look like. For forest management and decision-making, we have to accept that no exact forecast is possible. Rather, we have to base our planning on the projected trends including their uncertainty. At a global scale, the periodic reports by the Intergovernmental Panel on Climate Change (IPCC) summarize the state of the art of how scientists foresee the development of the future climate and the associated impacts on ecosystems, economy and society. Now, the 5th assessment report is approaching, and some comparisons to the last two reports are already possible. The 3rd Assessment Report (IPCC 2001) had assumed that the global climate might be warming by 1.4-5.9°C, with no probabilities given for different increases, and with extreme scenarios projecting even far higher temperature increases. The 4th assessment report (IPCC 2007) provided overall a more narrow range of the likely future of the global climate stating that temperatures will probably be between 2.0 and 4.5°C warmer than in the 1961-1990 period (with a likelihood of 66%), and it also said that temperature increases by more than 4.5°C are not excluded (see Rogelj et al. 2012). In summary, the most likely temperature increase by 2100 was said to be 3.0°C. First indications from global climate modelling studies for the 5th IPCC assessment report project an increase of 2.4-4.9°C as medians from three different scenarios of radiative forcing (following different emission scenarios that are similar to those used in earlier reports). A fourth scenario is added that assumes a more rigorous and rapid reduction of greenhouse gases than was ever used before, predicting a median temperature increase of only 1.1°C during the 21st Century. Overall, the model simulations for the 5th IPCC assessment report expect that the likelihood of having global temperature increase exceeding 4.9°C is 14%, thus also likely, but that the most likely warming scenario at the global scale is still 3.0°C. Thus, in general, the newest scenarios do project similar warming trends as we have seen in the 4th IPCC assessment report, although some scenarios point to somewhat higher warming trends than were calculated for the 4th report.

The global climate is simulated using so-called general circulation models (GCM), which project the climate future on physics-based processes and first-principles. For regional applications, such model outputs are not very useful, as the spatial resolution of GCMs is very coarse, usually in the range of 1°-2.5° Lat/Lon per model cell. For regions such as the Alps only very few cells are modelled and no variation in terrain elevation is considered. In order to obtain more realistic climate projections at a regional to local scale, two types of downscaling are often combined. First, so-called regional climate models (RCM) are applied to certain larger regions of the World (such as e.g. all or parts of Europe). These models contain the same mechanisms as the GCMs, are fed by GCM output, and then simulate the climate evolution within the study region by using the input of the GCMs from outside of the study region. The output of these models is thus very similar, providing a range of climate variables at high temporal and moderate spatial resolution, ranging typically between 15-50km per cell. This is a much better spatial representation of the climate in regions and the output is somewhat sensitive to mountains, their variation in elevation, and their effects on the climate system, though often the output is still too coarse for management and decision-making. Therefore, a statistics-based downscaling procedure is further applied (Gyalistras et al. 1994, Pielke & Wilby 2012, Meier et al. 2012) in order to scale the output from RCMs to finer spatial resolution ranging from e.g. 100m to 1km, which can be considered well-suited for management applications.

For the MANFRED project, we have used five different RCMs driven by four different GCMs resulting in six GCM/RCM combinations in order to study the impact of likely climate changes on forest species and ecosystems. Table 1 gives an overview of the models used, which originate mostly from the ENSEMBLES EU project, using GCM runs that were calculated for the 4th IPCC assessment report (IPCC 2007).

We downscaled basic RCM variables such as monthly temperature and precipitation to finer spatial resolution for the six models, typically to 1km or 100m cell size. The method used can be called the "anomaly-approach", where we scaled the deviation (also called anomaly) of the future compared to the current climate from coarser to finer resolution. This is an efficient method, since anomalies do not depend much on altitudinal lapse rates. Once downscaled, the

Model RCM/GCM Scenario:	A1Fi	A1B	A2	B1	B2
CLM/ECHAM5, run by MPI	–	x	x	x	–
RACMO2/ECHAM5, run by KNMI	–	x	–	–	–
HADRN3/HadCM3, run by HC	–	x	–	–	–
HIRHAM3/Arpège, run by DMI	–	x	–	–	–
RCA30/CCSM3, run by SMHI	–	x	x	–	x
RCA30/ECHAM5, run by SMHI	–	x	x	x	–

Table 1. Climate models used to assess the impact of climate change on forest ecosystems and tree species ranges in the MANFRED project. RCM models are labeled in bold face, while the GCMs used to feed the RCMs are typed in normal font.

anomalies are added to an existing high-resolution climate map such as those available from Worldclim (Hijmans et al. 2005) or from national mapping campaigns (e.g. Zimmermann & Kienast 1999). The most important step here was to generate anomalies appropriately. First, we needed to identify the reference period of the high-resolution climate maps. Worldclim is mapping e.g. average monthly values of the 1950-2000 period. Next, we generated the monthly climate anomalies for given periods in the future at the RCM output resolution. To calculate the anomaly of each projected future climate month of any RCM relative to the current climate, we used the simulated time series outputs for the period of 1950-2000 from each RCM. By this, we avoided projecting the modeled bias in RCMs should the recent past deviate from climate station measurements. We were only interested in projecting the relative differences between simulated recent past and simulated futures. Once anomalies were generated, we interpolated these anomalies to the high resolution of existing climate maps such as Worldclim and added them to these maps to project the future climate changes to the representations of the existing climate.

The development of climate anomalies was done by first averaging the monthly time series of minimum (*Tmin*), average (*Tave*), maximum (*Tmax*) temperature and precipitation (*Prcp*) over the period of 1951-2000 for each RCM run used, since these represent the same base period of Worldclim maps. Second, we then used monthly RCM outputs to calculate monthly anomalies relative to the 1950-2000 period means per month. We developed these monthly anomalies by: (a) subtracting current from future temperatures, and (b) dividing future by current climate for precipitation. The latter results in ratios of change, which avoids negative precipitation values that could else result after downscaling if the difference method (a) is used. All climate anomalies were first calculated at the spatial resolution of the RCM output, and were then scaled to an intermediate resolution of 1km by bilinear interpolation (and in a second interpolation step to 100m if necessary). All scaling analyses as described above were performed using "cdo"[1] and other tools applied to NetCDF data files. Figure 1 illustrates the projected climate change trend from the six used RCM simulations by the example of annual and seasonal (summer and winter half) means, and by the uncertainty in projected summer climates.

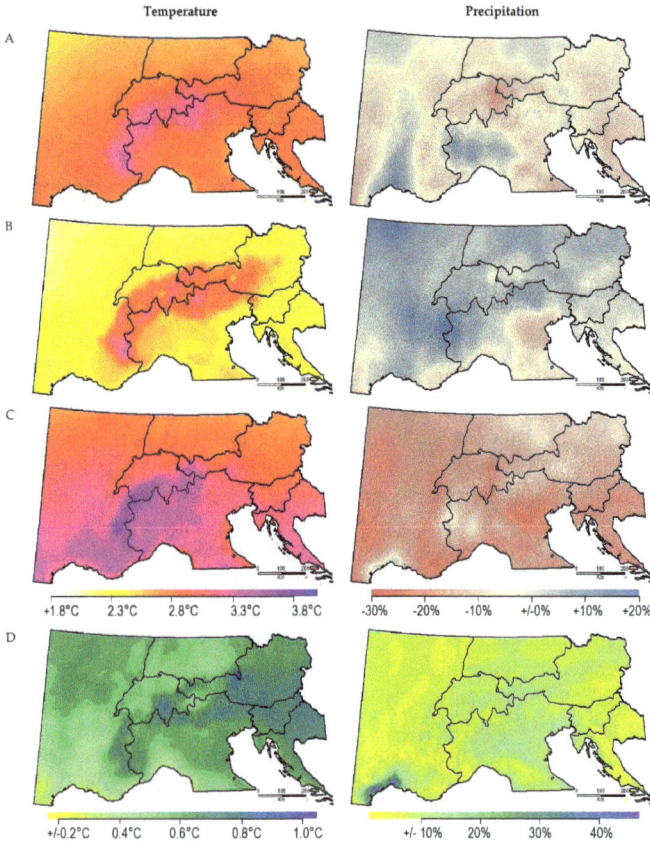

Figure 1. Climate anomalies for the A1B scenario by 2080 (deviations of the 2051-2080 period from the current, i.e. 1950-2000 climate) averaged over the six RCM models used to assess the impact of climate change on forest ecosystems and tree species ranges in the MANFRED project. A: Anomalies for annual temperature and precipitation; B: Anomalies for winter months (October-March): C: Anomalies for summer months (April-September); D: Uncertainties in summer anomalies among all 6 RCM models (calculated as the standard deviation among individual summer anomalies of the six models).

For temperature, we observe a general warming trend in the range of 1.8 to 4.0 °C over the Alps for the annual mean of the 2051-2080 period, with least warming in the winter half, and

1 Climate Data Operators (https://code.zmaw.de/projects/cdo)

highest warming in the summer months. The Alps generally face higher warming trends than the surrounding mainland, specifically in the winter months. In summer, the warming is more pronounced in the Western Alps and generally in the South of the Alps, while the northern ranges and lowlands will face a lower warming. Uncertainty among the six models is highest in the higher altitudes of the Alps, and generally increases towards the Eastern part of the Alps.

For precipitation, the annual trend is not very strong, with some regions South of the Alps obtaining a bit more, while most of the regions obtain a bit less precipitation annually. However, the seasonal differences are large. The summer half year is projected to obtain significantly less rainfall, with some regions in the Central Alps obtaining only 70% of the current summer rainfall amounts, and with only small regions in the Southwest and in the East of the Alps obtaining roughly the same amount as today. The winter half is projected to be wetter for most regions, especially the Southwestern Alps, with the Po plain and some Mediterranean regions obtaining less rainfall than today (-20%). The uncertainty among models is highest along the Mediterranean coast in the West, and is also comparably high in the Po plain and at higher elevation in the Alps, while in the plains north of the Alps, the six models show comparably high agreement.

The projected climate simulated differs quite significantly among the six models (Fig. 2). The HadRM3/HadCM3 model projects the highest (ca. +5° C), while the RCA30/CCSM3 model foresees the lowest (ca. +2.8° C) average summer temperature increase by 2100. With regards to precipitation, the HIRHAM/Arpege model projects the strongest (ca. -30%), while the RCA30/ECHAM5 model foresees the lowest (ca. -5%) reduction in summer precipitation over Europe. The year-to-year climate variability is significantly higher in the HIRHAM/Arpege and in the HadRM3/HadCM3 than in the CLM/ECHAM5 model. These differences indicate uncertainties with regards to climatic extremes we may face and with regards to the degree of climata change we will face in forest management decisions.

We also observe strong spatial variation in projected climate patterns, both among models (Fig. 1d) and throughout the projected time series (Figs. 3 & 4). The HadRM3/HadCM3 model reveals high spatial variation and additionally strong fluctuations around a warming trend until 2100 (Fig. 3), meaning that temperatures cannot be expected to gradually warm up. Early in the 21st century, regions North of the Alps are partly projected to show higher temperature increases than the south of the Alps, while after 2050, the South (and partly the West) of the Alps show clearly higher summer temperature increases. After 2030, 2055 and 2070, clear jumps to higher anomaly levels are observed in this model. The HIRHAM/Arpege model reveals a very high spatial variation and considerable temporal fluctuations in summer precipitation anomalies (Fig.4). This means that despite a general drying trend, some wet years are projected to occur, although with decreasing frequencies. On the other hand, such high fluctuations over time also indicate that very dry years are expected to occur increasingly more frequent. Some regions to the Southwest of the Alps show specifically high temporal fluctuations, with very wet years occuring infrequently.

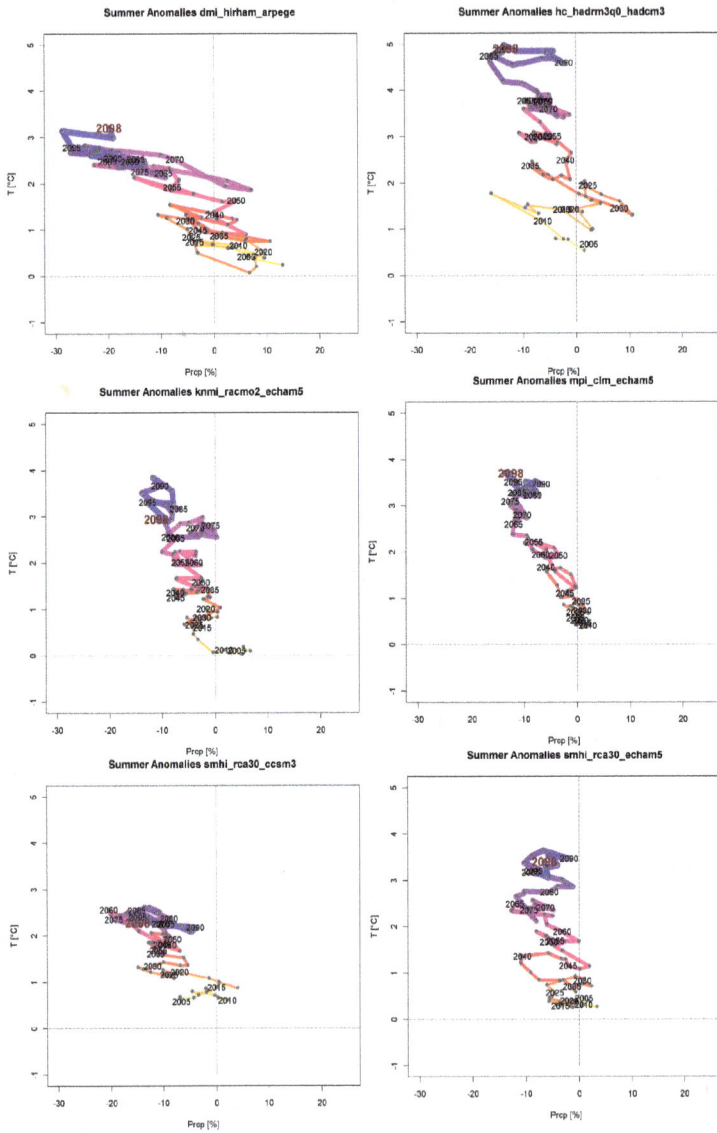

Figure 2. Time series of climate anomalies for the A1B scenario from 2001 until 2100 compared to the current climate (i.e. 1950-2000) over Europe for the six RCM models used in the MANFRED project. For each RCM, the anomalies are mapped as absolute (temperature) and relative (precipitation) values. The lines map the evolution of climate as 5-year averaged anomalies starting in 2005 (2001-2005). Every 5th year is labelled on each graph and grey dots represent the end year of the 5-year running average.

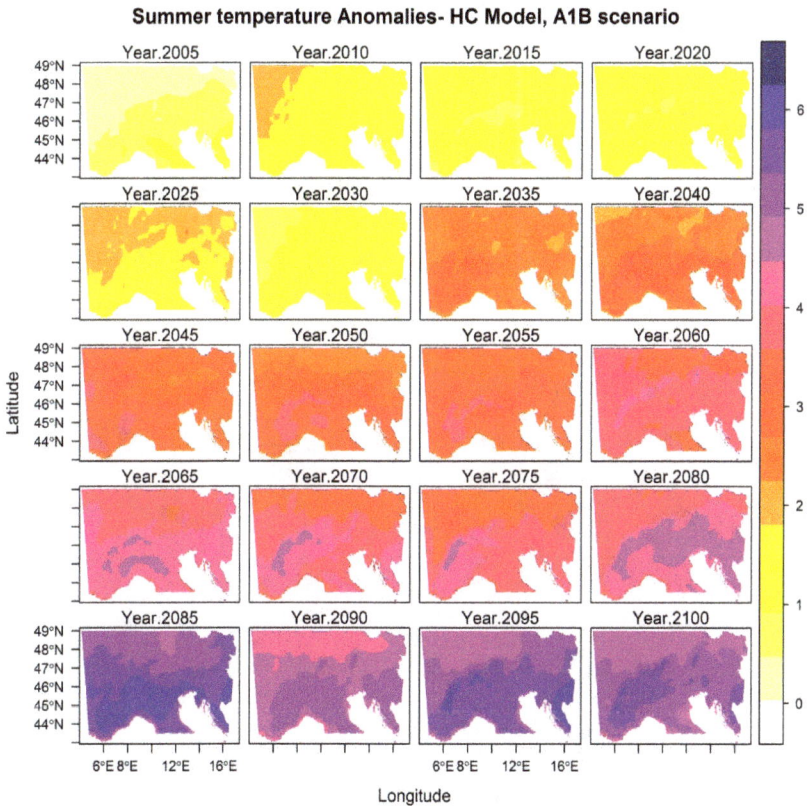

Figure 3. Spatial time series of absolute (°C) summer (April-September) temperature anomalies for the A1B scenario from 2001 until 2100 compared to the current climate (i.e. 1950-2000) over the Alps for the HadRM3/HadCM3 model used in the MANFRED project. Anomalies represent 5-year averages starting in 2005 for the 2001-2005 period.

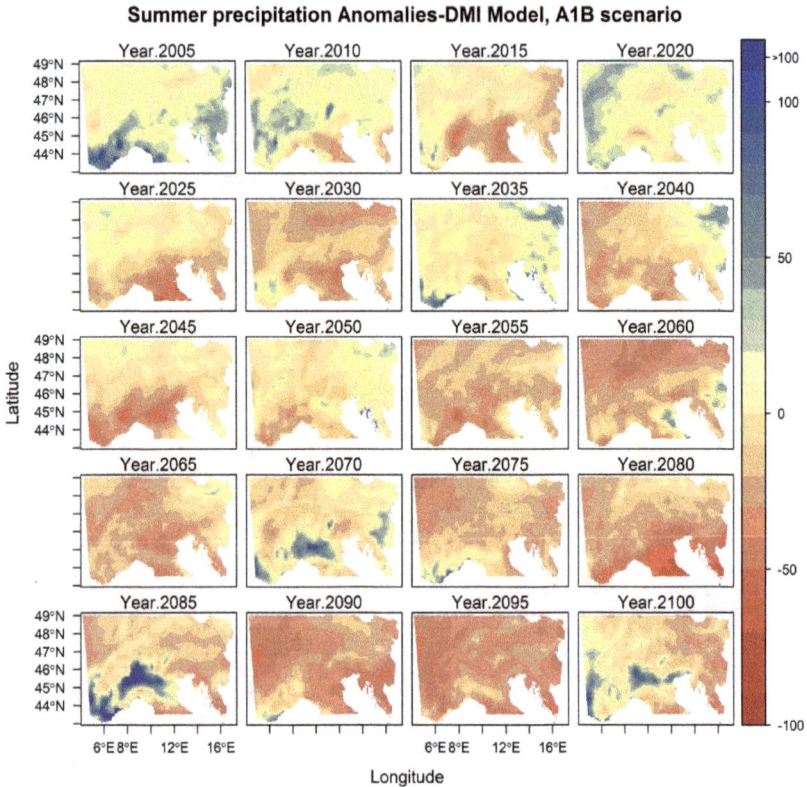

Figure 4. Spatial time series of relative (% change) summer (April-September) precipitation anomalies for the A1B scenario from 2001 until 2100 compared to the current climate (i.e. 1950-2000) over the Alps for the HIRHAM/Arpege model used in the MANFRED project. Anomalies represent 5-year averages starting in 2005 for the 2001-2005 period.

A significant change can also be expected from a change in seasonality (Fig. 5). The CLM model projects a "mediterranization" of the climate, by projecting significantly lower summer precipitations than today, and by simulating increased spring (March, April) and autumn (November) rainfall compared to today, and notably so after ca. 2050.

For forest management this means to be ready for clearly warmer, and at the same time also drier summers, which will have significant effects on some tree species, notably those with lower drought tolerance. This trend is particularly strong in the South of the Alps.

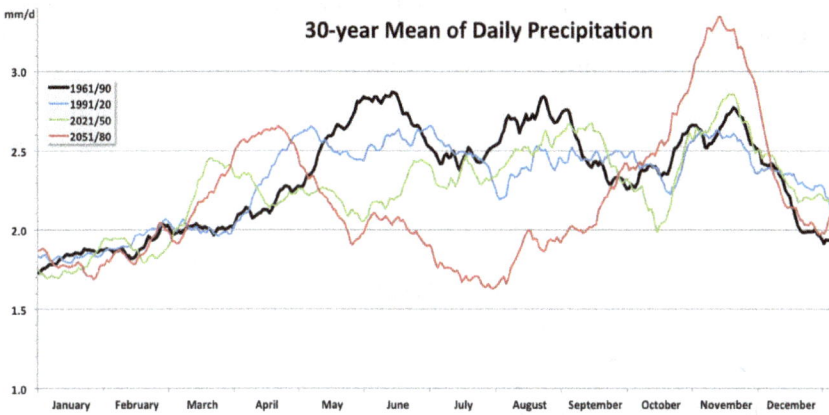

Figure 5. Change in precipitation seasonality across the European Alps as projected by the RCM model CLM that was driven by the ECHAM5 GCM. A trend to drier summer months and somewhat wetter spring (March/April) and autumn (November) months is apparent, specifically visible after 2050, when the RCM simulation projects a strong change to the precipitation regime.

Author details

Niklaus E. Zimmermann[1], Ernst Gebetsroither[2], Johann Züger[2], Dirk Schmatz[1] and Achilleas Psomas[1]

1 Swiss Federal Research Institute WSL, Birmensdorf, Switzerland

2 Austrian Institute of Technology, Vienna, Austria

References

[1] Gyalistras, D, & Storch, H. von, Fischlin A & Beniston M, (1994). Linking GCM generated climate scenarios to ecosystems: case studies of statistical downscaling in the Alps. Clim. Res. , 4, 167-189.

[2] Hijmans, R. J, Cameron, S. E, Parra, J. L, Jones, P. G, & Jarvis, A. (2005). Very high resolution interpolated climate surfaces for global land areas. International Journal of Climatology , 25, 1965-1978.

[3] IPCCClimate Change (2001). The Physical Science Basis. Contribution of Working Group I to the Third Assessment Report of the Intergovernmental Panel on Climate Change. Cambridge University Press, Cambridge, UK and New York, NY, USA

[4] IPCCClimate Change (2007). The Physical Science Basis. Contribution of Working Group I to the Fourth Assessment Report of the Intergovernmental Panel on Climate Change. Cambridge University Press, Cambridge, UK and New York, NY, USA

[5] Meier, E. S, & Lischke, H. Schmatz DR & Zimmermann NE, (2012). Climate, competition and connectivity affect future migration and ranges of European trees. Global Ecology and Biogeography , 21, 164-178.

[6] Pielke RA & Wilby RL (2012). Regional Climate Downscaling: What's the Point? EOS , 93(5), 52-53.

[7] Rebetez, M. (2006). Die Schweiz im Treibhaus. Bern, Haupt, 149pp.

[8] Rebetez, M, & Reinhard, M. (2008). Monthly air temperature trends in Switzerland 1901-2000 and 1975-2004. Theoretical and Applied Climatology , 91, 27-34.

[9] Rogelj, J. Meinshausen M & Knutti R, (2012). Global warming under old and new scenarios using IPCC climate sensitivity range estimates. Nature Climate Change: DOI:NCLIMATE1385

[10] Zimmermann, N. E, & Kienast, F. (1999). Predictive mapping of alpine grasslands in Switzerland: species versus community approach. Journal of Vegetation Science , 10, 469-482.

Chapter 3

Potential Future Ranges of Tree Species in the Alps

Niklaus E. Zimmermann, Robert Jandl,
Marc Hanewinkel, Georges Kunstler,
Christian Kölling, Patrizia Gasparini,
Andrej Breznikar, Eliane S. Meier, Signe Normand,
Ulrich Ulmer, Thomas Gschwandtner, Holger Veit,
Maria Naumann, Wolfgang Falk, Karl Mellert,
Maria Rizzo, Mitja Skudnik and Achilleas Psomas

Additional information is available at the end of the chapter

1. Introduction

Climate is one of the major drivers of plant and tree distribution, while soil variables or interspecific competition often are considered to primarily drive their local abundance. While the later is partly debated (Meier et al. 2010, 2011, 2012), the climate constraint to species ranges is generally accepted (Woodward 1987). The debate on climate change impacts on biodiversity and ecosystems is therefore specifically relevant to long-lived plants such as trees or shrubs, as these take many years to reach maturity and fecundity, and given their sessile growth strategy they are specifically vulnerable to rapid changes in climatic conditions. Also, forest management typically encompasses many decades, partly even reaching to the end of the 21st century, which highlights the challenge to manage such organisms successfully at such long planning horizons. This calls for careful and adaptive management strategies and for a good understanding of the uncertainties related to the expected changes and its impacts on trees and forest ecosystems.

Many approaches exist to project the impact of climate change on trees and forests. Yet, most of the approaches either can be applied only to comparably small regions, to few species only, or they need to be run at very coarse spatial resolutions in order to enable coverage of larger spatial extents. We distinguish the following five basic approaches of which we list major advantages and disadvantages. We then explain how we have used the last of these five approaches to project how species might respond in their habitat preference at the scale of the European Alps in response to projected climate change. Other approaches exist and often they glean from the five approaches described below:

1. Biogeochemistry-driven dynamic vegetation models (DVMs) are considered mechanistic with regards to physiological processes, and therefore represent growth and the fluxes of water, carbon (Sitch et al. 2003), and – in some cases – nitrogen cycling (Thornton et al. 2002). They are driven by input parameters such as daily precipitation, temperature, vapor pressure, or global radiation, and they model the water fluxes based on climatic input, on soil properties, and on biogeochemical processes of the vegetation. The behavior of vegetation is described in a larger set of parameters such as the C:N ratio of leaves, roots, stemwood, of values that decribe photosynthetic or hydraulic properties. Such parameters are time consuming to measure, and therefore such models simulate the behavior of plant functional types (or biomes) rather than individual species. Demography or structural details of the forest stand, interspecific competition or stochastic disturbances are handled less well (for an exception, see Hickler et al. 2012), and these models are most often operated at very coarse spatial resolution such as 10' to 0.5° Lat/Lon cells, and are therefore not easily applicable to forest management. This model family does usually not distinguish individual trees or cohorts, but rather biomass pools and translocation of carbon and energy between pools. They are useful when investigating coarse scale fluxes of carbon, water and nitrogen, and in studying the re-distribution of coarse biomes such as deciduous broadleaf, evergreen needleleaf, or C3 grassland vegetation belts in response to climate change.

2. Population dynamic models of vegetation such as the forest gap models (Bugmann 2001) represent a model family that is highly suited for forest management applications at regional spatial scale. These models are process-oriented – although they lack the physiological processes of DVMs – and include significant demographic and structural details. They simulate the fate of individual trees or cohorts, and they simulate both intra- and interspecies competition within a forest stand. Many of these models are not spatially explicit, and thus are operated at a single stand level. Few versions exist that are spatially explicit, and thus allow for regional assessment of management and climate change impacts. The model LandClim (Schumacher et al. 2006) is an example of such a spatially explicit forest dynamics model, while TreeMig (Lischke et al. 2006) is another. The latter additionally includes seed maturity, seed dispersal, and seed bank processes, and models the full regeneration process that is lacking in most other models. It therefore allows for simulating the natural migration of trees in a landscape following climate change. The model can be run at comparably fine spatial resolution, but regions such as the whole Alps or even all of Europe cannot easily be simulated due to computational demands. While forest gap models were the first such model type, new versions of demographic vegetation models that include other (than tree) plant functional types are currently being developed. While stand structural details are often well simulated in such population dynamic models, NPP, LAI and other ecosystem properties are less well simulated. Although, some models exist that are specifically oriented towards modelling such physiological details in forest stand

models. Such models can be applied to any forest, but due to the comparably high level of demographic detail, it is important to provide the necessary stand structural data as input to such models, in order to render them useful for forest management applications. Such input can be derived from forest inventories, although such information is not available yet at high spatial resolution. More research with regards to calibrating stand structure data from remotely sensed data sources is required to fill this gap.

3. Demographic range models (DRMs; Schurr et al. 2012) operate in a similar fashion as population dynamic models in that they also simulate the demography of species. They do so in a spatially explicit manner, but they usually lack inter-species competition, and only model one species at a time. This model type is thus best suited for treating dominant species. Habitat suitability is taken from species distribution models (see the fifth model family below), while demographic processes and spatial migration are modelled explic- itly. These models can be applied to larger spatial scales, but are moderately demanding to calibrate new species, as the model is very data hungry. It requires information on seed dynamics, on seed dispersal distances, and on the spatial population behavior. They can be seen as simplified versions of the second model family, although they are more data driven in their calibration, while forest gap models are more based on conceptual first principles.

4. Phenological (partly physiological) models such as PhenoFit (Chuine & Beaubien 2001, Chuine 2010) and conceptually similar models (Huey et al. 2012) attempt to calibrate those processes that directly affect – and moreover – constrain the demographic processes and life history, such as leaf unfolding, seed maturity, juvenile survival, etc. Such models therefore attempt to calibrate significant elements of the fundamental niche (Hutchinson 1957), while inter-species competition or demographic processes are not modelled. This model family can easily be applied to large spatial extents, but the calibration is very demanding for each species, so that only very few tree species have been calibrated to date.

5. Species distribution models (SDMs) are a simple but very efficient statistics-based method to map the spatial range of species and to project climate change impacts on species ranges (Zimmermann et al. 2010). The method is well matured from the statistical and conceptual end (Guisan & Thuiller 2005, Elith et al. 2006), and it is used for many different purposes including conservation management, theory testing in biogeography and ecology, species management, climate change impact assessment (Guisan & Zimmermann 2000). The method is based on calibrating statistical relationships between the observed spatial distribution of a species and climate and other spatial predictor variables that may explain this distribution. Sampling design is therefore very important, and the fitted spatial patterns represent the realized, not the fundamental niche. The method does not encom- pass any processes or details on transient responses following change, and no structural information is provided. Therefore, the method simply – but efficiently – provides an assessment of the suitability of any region for a species under current or projected future climate, and under the assumption that roughly the same species are available as potential competitors. SDMs are thus often used to assess, whether a given species has a future in a specific region or not, while the question whether it can reach a certain region and how

long it would take a species to get there are not handled by this method. SDMs therefore are best suited to assess habitat suitability and whether certain management options regarding species preference/selection are likely sustainable in the long run.

In the following, we present SDM simulations, often also termed climate envelope models (CEM), for major tree species of the European Alps in order to assess what the consequences of climate change on the habitat suitability of these tree species is. We used presence/absence information from forest inventories of France (Alps only), Northern Italy[1], Austria, Southern Germany Slovenia and Switzerland in order to build a database of tree species presences and absences across the Alps. We compiled data for ca. 50 tree species at a total of >80'000 inventory plots, although some countries did not distinguish all species at the same taxonomic level (some countries did e.g. not distinguish the different oaks or maples). We then compiled a series of climate maps under current and potential future climate from downscaled RCM models for future climates. We also compiled some topographic variables that may add to influence the spatial patterns of trees.

We finally used the following variables as predictors of species distribution in our models: (1) degree days with a 5.56°C threshold, (2) temperature seasonality (standard deviation of monthly values), (3) summer precipitation (sum of April to September monthly values), (4) winter precipitation (October to March), (5) potential yearly global radiation, (6) slope angle in degree, (7) topographic position (difference between the average elevation in a circular moving window applied to a 100m digital elevation model and the centre cell of the window) (Zimmermann et al. 2007), (8) aspect value (ranging from 0(south) to 100(north), and (9) distance to running waters. These parameters were then used to explain the spatial distribution of a species. The statistical model calibration then evaluated, what parameter has predictive power and what parameter can be discarded or downweighted because it did not significantly add to explain the spatial distribution of a tree species.

Potential future climate was taken from six different RCMs (see previous chapter), providing a range of potential climate futures. The use of several RCM models is meant to provide the mean trend that can be expected from climate change impacts on trees but also some measure of uncertainty associated with the projection of these trends (Araujo & New 2007, Thuiller et al. 2009). This was done because there is no consensus on which of the many climate models actually best describes the future faith of the climate. Rather, climatologists provide us with a range of models that are all considered equally likely to represent the climate future. Differences among climate models arise from divergent ways of parameterizing the complex dynamics of the global climate in such models. We therefore should not simply take one single model in order to assess the likely consequences of climate change on species range shifts. Our choice reflected differences in how much the different climate models simulated shifts in temperature and precipitation.

---

1 data from the National Inventory of Forests and Forest Carbon Pools -INFC 2005, National Forest Service of Italy - CFS

Several statistical models were used, since the choice of a statistical model has been shown to significantly contribute to uncertainty in projections (Buisson et al. 2010). More specifically, we used the following statistical models: (1) Classification and regression trees (CART), (2) Flexible discriminant analysis (FDA), (3) Generalized linear models (GLM), (4) Generalized additive models (GAM), (5) Artificial neural networks (ANN), and (6) Generalized boosted regression trees (GBM). Therefore, given the use of six statistical models and six future climate model runs, we model 36 different possible futures per species and time period. This allows for assessing the projection uncertainty from both the variability in climate models and the variability originating from the choice of statistical methods. Naturally, the different models perform with different average accuracies among all species, with GBM and GAM often outperforming other models such as CART or GLM. However, when projecting models to future climates, it is not necessarily the currently best performing models that will yield predictions with lowest uncertainties! We know e.g. from projections to regions that are outside of the calibration region (but with similar climate) that GLMs outperformed GAMs, although within the calibration regions GAMs were always better than GLMs (Randin et al. 2006). Therefore, we chose to use all models as potentially equally likely to provide future ranges of our modelled tree species, and we only discarded those models, which didn't meet minimal criteria in a cross-validation test.

We optimized each statistical model following procedures described in Thuiller et al. (2003, 2009). Some models produce probabilities, while other generate directly presence/absence output. Where the statistical models generated probabilities, we used the kappa statistic (Cohen 1960) to select the best threshold to split the probability maps into maps of species presence and absence values. We therefore had one presence/absence map per climate model/ statistical model combination available. We then built ensembles of these model projections and classified these as follows: (1) a species is unlikely to find a suitable habitat if less than 30% of the projections indicated presence of a species; (2) a species is moderately likely, associated with high uncertainty, if 30-60% of the projections suggested that the species is there; (3) a species is most likely present with rather low uncertainty under projected climates if in >60% of the 36 model projections presence of a species is simulated. Such a simple classification avoids an over-interpretation of the results from our modelling approach.

Figure 1 illustrates the potential future range shift in two species *Fagus sylvatica* L. (European beech) and *Picea abies* (L.) Karst. (Norway spruce) in eight panels, indicating the areas that are suitable for the two species under current and future climate conditions in three different time steps towards the end of the 21st century. Both species are expected to lose much terrain at low altitudes, and will retract to higher altitudes following climate change. Currently, Norway spruce is planted at lower altitudes than it occurs naturally. Obviously, these lower altitudes are still within the fundamental niche of the species, and the simulated maps also capture the extended range of the species to lower altitudes under both current and future climates. However, compared to beech, it extends to higher altitudes, reaching treeline in many parts of the Alps. This is specifically visible for the simulations of the 2051-2080 time period, where for Norway spruce clearly higher elevations are projected to be suitable than for beech. For both species, larger parts at low altitudes become unsuitable, while the habitat suitability in large areas in Southern Germany are projected to be highly uncertain for both species. This uncertainty arises from highly contradicting projections by both climate and SDM model combinations.

More than 50 tree species have been simulated for the range of the Alps, and only two species are displayed here. A more complete set of species can be checked out and downloaded under a special website[2]. It becomes obvious that mostly the more drought-tolerant, sub-mediterranean species such as *Quercus petraea* and *Quercus pubescens* can be expected to become more abundant at lower altitudes throughout the Alps, while other species such as *Acer pseudoplatanus, Tilia spp., Ulmus spp.* or *Abies alba* are likely further reduced in their ranges similar to beech and spruce (Fig. 2). Species from Mediterranean regions such as *Quercus ilex* or *Q. suber* are expected to extend their ranges to the North, but these species will not (or only barely) reach the areas currently suitable for beech by the end of the 21st century. Therefore, we can mostly expect sub-mediterranean, drought tolerant species to invade the spatial domain that is currently dominated by beech, namely: *Q. petraea, Q. pubescens, Ostrya carpinifolia* or *Fraxinus ornus* (see Fig.2). Different *Pinus* species are also expected to extend their ranges quite considerably. However, they will likely not extend to very fertile soils either, and some of the species like e.g. *P. sylvestris* might face indirect threats through insects and other pests, rather than direct threats from climate change alone. They may rather dominate at higher altitudes when summer precipitation is reduced, and larger parts of the Alps become considerably drier.

In fact, none of the individual SDM models is capable of projecting the effective fate of the different tree populations. The maps simply illustrate the habitat potential at certain time periods in the future. Species may still survive for quite a while at locations that become unsuitable. They will eventually face one or both of the following two threats: (1) physiological stress from a climate that they cannot tolerate, and (2) stronger competition by other, better adapted species or by antagonists such as insects, fungi, etc. that may profit in turn from a drier and/or hotter climate, and may infest trees that are less vigorous because of a combination of the two causes (1) and (2). Forest management can usually deal more or less well with the second type of threat if it is primarily due to changes in tree species competition. Dealing with changes in antagonists in forest management is more difficult, as the example of the Scots pine dieback in the Alps illustrates (Dobbertin et al. 2005, 2007, Bigler et al. 2006). Here, a rapid dieback at the lowest altitudes of the Scots pine distribution has been observed over the last 10 years, which is the area that is projected to eventually become unsuitable in the future from ensemble SDMs as well.

The presented SDM method has the advantage of rapidly indicating to managers and practitioners (1) the areas that become likely suitable soon, (2) the areas that are not expected to undergo significant changes, and (3) the areas that likely – and eventually – become unsuitable for maintaining viable tree populations naturally. The disadvantage of the method is that it cannot inform managers about the speed of the projected changes. Immigration into formerly unsuitable areas are not easy to predict, the invading species have to compete against the

---

2 http://www.wsl.ch/lud/manfred

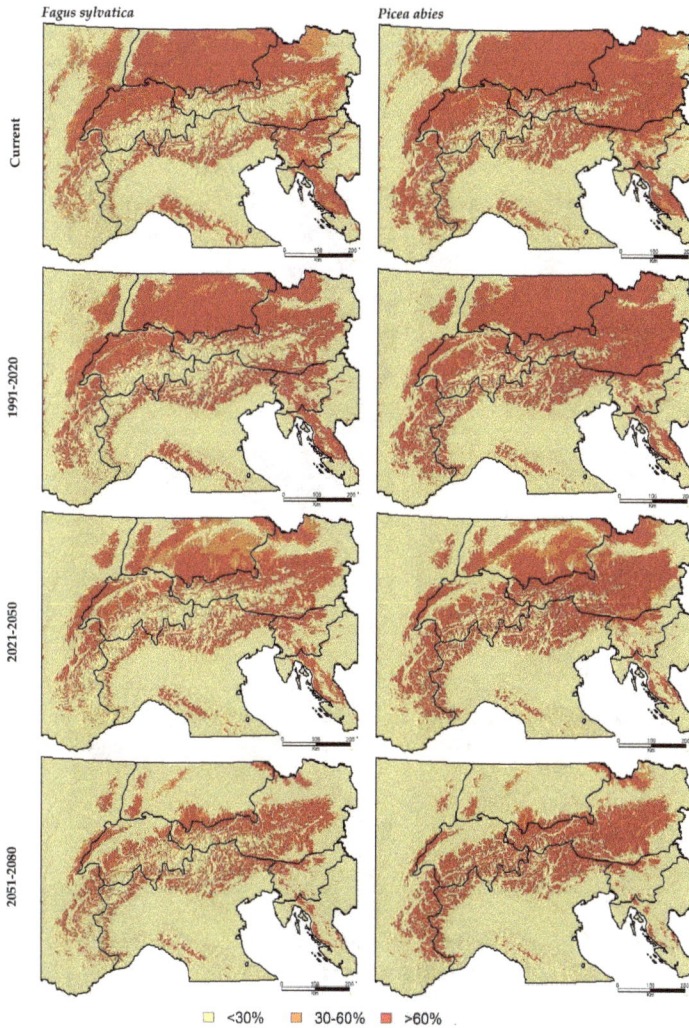

Figure 1. Projected future habitat suitability from ensemble SDM modeling of *Fagus sylvatica* and *Picea abies* for current and projected future climate in three time periods. Dark red colors represent high confidence of agreement of high habitat suitability from both six statistical and six climate models, while orange colors indicate high uncertainty in projected habitat suitability per time period.

already inhabiting species, and the replacement process might take a long time. Even more difficult is to forecast at what rate tree species can no longer survive, or render financial benefits when planted or managed under slowly deteriorating climates. In fact, most species grow better under warmer climates. However, when at the same time increasing drought levels

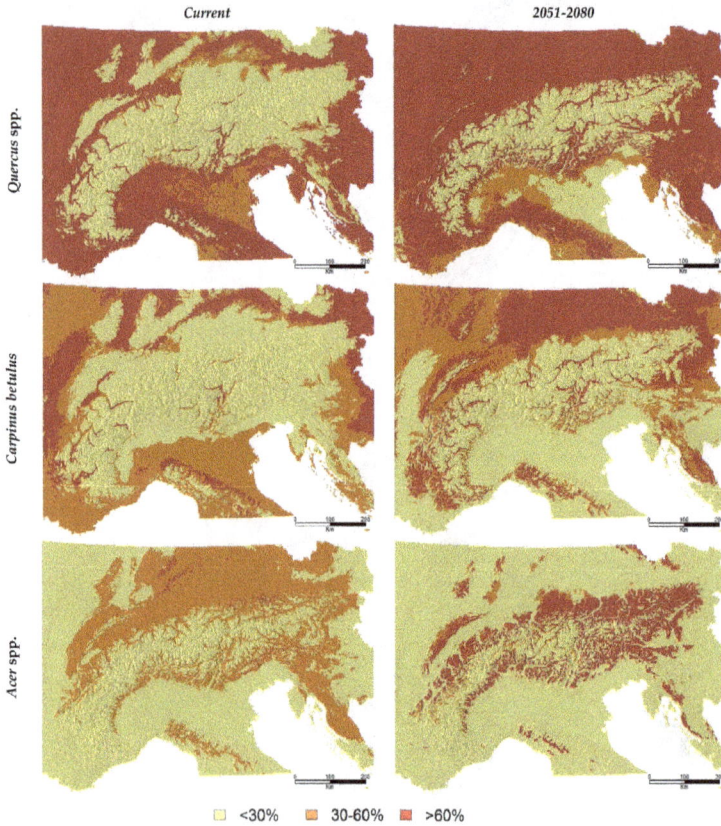

Figure 2. Projected future habitat suitability from ensemble SDM modeling of *Quercus* spp., *Carpinus betulus*, and *Acer* spp. for current and projected future climate in three time periods. Dark red colors represent high confidence of agreement of high habitat suitability from both six statistical and six climate models, while orange colors indicate high uncertainty in projected habitat suitability per time period.

exceed species-specific thresholds, then trees might rapidly decline in growth or in maintaining viable populations. This illustrates, how difficult the forecasting of rear edge populations of a species range is under changing climate conditions. The model results need to be combined with careful observations.

SDMs can be improved, by adding more informative predictors, such as soil information or site productivity estimates. However, the difficulty to obtain these predictors with identical units and sufficient quality over large areas spanning the entire range of a species renders this type of model improvement often very difficult. And doing it only for smaller areas often results in models that do not span the capacity of the species well, because a smaller area that represents only a portion of the ecological niche of a species can be calibrated from data of such spatial domains. Yet, model improvements have to be sought constantly, and novel GIS techniques combined with the increasing availability of ecologically meaningful remote sensing data allows constant refinement of model predictors.

Author details

Niklaus E. Zimmermann[1], Robert Jandl[2], Marc Hanewinkel[1], Georges Kunstler[4], Christian Kölling[5], Patrizia Gasparini[6], Andrej Breznikar[7], Eliane S. Meier[1], Signe Normand[1], Ulrich Ulmer[1], Thomas Gschwandtner[2], Holger Veit[3], Maria Naumann[5], Wolfgang Falk[5], Karl Mellert[5], Maria Rizzo[6], Mitja Skudnik[8] and Achilleas Psomas[1]

1 Swiss Federal Research Institute WSL, Birmensdorf, Switzerland

2 Federal Research and Training Center for Forests, Natural Hazzards and Landscape BLW, Vienna, Austria

3 Forstliche Versuchs- und Forschungsanstalt Baden-Württemberg FVA, Freiburg, Germany

4 National Research Institute of Science and Technology for Environment and Agriculture IRSTEA, Grenoble, France

5 Bayerische Landesanstalt für Wald und Forstwirtschaft LWF, Freising, Germany

6 Consiglio per la Ricerca e la sperimentazione in Agricoltura - Unità di Ricerca per il Monitoraggio e la Pianificazione Forestale, CRA-MPF, Trento, Italy

7 Slovenia Forest Service, Lubljana, Slovenia

8 Slovenian Forestry Institute, Lubljana, Slovenia

References

[1] Araujo, M. B, & New, M. (2007). Ensemble forecasting of species distributions. Trends in Ecology & Evolution , 22, 42-47.

[2] Bigler, C, Bräker, O. U, Bugmann, H, Dobbertin, M, & Rigling, A. (2006). Drought as an inciting mortality factor in Scots pine stands of the Valais, Switzerland. Ecosystems , 9, 330-343.

[3] Dobbertin, M, et al. (2005). The decline of Pinus sylvestris L. forests in the swiss Rhone Valley- a result of drought stress? Phyton-Annales Rei Botanicae , 45, 153-156.

[4] Dobbertin, M, et al. (2007). Linking increasing drought stress to Scots pine mortality and bark beetle infestations. Thescientificworldjournal , 7, 231-239.

[5] Bugmann, H. (2001). A review of forest gap models. Climatic Change , 51, 259-305.

[6] Buisson, L, Thuiller, W, & Casajus, N. Lek S & Grenouillet G, (2010). Uncertainty in ensemble forecasting of species distribution. Global Change Biology , 16, 1145-1157.

[7] Chuine, I, & Beaubien, E. G. (2001). Phenology is a major determinant of tree species range. Ecology Letters , 4, 500-510.

[8] Chuine, I. (2010). Why does phenology drive species distribution? Philosophical Transactions of the Royal Society B-Biological Sciences , 365, 3149-3160.

[9] Elith, J, et al. (2006). Novel methods improve prediction of species' distributions from occurrence data. Ecography , 29, 129-151.

[10] Guisan, A, & Thuiller, W. (2005). Predicting species distribution: offering more than simple habitat models. Ecology Letters , 8, 993-1009.

[11] Guisan, A, & Zimmermann, N. E. (2000). Predictive habitat distribution models in ecology. Ecological Modelling , 135, 147-186.

[12] Hickler, T, et al. (2012). Projecting the future distribution of European potential natural vegetation zones with a generalized, tree species-based dynamic vegetation model. Global Ecology and Biogeography , 21, 50-63.

[13] Huey, R. B, Kearney, M. R, & Krockenberger, A. Holtum JAM, Jess M, Williams SE, (2012). Predicting organismal vulnerability to climate warming: roles of behaviour, physiology and adaptation. Philosophical Transactions of the Royal Society B-Biological Sciences , 367, 1665-1679.

[14] Hutchinson, G. E. (1957). Concluding remarks. Cold Spring Harbour Symposium on Quantitative Biology , 22, 415-427.

[15] Lischke, H, Zimmermann, N. E, Bolliger, J, Rickebusch, S, & Löffler, T. J. (2006). TreeMig: A forest-landscape model for simulating spatio-temporal patterns from stand to landscape scale. Ecological Modelling , 199, 409-420.

[16] Meier, E. S, Edwards, T. C, Kienast, F, Dobbertin, M, & Zimmermann, N. E. (2011). Co-occurrence patterns of trees along macro-climatic gradients and their potential influence on the present and future distribution of Fagus sylvatica L. Journal of Biogeography , 38, 371-382.

[17] Meier, E. S, et al. (2010). Biotic and abiotic variables show little redundancy in explaining tree species distributions. Ecography , 33, 1038-1048.

[18] Meier, E. S, Lischke, H, Schmatz, D. R, & Zimmermann, N. E. (2012). Climate, competition and connectivity affect future migration and ranges of European trees. Global Ecology and Biogeography , 21, 164-178.

[19] Randin, C. F, Dirnböck, T, Dullinger, S, & Zimmermann, N. E. Zappa M & Guisan A, (2006). Are species distribution models transferable in space? Journal of Biogeography , 33, 1689-1703.

[20] Schumacher, S, & Bugmann, H. (2006). The relative importance of climatic effects, wildfires and management for future forest landscape dynamics in the Swiss Alps. Global Change Biology , 12, 1435-1450.

[21] Schurr, F. M, Pagel, J, Cabral, J. S, Groeneveld, J, Bykova, O, Hara, O, Kissling, R. B, Linder, W. D, Midgley, H. P, Schröder, G. F, Singer, B, & Zimmermann, A. NE, in press. How to understand species niches and range dynamics: a demographic research agenda for biogeography. Journal of Biogeography.

[22] Sitch, S, et al. (2003). Evaluation of ecosystem dynamics, plant geography and terrestrial carbon cycling in the LPJ dynamic global vegetation model. Global Change Biology , 9, 161-185.

[23] Thornton, P. E, et al. (2002). Modeling and measuring the effects of disturbance history and climate on carbon and water budgets in evergreen needleleaf forests. Agricultural and Forest Meteorology , 113, 185-222.

[24] Thuiller, W. (2003). BIOMOD- optimizing predictions of species distributions and projecting potential future shifts under global change. Global Change Biology , 9, 1353-1362.

[25] Thuiller, W, Lafourcade, B, Engler, R, & Araujo, M. B. (2009). BIOMOD- a platform for ensemble forecasting of species distributions. Ecography , 32, 369-373.

[26] Woodward, F. I. (1987). Climate and plant distribution. Cambridge University Press, Cambridge

[27] Zimmermann, N. E, Edwards, T. C, Moisen, G. G, Frescino, T. S, & Blackard, J. A. (2007). Remote sensing-based predictors improve distribution models of rare, early successional and broadleaf tree species in Utah. Journal of Applied Ecology , 44, 1057-1067.

[28] Zimmermann, N. E, et al. (2009). Climatic extremes improve predictions of spatial patterns of tree species. Proceedings of the National Academy of Sciences of the United States of America , 106, 19723-19728.

[29] Zimmermann, N. E, Edwards, T. C, Graham, C. H, Pearman, P. B, & Svenning, J. C. (2010). New trends in species distribution modelling. Ecography , 33, 985-989.

Chapter 4

Developing a Background for Forest Adaptation Strategies in the Alps: A Perspective for Policy Building

Luca Cetara and Federico Mannoni

Additional information is available at the end of the chapter

1. Introduction

Over the last two decades, since the Rio de Janeiro's UN Conference on Environment and Development (UNCED) in 1992, a growing demand for policy making in the forest sector has emerged at the global level in connection to the challenge of climate change.

The UN-declared 2011 "International Year of Forests" ended with some important steps forward achieved at the Durban's 17th Conference of the Parties of the United Nations Framework Convention on Climate Change [Perugini et al., 2012].

Meanwhile, the evolution of this process has progressively resulted in an articulated international regulatory framework based on a plurality of agreements and policy initiatives coping with the complex issue of protecting the forest ecosystems and managing them in accordance to the principles of sustainability.

Historically, among the landmark steps in the development of regulatory principles and schemes addressing the forest sector and recognizing the inner multi-functionality of forest ecosystems in the international environmental law, it is worth mentioning, for instance, the Convention on Biological Diversity (CBD), that dedicates a detailed work programme to protect the biodiversity stored in forests, and the United Nations Convention to Combat Desertification (UNCCD), that recognizes the relevance of the role of forests in fighting the process of desertification connected with global warming [EUROPEAN COMMISSION, 2010].

In any case, as far as international schemes are concerned, and given the intrinsic transboundary dimension of the multiple, interconnected ecological, economic and social functions performed by forests as well as of the potential effects of climate change dynamics on them, the implementation of rational policy making in the sector requires forms of supranational endorsement flexible enough to facilitate sustainable forest management decisions that can be suitable to the different characteristics and conditions of the various forest sites existing on the Earth.

At the European level, Sustainable Forest Management (SFM) was adopted by the EU and its Member States as the central approach to forestry since 1993, with the adoption of Helsinki Resolution 1 the Ministerial Conference on the Protection of Forests in Europe.

Since then, European institutions refer to SFM as "the stewardship and use of forest lands in a way, and at a rate, that maintains their biodiversity, productivity, regeneration capacity, vitality and their potential to fulfill, now and in the future, relevant ecological, economic and social functions, at local, national and global levels, and that does not cause damage to other ecosystems" [MCPFE, 1993].

As the policy principles are defined, when aiming at implementing evidence-based sustainable forest management, taking account of the potential trade-offs and mutual benefits among the functions of mitigation and adaptation to climate change, as well as between the many other forest functions, becomes a crucial task to be undertaken [Spittlehouse and Stewart, 2003, FOREST EUROPE 2011].

Though, since the UNFCCC recognized the role of forests in the balance of global greenhouse gas emissions, a broader emphasis was traditionally put on their potential for climate change mitigation, either by reducing emission as a source for potentially wood-based renewable energy and as a sink for carbon sequestration [Guariguata et al., 2008, Campbell et al., 2009, FOREST EUROPE 2011].

It is nevertheless since 1990s that evidence for Europe continues to show that, according to current climate scenarios, forest ecosystems in this part of the world are expected to be particularly vulnerable to varying climate conditions.

In this context, the adaptation of the European forest sector to climate change represents a priority for ensuring that the provision of goods and services from forests can be maintained [Lindner, M. Kolström 2008, FAO 2011].

However, the development of adaptation measures is bound to a number of challenges to be dealt with in a context of scientific uncertainty. On the one hand, there is yet limited knowledge about the vulnerability of ecosystems and species as well as climate change impacts on the functional characteristics of forests in different bio-climatic zones (inherent adaptive capacity). On the other, adaptation measures also depends on a range of socio-economic conditions, whose future adjustments are uncertain [Burton et al., 2002, Spittlehouse and Stewart, 2003, Lindner, Kolström, 2008].

2. EU and Alpine legal framework for mitigation and adaptation in the forest sector

At the EU level, work is in progress and actions have been proposed to support and enhance sustainable forest management and the role of forest multi-functionality in the contribution to the Lisbon EU 2020 Strategy and the Gothenburg Agenda on sustainable development.

Remarkably forerunning to the processes at stake, the EU Forest Action Plan [European Commission, 2006], building on the Council Resolution of 15 December 1998 on a forestry strategy for the European Union, have identified four main guiding objectives commonly on which to ground the implementation of a coordinated forest policy by the EU and Member States that need to be address the concepts of multi-functionality and sustainable forest management:

i.   improving long-term competitiveness;

ii.   improving and protecting environment;

iii.   contributing to quality of life;

iv.   fostering communication and coordination

More specifically, in order to implement these objectives, the Action Plan outlines eighteen Key Actions.

All these actions, in view of their intrinsic interrelation, have a relevance to be taken into account when devising adaptation strategies in the forest sector. Some of them, however, together with the measures outlined for their practical implementation, can be intrinsically seen as directly referable to policy-building for adaptation to climate change:

- the support of research and studies on climate change impacts and adaptation measures (Key Action 6);

- the enhancement of EU forests protection from biotic and abiotic hazards also by encouraging cooperation between Member States to study particular regional problems with the condition of forests (Key Action 9);

- maintain and enhance the protective functions of forests against the increasing threats of natural disasters and extreme events also through coordinated monitoring and planning, awareness raising and knowledge transfer on natural hazard and risk management, with a focus on mountain areas (Key Action 11)

A report on the state of the art of its implementation of the EU Forest Action Plan 2007 – 2011 will be presented to the Council and the European Parliament during 2012, in order to evaluate potential further developments.

Relevant to the analytic scope of the present paper, the EU Forest Action Plan has aimed at mainstreaming adaptation of European forests to climate change as a clear policy objective in a context where, in compliance with the subsidiarity principle, competence in forest policy lies primarily with the Member States. The role of the EU in forest policy, in fact, technically applies according to the principle of vertical subsidiarity (art. 5 of the EU Treaty), consequently meaning that the EU competencies are expected to simply perform in those cases where action by Member States is not sufficient and can be better achieved at the Union level.

With regard to forest policies, in particular, the EU role was historically limited to monitor and report on the state of EU forests; anticipating and stimulating attention of Member States on emerging global challenges and trends; proposing and coordinating or supporting early action at the EU level.

Moving in this complex, multi-dimensional institutional framework, innovative policy approaches have to be conceived to address the challenges posed by the need to ensure sustainability and adaptation of those forest resources that, due to their peculiar vulnerability and trans-boundary relevance, require apt regional policy-making strategies.

Not for a case, it has been recently affirmed by the FAO [2011] that there is a need for measures to protect mountain forests to be based on enhanced coordination at international and national levels, in consideration of local specificities and by integrating forest issues into broader policies and programmes.

Here, we focus on the Alpine forests and their specific functions, with the aim of analyzing a potential action framework for an effective and evidence-based policy building that can devise adaptive forest management strategies suitable to the peculiar adaptation needs of the area, representing at the same time an integrated element of the guiding measures and principles set out at the multiple level of decision making (EU, Alpine, national, local) co-existing in a multinational model of governance and cooperation.

Alpine forests retain the characteristics to represent an important pilot area for implementing adaptive strategies under consideration here, for a range of institutional, ecological and socio-economic reasons.

Under an institutional perspective, the Alpine Convention, quite peculiarly in international environmental law with regard to this sector, provides a regulatory framework for implementation of transboundary cooperation in sustainable forest management: the Alpine Convention was in fact the first framework agreement at the level of international law on the protection and sustainable development of a transboundary mountain region.

More precisely, the Alpine Convention is an international treaty between the Alpine countries (Austria, France, Germany, Italy, Liechtenstein, Monaco, Slovenia and Switzerland) as well as the EU open to signature in Salzburg (Austria) on 7 November 1991, aimed at promoting sustainable development in the Alpine area and at protecting the interests of the people living within it, embracing the environmental, social, economic and cultural dimensions.

In any case, as commonly designed within the international environmental law due to the technical and scientific complexity of the provisions' contents, the implementation of the general principles of the framework agreement is to be further delegated to the Protocols [Munari, 2006].

For implementing the Alpine Convention's goals for the forest sector, since 1996, the "Mountain Forests" Protocol - done at Brdo (Slovenia) on 27th February of that year - identifies a set of specific functions of Alpine mountain forests whose conservation and enhancement is to be considered as a priority in the overall regional sustainable development policy.

By the attribution of a legal status to these functions, the Contracting Parties have therefore designed a platform for international cooperation (art. 4) at multiple levels of decision-making (art. 3) to pursue the preservation and improvement of forest assets even in a cross-sectoral perspective with other policy fields, also by means of cooperation in research, education and information (Chapter III).

Particularly relevant in a perspective of climate change adaptation, the "Mountain Forest" Protocol highlights the relevance of the protective function (art. 6), the economic function (art. 7), and of the functions of social and ecological character such as, among the others, water resources, clean air, biodiversity and recreation (art. 9).

3. Major impacts of climate change on Alpine forests and their functions

Climate change is expected to determine major ecological, social and economic effects on Alpine forests: as a result of the combined effects of land abandonment and decrease of limiting tree growth factors at higher altitudes because of higher temperatures, the process of forest cover expansion in the Alps - currently around 7.5 million ha, the 43% of the total Alpine land surface [EEA, 2009] - which is ongoing in the last decades is expected to continue, probably at higher rates.

Climate change may also affect tree growth patterns and species distribution, as well as increase the spread and intensity of, respectively, abiotic (fires, storms, drought, precipitations, atmospheric pollution) and biotic (pests and diseases) disturbances. These climate change impacts may therefore trigger plural interdependent effects that could alter a range of socially and economically critical Alpine-specific forest functions.

Nevertheless, knowledge still lacks on the magnitude and trends of these impacts and on how they may affect, positively or negatively, the livelihood of highland and lowland communities in a socio-economic perspective [IPCC, 2007, Lindner, Kolström et al. 2008, FAO, 2011].

In this line, the mentioned research outcomes, policy making guidelines and experience in Alpine forest adaptive management have suggested to place a particular focus on:

- the *protective function* for human infrastructures from natural hazards like flooding, debris flows, landslides, avalanches and rockfalls, is one of the most important function played by mountain forests: for instance, 20% of Austria's forests is considered as having a protective role, meanwhile 63% of Bavarian forests are declared as having a protective function against soil erosion and 43% against avalanches, in Germany. Concerning the economic and social value related to mountain forests' protective function in Alpine territories, a telling example comes from the Swiss Alps where it has been estimated that, in their absence, the cost of ensuring protection against avalanches using permanent *ad hoc* defense structures would be around some 89 billion Euros, 5-20 times more than that of maintaining healthy protection forests [FAO, 2011]. However, protection forests are sensitive to varying climate dynamic, even though it is difficult to assess the complexity of interaction, as it depends on many structural and compositional properties. In theory, diverse species composition, sufficient natural regeneration and an optimal structure favour the forests' protective function. As climate change could significantly alter the plant's phenology, physiology and distribution, strategies have to be implemented in the Alps to deal with the potentially increased silvicultural challenges already at play - poor regeneration, low proportion of medium-aged trees, exposure to intensified natural disturbances [Lindner, Kolström et al., 2008, FAO, 2011]. In this regard, some significant actions have already been undertaken at the pan-alpine level. In particular, the MANFRED Project "Management Strategies to adapt Alpine Space forests to climate change risks", launched in the context of the European Territorial Cooperation Alpine Space Programme 2007 - 2013 has developed a framework to undertake cooperative action in research and policy-making with regard to assessing the impacts of climate change on forests all over the Alpine arc, with a special focus on the interrelation among climate change scenarios and biotic and abiotic hazard factors dynamics. The expected results goes precisely in the direction of improving knowledge-based transnational cooperation and experience sharing in forest adaptive management at the Alpine level in order to address, among the others, the capacity of forests in the Alps of enduring their protective function also in a context of varied climatic conditions.

• The *productive function* of wood and non-wood forest goods and services may be consistently altered by climate change with both *a*) potential pros and cons as well as with *b*) trade-offs between the different potential economic uses of the same forest parcel. In the first case, for instance, wood productivity is expected to increase at higher elevations as a consequence of climatic changes as far as the water supply is sufficient, meanwhile at lower elevations changes species competition and adaptive capacity to natural disturbances may lead to decreasing productivity, also affecting the economic profitability patterns and their differences existing between highland and lowland forestry [Lindner, Kolström et al., 2008, FAO, 2011]. Instead, referring to different potential forested land uses, by the way of example, trade-offs exist between use of forest as carbon sinks or for biomass production, which are determined by both socio-economic conditions such as market prices and demand for biomass energy from forests as well as climate change-related effects [Lindner, Kolström et al., 2008]. In fact, studies suggest that the Alps are expected to maintain their carbon sink capacity at least for the first half of the 21st century, while increasing respiration rates and natural disturbances at low elevation sites may decrease thus making Alpine forests a carbon source [Karjalainen et al., 2002, Thürig et al., 2005, Zierl and Bugmann, 2007].

• The *social and economic functions* of forests include other forest services such as biodiversity, fresh and clean air water supply, or landscape that are finally relevant for the provision of other cultural and tourist-related economic ecosystem services potentially bearing site-specific positive externalities. The value of these positive externalities is often not disclosed on markets as they are generally public goods, but their importance is crucial for the livelihood of communities inside and outside the mountain regions [TEEB, 2010; FAO, 2011]. Hence, their assessment is fundamental for the implementation of rational public choices in forest management adaptation planning.

4. Adaptation strategies for the Alpine forest sector

The Alpine region as a whole can be considered as a particularly vulnerable area to the impacts of climate change, both physically and socially, with a clear need to build resilience through improved adaptation capacity. A strong need for adaptation has emerged from several scientific and policy-oriented sources (including the Climate Action Plan adopted by the Alpine Convention in Evian, 2009), experiences conducted in the Alpine countries and in other literature (EEA 2009, OECD 2007). In particular the necessity to focus and undertake cooperative research on the adaptation side of mountain forests in the Alps has been increasingly recognized as a central strategic topic in the framework of the Territorial Cooperation goals of the "Alpine Space Programme". As recalled above, the MANFRED Project, for example, directly addresses the topic of adaptation in Alpine forests to the impacts of climate change. In short, the project was focused on the assessment of potential future climate and land use dynamics' patterns, on the related impacts of them on abiotic and biotic hazard factors, on the historical frequency of and statistics on extreme events and related management practices in the Alpine forests, and on the sharing of knowledge about protection forests engineering in the Alps. All these fields of activity are expected to contribute to the definition of shared guidelines for sustainable forest adaptive management at the Alpine level, also based on the direct involvement of relevant stakeholders and the sound consideration of their preferences during the multiple stages of the project implementation.

Some forest-based adaptation activities have been recently identified as including monitoring and maintaining forest health, vitality and diversity; implementing integrated forest fire management; enhancing landscape connectivity and reducing forest fragmentation; monitoring and removing invasive species and addressing pest and disease threats; implementing reduced-impact logging; selecting appropriate species for use in planted forests; undertaking forest restoration and rehabilitation, particularly on slopes, through specific measures and tools that can facilitate the adaptation process [Ciccarese et al. 2012].

However, the concrete design and implementation of climate change adaptive strategies as part of sustainable forest management requires making policy choices between sets of different adaptation options often implying potential co-benefits and trade-offs between environmental, social and economic functions [Spittlehouse and Stewart, 2003, FOREST EUROPE 2011].

Moreover, choices between adaptation measures have to be taken in a long-term perspective - as the long life-span and growing conditions of trees does not allow for rapid adaptation to environmental change in a context of temperature increase in the Alps over the last century that is already twice the global average (about 1.5°C), with a slightly growing tendency at higher altitudes [Houghton et al., 2001, Casty et al., 2005, IPCC, 2007]

Nevertheless, there is still a high degree of scientific uncertainty about regional future climate conditions and how they will impact on forests health and growing patterns [Roetzer, 2005; IPCC, 2007, Lindner, Kolström 2008]. For forest sector policy-making to be effective, reliable information has to be available: collaborative, interdisciplinary research has to be further developed in order to provide policy-makers and relevant stakeholders inside and outside the forest sector with the necessary evidence-based tool for rational decision-making [FOREST EUROPE 2011, FAO 2011]. Furthermore, research demonstrated the ability to develop sensible decision support systems for different sectors, that can be applied in different regions (some examples for the Alps: Climalptour, SHARE, etc.).

5. Policy framework and perspectives for policy building

The actual use of forests can be socially and historically rooted in each considered area [Agnoletti and Anderson a) and b), 2000, Williams, 2006]. Different forest uses and policies can bring different contributions to regional economic development trends, by producing variable degrees of income and wealth from forest marketed and non-marketed services [FOREST EUROPE 2011].

The historical persistence of some forest uses observed locally and the priorities set by policy makers at different levels (e.g. increasing wood supply and mobilization are common to several European, Alpine and non-Alpine regions' forest policy) suggest to assign different relative weights to the contribution that each forest function supplies to the economy and society in different countries and sub-regions.

In turn, this discretionary and site-specific weighting is likely to impact on the actual mix of adaptation measures that regional governments will support, incentivize and implement.

Thus, when selecting sector-specific adaptation measures, decision makers are called to inspect:

• the physical and ecological features of the region that participate in determining its inherent adaptive capacity [Lindner and Kolström 2008, EEA 2009],

• the composition and historical roots of the regional economy – where some economic indicators for the profitability of forestry have been calculated for EU forests [Kovalcik 2011],

• the priorities set (or to be set) at different political levels in the region under inquiry [Lindner and Kolström 2008].

From this situation, a few guiding principles for the definition of suitable adaptation policies can be extracted:

• climate change impacts on the forest functions can assume a variable social and economic weight, on the basis of some situational variables, which may include the structure of the economy, the forest sector contribution to regional GDP, the regional labor market structure, the social consequences of policies and trends observed in the forest sector, the functions and services provided by forest ecosystems, and their formal understanding from the scientific community;

• public awareness existing on climate change impacts at the local level is a key factor to be taken into account when defining adaptation policies. In particular the perceptions of policy makers, of the scientific community, of qualified stakeholders and of the general public on the relative weights and role of the climate change-affected forest functions at the regional level should be attributed a specific policy focus. Namely, particularly relevant is the dissemination of the available information to user groups and forest owners [Lindner and Kolström 2008, FAO 2011];

• considerations on, and choices about trade-offs and co-benefits eventually deriving from adaptation measures in the forest sector are likely to be context-specific, i.e. dependent on the framework conditions observed in the region at the economic, social, ecological and policy-level.

Current figures reveal that in Europe the average share of forestry in the GDP is modest and corresponds on average to 0,31%, with slightly higher values in Austria and Slovenia. [FOREST EUROPE 2011]. At the same time, entrepreneurial income per hectare shows very variable values across EU [Kovalcik, 2011].

In such a context, the regions where forestry and forest policies have only marginally developed, as it is the case also with some Alpine countries, phenomena of policy change and "paradigm shift" towards a comprehensive adaptation strategy to climate change and a concomitant sustainable forest management may be triggered only by a composite set of factors.

Among them, the following can be recalled: the results of scientific research, the increased understanding of the physical mechanisms behind forest functions, the novel knowledge available on the possible ecological, economic and social impacts of climate change, and the growth opportunities for the forest sector which have been identified at the economic and policy-analysis level, also in the framework of the 2012 United Nations Conference on Sustainable Development (UNCSD) in Rio de Janeiro, that explicitly addressed the theme of "Green economy in the context of sustainable development and poverty eradication [OECD 2011, UNEP 2011, UNECE 2012]. Conversely, the process of change in Alpine forest management may be hindered by the persistence of a high degree of regional vulnerability.

6. Conclusions

Building resilience to climate change in the Alpine forest sector by implementing suitable adaptation policies can therefore facilitate the sector's innovation and change. The process of policy change in addressing the challenges of the forest sector can be read as the consequence of an iterative *feedback* impacting on the framework conditions of the sector and the regional economy at large. This may stimulate a learning process consisting in information feedbacks being able to revise the behavioral model of decision making [Sterman, 1994], and potentially inducing increasing modifications in forest management.

In conclusion, such a feedback effect can be supported by several factors and conditions, that participate in determining the intensity and potential for change of the forest policies, including:

- the presence (or absence) of already openly defined policy goals (policy stability of the forest sector);

- the existing stakeholders' claims and expectations;

- the expected impacts of climate change on the regional economy and society, and their relative weights;

- the resilience of Alpine forests to climate change and their actual and potential contribution to societal adaptation [Innes et al., 2009];

- the possible social and economic benefits that can derive from a more efficient forest management and target-oriented expenditure (e.g. reflecting in higher entrepreneurial revenues and employment levels, differently from the present situation) [FOREST EUROPE, 2011];

- the resulting attractiveness (in terms of expected economic, ecological and societal benefits) of the "paradigm shift" itself for key regional stakeholders and policy makers.

Author details

Luca Cetara[1] and Federico Mannoni[2]

1 Accademia Europea di Bolzano, Bolzano, Italy

2 Istituto per le Piante da Legno e l'Ambiente – IPLA, Torino, Italy

References

[1] Agnoletti, M, & Anderson, S. eds.). Forest History: International Studies on Socioeconomic and Forest Ecosystem Change. In: IUFRO Research Series #2. New York, NY: CABI Publishing. (2000a). , 418.

[2] Agnoletti, M, & Anderson, S. eds.). Methods and Approaches in Forest History. In: IUFRO Research Series #3. New York, NY: CABI Publishing. (2000b). , 281.

[3] Alpine ConventionProtocol on the Implementation of the Alpine Convention relating to Mountain Forests: "Mountain Forest" Protocol. (1996). http://www.alpenkonvention.org/en/convention/protocols/Documents/protokoll_bergwaldGB.pdfaccessed 02 May 2012)

[4] Alpine ConventionAction Plan on Climate Change in the Alps: AC_X_B6_fr. ; 2009 http://www.alpconv.org/en/ClimatePortal/actionplan/Documents/ AC_X_B6_en_new_fin.pdfaccessed 24 April (2012).

[5] Brang, P, Schönenberg, W, Frehner, M, Schwitter, R, Thormann, J, & Wasser, B. Man-agement of protection forests in the European Alps: an overview. Forest, Snow and Landscape Research (2006).

[6] Burton, I, Huq, S, Lim, B, Pilifosova, O, & Schipper, E. L. From impacts assessment to adaptation priorities: the shaping of adaptation policy. Climate Policy (2002).

[7] Campbell, A, Kapos, V, Scharlemann, J. P. C, et al. Review of the literature on the links between biodiversity and climate change: impacts, adaptation and mitigation,. Secretariat of the Convention on Biological Diversity. Montreal: Technical Series n. 124; (2009).

[8] Casty, C, Wanner, H, Luterbacher, J, Esper, J, & Böhm, R. Temperature and Precipita-tion Variability in the European Alps Since 1500. International Journal of Climatolo-gy (2005)., 1855-1880.

[9] Ciccarese, L, Mattsson, A, & Pettenella, D. Ecosystem services from forest restoration: thinking ahead. New Forests International Journal on the Biology, Biotechnology, and Management of Afforestation and Reforestation 2012. http://www.scribd.com/doc/54145282/European-Forestsaccessed 3 June (2012).

[10] European CommissionGreen Paper on Forest Protection and Information in the EU: Preparing forests for climate change. COM ((2010). final.

[11] European CommissionCommunication from the Commission to the Council and the European Parliament on an EU Forest Action Plan. COM((2006).

[12] European Environment AgencyRegional climate change and adaptation. The Alps facing the challenge of changing water resources. Copenhagen: EEA Report (2009). and 51-53.(8), 18-19.

[13] FAOThe state of the world's forests report, Rome: FAO; (2011).

[14] FOREST EUROPEUNECE, FAO. State of Europe's forests (2011). status and trends in sustainable forest management in Europe: conference proceedings. 14.16 June. 2011, Oslo, Ministerial Conference on the Protection of Forests in Europe (FOREST EU-ROPE)

[15] Guariguata, M. R, Cornelius, J. P, Locatelli, B, Forner, C, & Sanchez-azofeifa, G. Miti-gation needs adaption: Tropical Forestry and climate change. Mitigating Adaptive Strategies for Global Change (2008).

[16] Hemmati, M. Multi-Stakeholder Processes for Governance and Sustainability, Be-yond Deadlock and Conflict, London: Earthscan; (2002).

[17] Houghton, J. T, Ding, Y, Griggs, D. J, Noguer, M, Van Der Linden, P. J, Dai, X, Mas-kell, K, & Johnson, C. A. Climate change 2001: The scientific basis. Cambridge: Cam-bridge University Press; 2001. http://www.grida.no/climate/ipcc_tar/wg1/index.htmaccessed 18 May (2012).

[18] Innes, J, Joyce, L, Kellomaki, M, et al. Management for adaptation. In: Seppälä R., Buck A., Katila P. (eds). Adaptation of forests and people to climate change. Vienna: International Union of Forest Research Organizations IUFRO world series: (2009). , 135-169.

[19] IPCC- Intergovernmental Panel on Climate ChangeSummary for Policymakers.. In: Parry M.L., Canziani O.F., Palutikof J.P., van der Linden P.J. and Hanson C.E. (Eds.). Climate Change 2007. Contribution of Working Group II to the Fourth Assessment Report of the Intergovernmental Panel on Climate Change. Impacts, Adaptation and Vulnerability. Cambridge: Cambridge University Press; (2007). , 7-22.

[20] IPCC- Intergovernmental Panel on Climate ChangeWorking Group II Report: Impacts, adaptation and vulnerability: food, fibre and forest products. Cambridge: University Press; (2007).

[21] Karjalainen, T, Pussinen, A, Liski, J, Nabuurs, G. J, Erhard, M, Eggers, T, Sonntag, M, & Mohren, G. M. J. An approach towards an estimate of the impact of forest management and climate change on the European forest sector carbon budget: Germany as a case study. Forest Ecology and Management: (2002). , 87-103.

[22] Kovalcic, M. Profitability and competitiveness of forestry in European countries. Journal of Forest Science (2011). , 369-376.

[23] Lindner, M, & Kolström, M. Impacts of climate change in European forests and options for adaptation, Report to the European Commission Directorate-General for agriculture and Rural Development; 2008. http://ec.europa.eu/agriculture/analysis/external/euro_forests/full_report_en.pdfaccessed 3 March (2012).

[24] Munari, F. Tutela Internazionale dell'Ambiente. In: Carbone S.M., Luzzato R. Santa Maria L. Istituzioni di Diritto Internazionale. Torino: Giappichelli; (2006). , 406-445.

[25] OECDTowards Green Growth. Paris: (2011). http://www.oecd.org/dataoecd/37/34/48224539.pdfaccessed 26 February 2012).

[26] Perugini, L, Vespertino, D, & Valentini, R. Conferenza di Durban sul clima: nuove prospettive per il mondo forestale. Forest@ 2012; http://www.sisef.it/forest@/contents/?id=688accessed 5 April (2012). , 9(1), 1-7.

[27] Price, M. F, et al. Mountain Forests in a Changing World- Realizing Values, addressing challenges. Rome: FAO/MPS and SDC; (2011).

[28] Roetzer, T. Climate change, stand structure and the growth of forest stands. Annalen der Meteorologie (2005). , 40-43.

[29] Somlai, I. G. Identifying stakeholders: Approach to Social Forestry Conflicts. International Journal of Social Forestry (2008). , 83-95.

[30] Spittlehouse, D. L, & Stewart, R. B. Adaptation to climate change in forest management. BC Journal of Ecosystems and Management (2003). , 1-11.

[31] Sterman, J. D. Learning in and about complex systems. System Dynamics Review (1994). , 291-330.

[32] TEEBThe Economics of Ecosystems and Biodiversity: Mainstreaming the Economics of Nature: A synthesis of the approach, conclusions and recommendations of TEEB, 2010.   http://www.teebweb.org/Portals/25/TEEB%20Synthesis/TEEB_SynthReport_09_2010_online.pdfaccessed 15 April (2012).

[33] Thuerig, E, Palosuo, T, Bucher, J, & Kaufmann, E. The impact of windthrow on carbon sequestration in Switzerland: a model-based assessment. Forest Ecology and Management (2005). , 337-350.

[34] UNCEDReport of the United Nations Conference on Environment and Development 1992: Conference proceedings, June Rio de Janeiro, Annex III, 2b. http://www.un.org/documents/ga/conf151/aconf15126-3annex3.htmaccessed 20 April (2012). , 3-14.

[35] UNEPTowards a Green Economy: Pathways to Sustainable Development and Poverty Eradication. Nairobi: United Nations Environment Programme; 2011. www.unep.org/greeneconomyaccessed 15 April (2012).

[36] Williams, M. Deforesting the Earth: From Prehistory to Global Crisis, An Abridgment. Chicago: University of Chicago Press; (2006). p xviii and 543.

[37] Zierl, B, & Bugmann, H. Sensitivity of carbon cycling in the European Alps to changes of climate and land cover. Climatic Change (2007). , 195-212.

Chapter 5

Risk Assessment for Biotic Pests Under Prospective Climate Conditions

Holger Griess, Holger Veit and Ralf Petercord

Additional information is available at the end of the chapter

1. Introduction

Mountain ecosystems are among the most species-rich sites on the planet and are particularly sensitive to climatic changes. Although the Alps are certainly the best-studied mountain region in the world, long-term studies of species distribution are still hard to find. In particular, there are no comparative studies covering the Alpine region as a whole. Climate change will affect the mountain forests, on the one hand through direct impacts such as longer drought periods, increased storm damage risk, and on the other hand, through indirect effects such as the development of forest pests. Temperatures in the European Alps have increased twice as much as the global average temperature since the late nineteenth century and are predicted to rise by an average of 0.3-0.5°C per decade in the next century. The effects of a rising temperature will influence the occurrence and the impact of biotic risks. A suite of biotic and abiotic factors, ranging from climatic variables and edaphic conditions, to competition, predation and resource distribution, have been invoked to explain the limits in species distributions. Still, our understanding of the spatial dynamics at range limits and their underlying factors is far from complete. Our ability to predict the ecological and evolutionary responses of organisms to climate change requires understanding of the mechanistic links between climate and geographical range limits. The consequences of climate change may involve expansion, contraction, or shifts in species range, and some species may migrate to new areas and exploit previously unutilized resources. The latter situation can only occur if conditions in the new part of the range have been or have become suitable; for example, global warming can allow temperature-limited species to colonize latitudes or altitudes beyond their original distribution. Within this Project we concentrated our efforts on the following two questions:

Will the effects of climate change promote a migration of forest pests from south to more northerly alpine regions?

Does the changing climate increase the habitat suitability for quarantine organisms?

2. Forest pest migration — The pine processionary moth

An increasing number of taxa are undergoing significant range shifts in response to human-assisted dispersal and changes in environmental factors, notably. Often these range shifts are into novel environmental space, from both biotic and abiotic perspectives.

Over the last century, climatic isotherms over Europe have been displaced by an average of 120 km northwards. A latitudinal and altitudinal expansion has been documented in several important forest pest species in the northern hemisphere, for instance Thaumetopoea pityocampa (pine processionary moth) in central Europe. The pine processionary moth is a monovoltine conifer defoliator, widespread in the Mediterranean region. Adults emerge in the summer, mate and lay eggs within 2–4 days. The gregarious larvae hatch after ca. 1 month and their development progresses until late winter, with the larvae sheltering inside self-spawn silk nests. Pupation takes place in the soil, where the pupae undergo an obligatory diapause. *T. pityocampa* exhibits a variable phenology across its range of distribution, with late adult emergence (August–September) and early pupation (January–February) occurring in the warmest regions, whereas early adult emergence (June– July) coupled with late pupation (March–April), is observed in populations colonizing the northernmost, or high-altitude areas.

In the last three decades, substantial expansion of the range of this economically important pine defoliator has taken place both latitudinal and altitudinal. In the Paris Basin the range boundary shifted by 87 km northwards between 1972 and 2004; in the Italian alpine region, an altitudinal shift of 110 – 230 m upwards occurred between 1975 and 2004.

3. Increase habitat suitability — The emerald ash borer

The emerald ash borer, *Agrilus planipennis*, is one of the most feared beetles. It has the highest quarantine status. National Plant Protection Organizations (NPPOs) around the world continuously inspect their indigenous *Fraxinus* trees and imported wood products on the presence of this beetle because its presence usually results in the destruction of *Fraxinus* trees.

Agrilus planipennis belongs to the family of the Buprestidae, also called metallic wood-boring beetles. Buprestids are relatively small, elongated beetles. The emerald ash borer measures approximately 7.5-15 mm. Its larvae measures about 15-30 mm. Females lay eggs on the bark of the host trees in May-June. One female lays about 65-90 eggs during her lifetime. Eggs hatch and the larvae bore to the phloem area just behind the bark, where they create serpentine tunnels. The development through four larval stages takes about 1-2 years depending on the vigor of the host trees in temperate zones. Fourth instar larvae excavate chambers either in the bark or slightly in the sapwood, where they become prepupa in September-October. Most of the population overwinters in this stage, although some individuals overwinter as earlier-instar larvae. In these chambers, they develop into pupae during spring, and emerge as adults in May/June. The adults live approximately 2-3 weeks, during which they disperse no more than several kilometers. Host plants of *A. planipennis* are almost exclusively *Fraxinus* trees: in northern America they include *F. americana, F. chinensis, F. japonica, F. lanuginose,* and *F. nigra,* but the emerald ash borer was also found in *Ulmus davidiana* var. *japonica, Ulmus propinqua, Juglans mandshurica* var. *sieboldiana,* and *Pterocarya rhoifolia*. In Europe host trees are *F. excelsior, F. pennsylvanica,* and possibly *F. angustifolia.*

The native distribution of *A. planipennis* includes Mongolia, central, eastern, and north-eastern China, Taiwan, both Koreas and Japan. In July 2002 the species was found in Northern America: specimens were identified in south-eastern Michigan, USA. By March 2009, emerald ash borer infestations had been found in a total of ten states and two Canadian provinces. Estimates indicate that more than 53 million native ash trees had been killed by *A. planipennis* in those states by 2007.

It is believed that the species entered the USA at Detroit, in freight from cargo ships. In Europe, the species was found in Moscow, Russia in 2005 and seems to be spreading. In city squares or along railway tracks many *Fraxinus* trees are declining or dying. In some places, 70-80% of the *Fraxinus* trees have lost most of their foliage. The pathway of this introduction is not known. The species has not been found in other areas of Europe.

The species damages infested trees because the larvae bore serpentine tunnels in the cambial layer and the inner bark (phloem), disrupting, or even completely cutting off the nutrient flow from the leaves to the roots. During the early stage of an infestation, when *A. planipennis* population is low, the initial damage is low. However, after 2 to 3 years of continuous infestation, the population builds up, and eventually the tree's nutrient and water transport system is disrupted, culminating in wilting and eventual tree mortality. *A. planipennis* will kill apparently healthy trees during high beetle population levels which are probably triggered by a few years of hot and dry climatic conditions. *A. planipennis* can cause severe damage to ash stands over eight years of age that are not crown-closed, with good sun light penetration, and that are comprised of trees with bark fractures. After one to two years of infestation, the bark often falls off in pieces from damaged trees thereby exposing the tunnel-ridden sapwood. In the most severe cases, entire stands may be destroyed. In its natural area, *A. planipennis* population dynamics are balanced due to natural enemies: parasitoids, predators, and entomopathogenic fungi. Apparently, the absence of most of these natural enemies in the 'new' areas has an enormous impact on the population dynamics, allowing build-up of *A. planipennis* populations to such high levels as to become a true pest.

Other symptoms that may indicate an *A. planipennis* infestation and may be seen from the outside, include crown dieback/chlorosis, epicormic shoots, increased woodpecker and squirrel feeding, bark deformities, foliage feeding, and exit holes. The exit holes are D-shaped of 3.5-4.1 mm in diameter. A clear sign of an *Agrilus* sp. infestation that cannot be seen from the outside is the presence of typically serpentine shaped larval galleries, mostly filled with shavings. They can easily be found by peeling away the bark. Once the signs and symptoms become apparent, the trees are often severely infested. The infestation may have spread to surrounding areas where signs are not yet apparent. Based on the infestation symptoms, examination techniques applied by inspectors should include visual inspection, crown survey, and bark peeling. The most effective way to control and eradicate an outbreak of *A. planipennis* is by felling and chipping or burning infested trees. This should be done in the infested and as a preventive measure in the surrounding area.

The fact that ashes are very common in the alpine region and that the species has already entered Eastern Europe poses a serious phytosanitary risk.

4. Species distribution model

Global climate change is ever more evident. Consequently, geographic areas corresponding to biomes, ecosystems and species' ecological niches are changing, which is likely to affect the distribution of many species. Species distribution modeling can be used to provide a rapid evaluation of the potential impact of climate change on the distribution of ecosystems and the species that inhabit them. The process consists of detecting changes in species distribution by comparing the potential distribution areas in the current climate with the potential distribution areas, based on a species' current climate preferences, under future climatic conditions. Future potential distribution areas of occurrence are identified using climate layers based on the projections of General Circulation Models (GCM). Climate research institutions from various countries generate these which predict future climatic conditions under different emission scenarios developed by the Intergovernmental Panel on Climate Change (IPCC).

The environmental data for the global climatic conditions was obtained from the WorldClim database, which provides 19 layers of bioclimatic variables (Table 1) compiled from monthly data. The data for the present day conditions was collected from 1950 to 2000 with a spatial resolution of 30 arcsec (1 km^2). For the niche models under future conditions WorldClim data was obtained for the years 2020, 2050 and 2080. Prior to use, the WorldClim grids were fitted in Arc-GIS to the extent of the natural and invaded range. This allows the projection of the models into the area of interest. In addition to the climatic data we added altitude data for the whole range.

A commonly employed method for risk assessment with regard to range shifts and range extension of invasive species is climate-based ecological niche modeling. It has been used to predict potential ranges for recent forest invaders in North America e.g. the Asian longhorn beetle However, despite its possible ecological and economic impact there is no climate-based model of the potential large-scale distribution of the emerald ash borer in the alpine region.

To determine the potential geographic distribution for *T. pityocampa* and *A. planipennis* in the alpine region we used ecological niche modeling procedures. Ecological niche modeling involves generating a set of predictive rules that describe relationships between a species georeferenced occurrence and a suite of spatially explicit environmental variables, most commonly bioclimatic factors such as temperature and precipitation.

The basis for spatial analysis of species distribution is occurrence data. Occurrence data is a snapshot of species in time and space. To analyze occurrence data, we used the data provided

Bioclimatic Variable	Description
BIO1	Annual Mean Temperature
BIO2	Mean Diurnal Range (Mean of monthly (max temp - min temp))
BIO3	Isothermality (BIO2/BIO7) (* 100)
BIO4	Temperature Seasonality (standard deviation *100)
BIO5	Max Temperature of Warmest Month
BIO6	Min Temperature of Coldest Month
BIO7	Temperature Annual Range (BIO5-BIO6)
BIO8	Mean Temperature of Wettest Quarter
BIO9	Mean Temperature of Driest Quarter
BIO10	Mean Temperature of Warmest Quarter
BIO11	Mean Temperature of Coldest Quarter
BIO12	Annual Precipitation
BIO13	Precipitation of Wettest Month
BIO14	Precipitation of Driest Month
BIO15	Precipitation Seasonality (Coefficient of Variation)
BIO16	Precipitation of Wettest Quarter
BIO17	Precipitation of Driest Quarter
BIO18	Precipitation of Warmest Quarter
BIO19	Precipitation of Coldest Quarter

Table 1. Structure of bioclimatic data used for modeling

by our project partners and publicly available data provided through international platforms. Much of the available data are historical observations which may not reflect the current presence of taxa due to recent ecological processes or human interventions, such as forest conversion to agriculture and other changes in land use. Nonetheless, such data is still useful to gain insights into the ecological processes behind the geographical distribution of species distribution. Observation data must be organized following the format specified by the applied niche modeling software. The analyses can be based on presence or absence data or a combination of both. Basic data consists of a taxonomic identification of an observed individual and the location of the collection or observation site. Presence points may be associated with additional data describing the date of collection, source of coordinates. Information about the collection date provides a time dimension to the analyses and can be used to analyze trends in species distribution.

The opposite of presence points is absence data. While absence data can also be relevant in spatial analyses, e.g. monitoring trends in species distribution, it is often challenging to understand concrete reasons for the absence of taxa in a geographic unit, complicating the use of this data in ecological analyses. Therefore, the ecological niche modeling for this project is based on presence points only.

Often, the available set of presence data does not cover the entire range of a species' natural distribution. Building a reliable ecological niche model can be problematic, since only limited information is available on the native range and out of range occurrence of the species. Sources claim that e.g. the emerald ash borer may be native to northeastern China, however documented observations and museum specimens are scarce. It was possible to identify 496 georeferenced occurrence records for *A. planipennis* and 169 for *T. pityocampa* from various sources. Species distribution modeling programs such as Maxent (PHILLIPS et al. 2004, 2006) enable us to approximate the full distribution range. These programs are practical tools to identify those areas where a species is likely to occur. The results of the species distribution modeling analysis can be used for different combined spatial analyses, e.g. evaluating the impact of climate change on the future distribution of species. Species distribution modeling programs identify sites with similar environments to those where a species has already been observed as potential occurrence areas. The data required to identify these potential distribution areas include species presence points as well as the raster data of environmental variables covering the study area. First, a niche is defined based on the environmental values that correspond to the presence points used in the analysis. Then, the similarities between the environmental values at a specific cell in the data raster and those of the niche of the modeled species are calculated for each raster cell in the study area. With this information, the model calculates the probability of a species occurrence in each raster cell. Even though the next analysis is based on climate data, Maxent also allow us to include other types of variables in the model such as altitude.

Maxent calculates the species' realized niche and probability of occurrence using an algorithm for maximum entropy. Since Maxent has fared well in evaluations, in comparison to other niche modeling programs, it is the program of choice in this project.

Maxent is a general-purpose method for making predictions or inferences from incomplete information. Its origins lie in statistical mechanics, and it remains an active area of research with an Annual Conference, Maximum Entropy and Bayesian Methods, that explores applications in diverse areas. The idea of Maxent is to estimate a target probability distribution by finding the probability distribution of maximum entropy (i.e., that is most spread out, or closest to uniform), subject to a set of constraints that represent our incomplete information about the target distribution. The information available about the target distribution often presents itself as a set of real-valued variables, called "features", and the constraints are that the expected value of each feature should match its empirical average (average value for a set of sample points taken from the target distribution). When Maxent is applied to presence-only species distribution modeling, the pixels of the study area make up the space on which the Maxent probability distribution is defined, pixels with known species occurrence records constitute the sample points, and the features are climatic variables, elevation and functions thereof. Maxent offers many advantages. It requires only presence data, together with environmental information for the whole study area. It can utilize both continuous and categorical data, and can incorporate interactions between different variables. Efficient deterministic algorithms have been developed that are guaranteed to converge to the optimal, maximum entropy, probability distribution. The Maxent probability distribution has a concise mathematical definition, and is therefore amenable to analysis. For example, as with generalized linear and generalized additive models (GLM and GAM), in the absence of interactions between variables, additivity of the model makes it possible to interpret how each environmental variable relates to suitability. Over-fitting can be avoided by using regularization. The output is continuous, allowing fine distinctions to be made between the modeled suitability of different areas. If binary predictions are desired, this allows great flexibility in the choice of threshold. Maxent is a generative approach, rather than discriminative, which can be an inherent advantage when the amount of training data is limited.

It is important to realize that when a geographical area shows environmental conditions favorable for a species, this does not necessarily mean that the species actually occurs in this area.

5. Results

The range of the pine processionary moth under present day climatic conditions in the Alpine region is illustrated in Figure 1. The habitat suitability is very low across the Alpine arc. Only two departments in France show a higher suitability. In the department of Rhône-Alps areas with low habitat suitability are evident the westernmost parts. A low to medium habitat suitability can be observed in the southern part of the department of Provence-Alpes-Côte D 'Azur.

The habitat suitability under possible climate conditions in the future is pictured in Figure 2. The overall habitat suitability is still very low across the Alpine arc. The area with a low suitability has increased in the departments of Rhône-Alps and Provence-Alpes-Côte D 'Azur. Yet for the latter department the suitability decreased from medium-low to low.

Under present day climate conditions our models show very low habitat suitability across the Alpine region for the emerald ash borer. The exception is a small area in the east of the department Rhône-Alps where low habitat suitability can be observed (Figure 3).

The results of our model for the habitat suitability of the emerald ash borer under conditions influenced by climate change are illustrated in Figure 4. In comparison to the present day condition we can observe a significant increase of area with low and medium habitat suitability. This increase in area is concentrated in the German and Austrian part of the Alps.

6. Discussion

With the results of our models we can now return to the beginning and answer the questions.

- Will the effects of climate change promote a migration of forest pests from south to more northerly alpine regions?

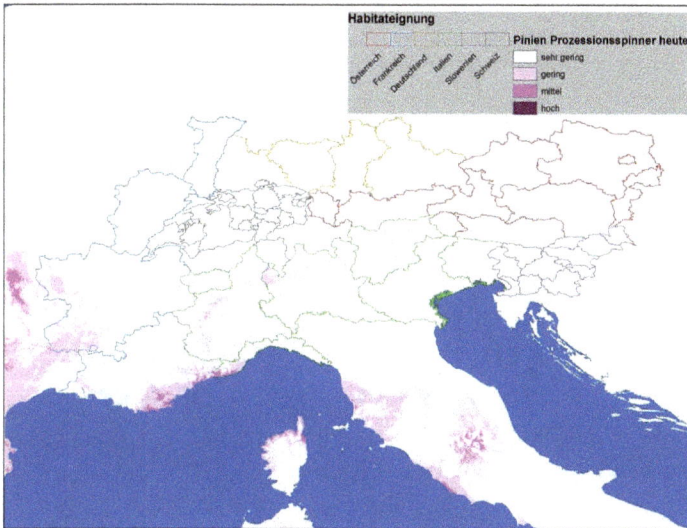

Figure 1. Potential habitat suitability Pine Processionary Moth in the Alpine region under present climatic conditions

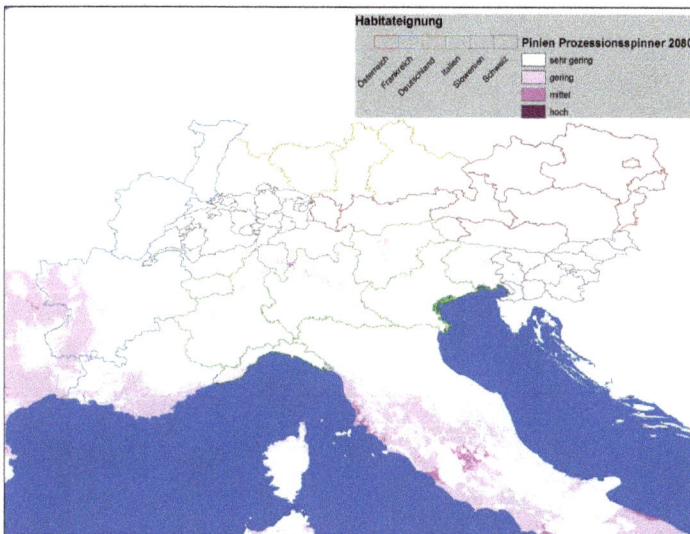

Figure 2. Potential habitat suitability Pine Processionary Moth in the Alpine region under possible climatic conditions in 2080

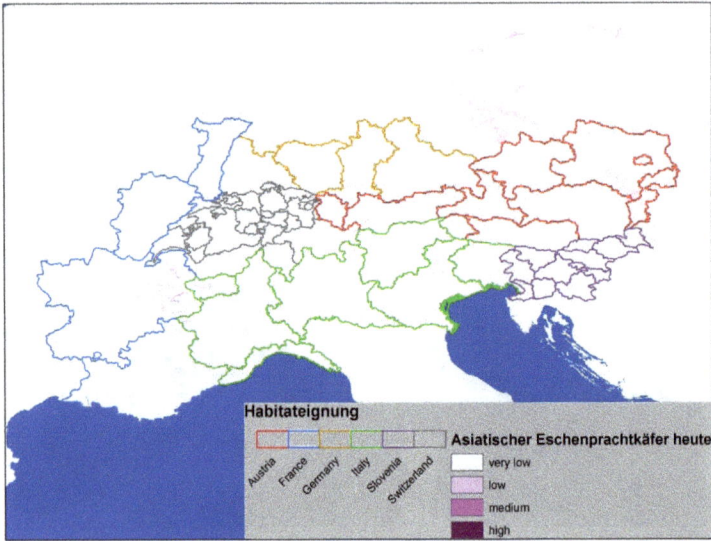

Figure 3. Potential habitat suitability Emerald Ash Borer in the Alpine region under present climatic conditions

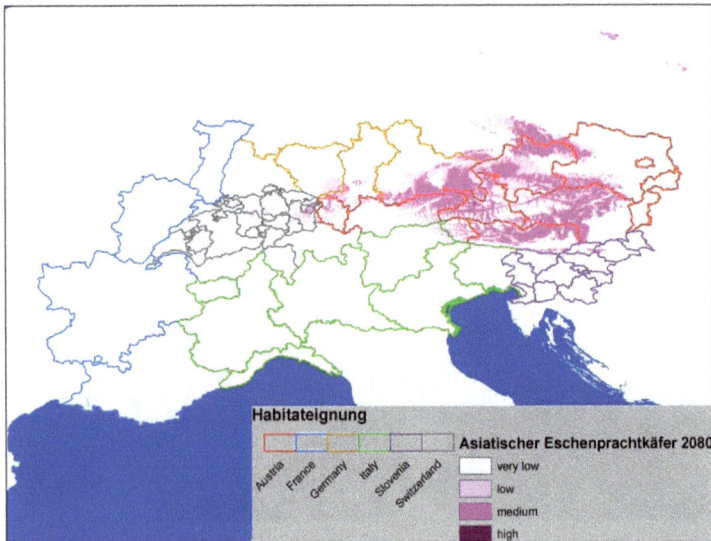

Figure 4. Potential habitat suitability Emerald Ash Borer in the Alpine region under potential climatic conditions 2080

• Does the changing climate increase the habitat suitability for quarantine organisms?

In case of the pine processionary moth as an example for a forest pest with potential to migrate from south to north, our models show that a migration is possible under changing climate conditions in the future. Yet the range enlargement is not as big as other studies, which base their models solely on a rise in temperature, imply. Our models show that the main factors for a possible range enlargement for *T. pityocampa* are the temperature seasonality, the annual range of temperature and the mean temperature in the coldest quarter. To better understand why these factors are important in the habitat suitability of *T. pityocampa* we need to cross validate them biology of *T. pityocampa*. The pine processionary moth has been worked on in many scientific papers, the knowledge of the biology of it is still limited. We need to increase that knowledge in order to make the results for the species distribution modeling for the pine processionary moth more precise.

Even if we can't see a dramatic change according to our model, a higher impact of thaumetopoea processionea is still possible. For sites with less favorable climate conditions, like southern slopes in the alpine region, an infestation with *T. pityocampa* might have severe consequences. These sites are usually so dry that water is a limiting factor for tree growth. Pines are the only trees which are able to compete under these conditions. The pines are under constant water stress which makes the susceptible to pests. An infestation with T. pitycampa might lead to a destabilization of the whole stand and may cause the loss of forest functions in that area.

The emerald ash borer is a feared quarantine pest organism in Europe. An occurrence and spread can have serious ecological and economic consequences. Our models show that for the northern part of the Alps a changing climate will lead to a more suitable habitat. A more suitable habitat for the emerald ash borer increases the risk of an outbreak after an initial infection with this quarantine organism.

Due to the inner European trade routes and tourism the alpine region has a high road density. The number of vehicles which are traveling across the whole alpine space is increasing every year. Traffic is a possible pathway for an infection with the emerald ash borer. An increase in traffic in the future combined with higher habitat suitability due to climate change will lead to high possibility of the occurrence and outbreak of the emerald ash borer in the alpine region.

In order to decrease the chance of a possible outbreak a better monitoring system in the areas with a higher suitability is needed. Within the Manfred-Project an internet based monitoring platform has been developed. This platform allows a quick exchange of information between the forest protection specialists in the alpine region. This helps to spread the information about a detected infection in the earliest possible stage and minimizes the time to take counteractions against the thread.

Author details

Holger Griess[1], Holger Veit[2] and Ralf Petercord[1]

1 Landesanstalt für Wald und Forstwirtschaft Bayern (LWF), Freising, Germany

2 Forstliche Versuchs- und Forschungsanstalt Baden-Württemberg (FVA), Department of
Biometrics, Freiburg im Breisgau, Germany

References

[1] Phillips, S. J., Dudik, M. & Schapire, R.E. 2004. A maximum entropy approach to species
    distribution modeling. Pages 655-662 in Proceedings of the 21st International Confer-
    ence on Machine Learning. ACM Press, New York

[2] Phillips, S. J., R. P. Anderson, and R. E. Schapire. 2006. Maximum entropy modeling of
    species geographic distributions. Ecological Modelling, 190:231-259.

Chapter 6

Abiotic Stressor: Storms

Bin You and Mitja Skudnik

Additional information is available at the end of the chapter

1. Introduction

Storms have been the predominating natural disturbance that seriously affected the forest ecosystems in Europe. Storms tend to occur more frequently and intensively, leading to more serious damages in recent decades. Over 130 separate storm events that induced noticeable damages to forests were identified during 1950 – 2010 [1]. Based on the percentage of growing stock that were damaged, 11 most catastrophic windthrows were identified: January 1953, September 1967, September 1969, November 1972, October 1987, January-March 1990, December 1999, November 2004, January 2005, January 2007, January 2009. The total damage by various disturbances to Europe forests is illustrated in Figure 1[3], from which it can be easily indentified that windthrows is the most significant nature disturbances. As regard to seasonal distribution of storm events, winter storms are more common than summer storms. The overwhelmingly majority of storm damages were found between November and January. It is also important to notice that some storm events might not make significant damage in large areas, but could also be extremely devastating at local level, causing catastrophic losses to local forest owners at a certain region.

2. Historical storm damage in Alpine space member states

Data on timber harvest records from four countries (Austria, Germany (Baden-Württemberg), Italy (Bolzano) and Slovenia) were used in order to determine how in a recent history the storms affected forests of the Alpine region. Available data dates back to year 2000 till 2010 and includes information on volume of trees that were logged during the specific year because of storm (wind) damages (LIT: see regional reports for details). Those data are usually related with forest management units but for this analysis the data were summarized to NUTS 3 level [2]. The common time span for all four countries was from 2002 to 2008. The data presented

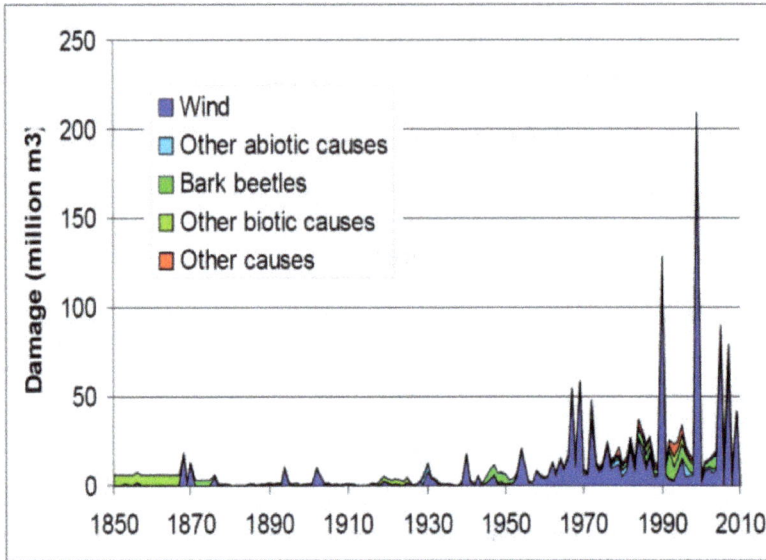

Figure 1. Total damamges to European forests (Retrieved from Schelhaas 2008a)

in the graph (Figure 2) shows the volume of trees that were cut because of storm damage. The volume is strongly influenced by the area of forests so the main purpose of the graph is just to show some extreme years. In Austria those were 2002, 2007 and 2008; in Baden-Württemberg 2000, 2001, 2003 and 2007, in Bolzano 2003 and 2005; in Slovenia 2008. All together in eight years (from 2002 till 2008) for those regions more than 30 million m3 of wood (trees) was registered to be cut because of storm damages.

The forest area at each NUTS 3 region differs greatly. Total volume of cuts per unit (NUTS 3 region, country etc.) was divided by its corresponding area of forest to avoid the influence of area on the volume of logged trees. With unit value per hectare ("m3/ha of forest" or with multiplying it with 100 to "m3/100 ha of forest") it is possible to compare timber data between different NUTS 3 regions and between years.

Forest areas were calculated from CORINE Land Cover maps, which were produced in 2000 [3] and then updated in 2006 [4]. For selected NUTS 3 regions from year 2000 to 2006 the area of mixed forest has decreased from 2,291,285 ha in year 2000 to 2,150,907 ha in 2006 which represented a decline of 140,378 ha. On the contrary the area of coniferous and broadleaved forests has increased by 39,213 ha (from 3,412,648 to 3,451,861 ha) and by 66,587 ha (1,097,571 to 1,164,158 ha). All together the area of forests in selected countries has decreased by 34,578 ha in the period between 2000 and 2006. With the interpolation of those changes between year 2000 and 2006 the interpolation for each NUTS 3 region for period 2000 to 2010 was done. The aim was to get some robust information on the area of forest per NUTS 3 region in each year.

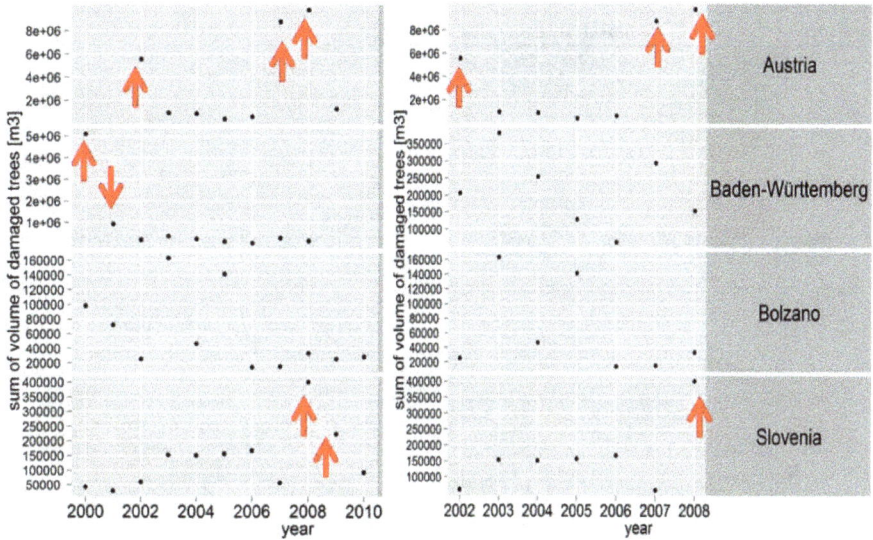

Figure 2. Sum of m3 of timber harvested because of storm damages per year for all available years (left) and for the jointly time span (right). The values on the y axis are due to big differences between countries not the same.

For the period from 2000 to 2010 the annual mean value of damages because of storm on NUTS 3 level was 66 m3 on 100 ha of forest. The median was much lower 11.45 m3 on 100 ha of forest which shows that in this time span there were some very big damaging events in some of the forests in Alpine Space countries. From the maps below it is possible to recognize which NUTS 3 regions were more exposed to storm damages in the forests during the period 2000 - 2010. If the country is not coloured there was no data for selected year. The used thresholds (see Table 1) were calculated from the whole dataset (all regions and all years) and then divided to four quantiles. The 4th quantile was further divided on the 90th percentile and 99th percentile. So the 99th percentile would mean that 99 % of all data on cuts in region from 2000 to 2010 were smaller than 1106.778 m3 per 100 ha of forest. With these subdivisions it is possible to identify also those regions with extremely large amount of damaged trees in selected year in comparison to other regions.

From the maps (Figure 3) it is possible to see some spatial and time patterns how the forests in selected countries of Alpine Space were affected by storms in the period from 2000 to 2010. From the maps and graphs (Figure 4 and Figure 5) it is possible to identify two peaks: first one during years 2000 to 2002 (when a big influence of windstorm Lothar especially in region Baden-Württemberg was noticed) and second one in 2007 and 2008 when forests in central Austria and north Slovenia suffered more severe storm damages. In those two years in all regions (Baden-Württemberg, Bolzano, Austria and Slovenia) together almost 20 million m3 of trees were cut because of storm damages.

Rang	Colour	Calculated threshold	Damaged timber [m³/100 ha of forest]
1		25 %	0 – 3.785
2		50 % (median)	3.786 – 11.450
3		75 %	11.451 – 31.904
4		90 %	31.905 – 111.982
5		99 %	111.983 – 1106.777
6			> 1106.778

Table 1. Legend: Map of NUTS 3 regions with calculated cuts on 100 ha of forest per year.

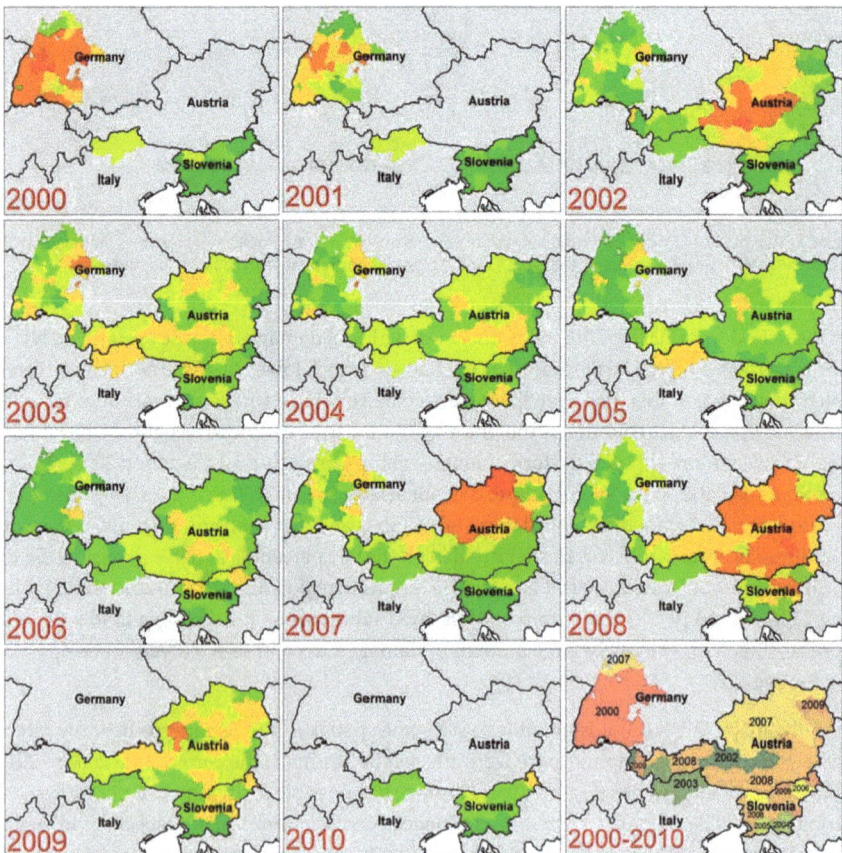

Figure 3. Maps of NUTS 3 regions with the information on volume of cut trees that were damaged by storms [m3/100 ha of forest]

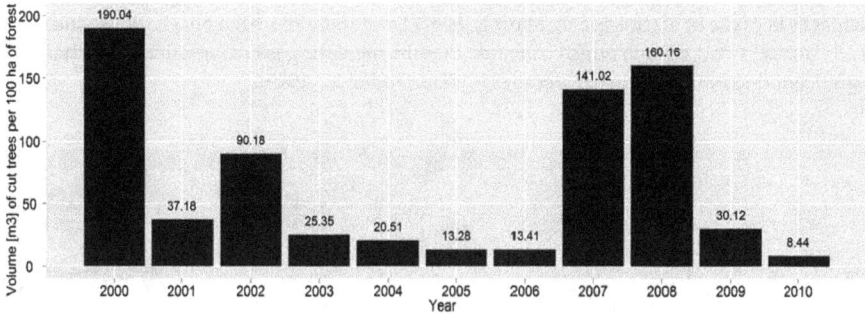

Figure 4. Volume of trees that were logged because of storm damage in the period from 2000 to 2010. With the goal to make data comparable between years. For each year the sum of cut trees was divided with the area of forest for same year and multiplied with 100, so the values are in m3 per 100 ha of forest.

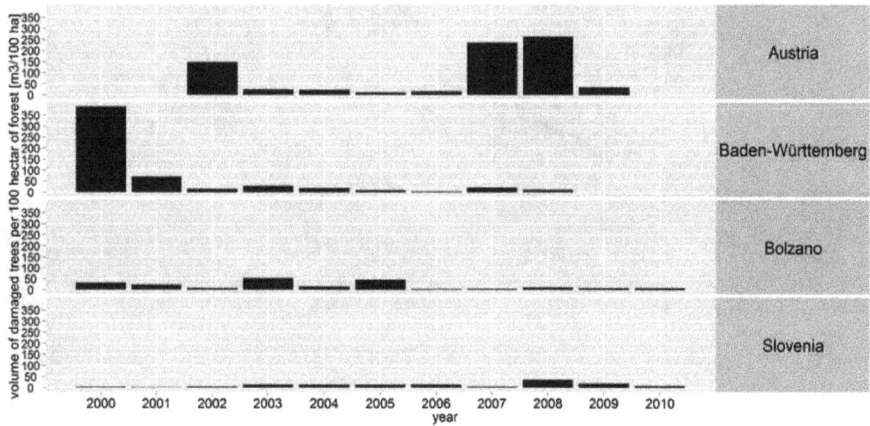

Figure 5. Volume of trees that were logged because of storm damage in the period from 2000 till 2010 divided on regions (country)

During the period between 2000 and 2010 there was eight NUTS 3 regions in which, because of storm damages, more than 1,100 m3 of trees per 100 ha of forest were cut. All of them are located in Germany and in Austria and only one is located directly in the Alps (Figure 6, Table 1). Considering common time span for all selected regions (from 2002 to 2008) it shows that because of storm damages in recent years (2007 and 2008) almost 65 % of all volume which has been recorded in timber harvest records was cut. The available data set illustrated that there were two extremely stormy periods (2000 and 2007/08) with high volumes of storm-damaged trees in harvest records, but the period of available data in this analysis is too short to predict if the storm damages in the forests of selected countries are increasing. Also reviewed literature on historical storm damages in EU show that some of the authors conclude that the forest damages because of storms are increasing [5, 6, 7] and there has been also various studies of the historical wind climate which conclude that the increasing is not significant but there are significant fluctuations between decades [8] (Dorland et al., 1999).

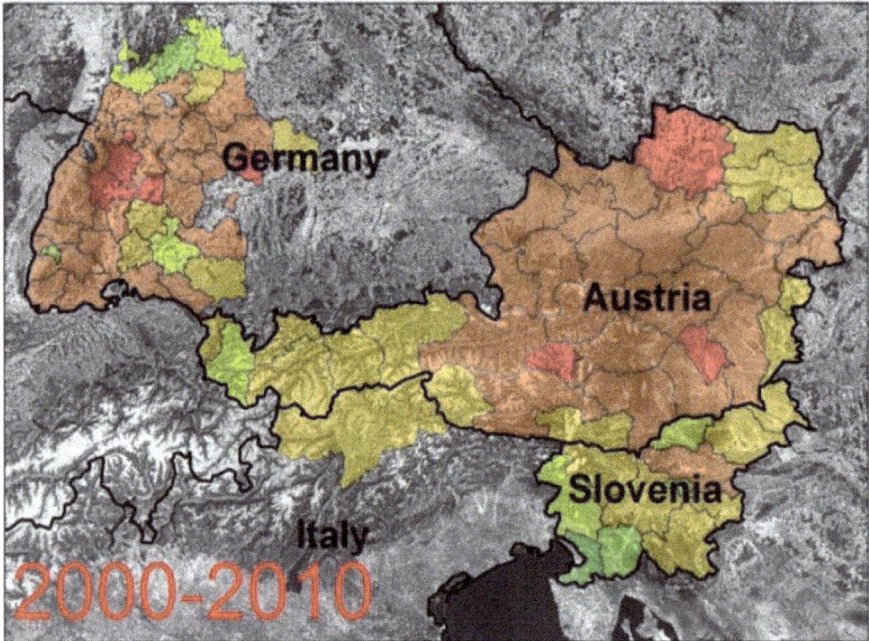

Figure 6. NUTS 3 regions with maximum volumes of cut trees [m³/100 ha of forest] because of storm damage between years 2000 and 201

3. Contributing factors

It is clear that forest storm damages are triggered by many factors. This study tried to analyze some of the most contributing factors under commonly recognized forest and climatic conditions, aiming at provoking actions and plans to reduce storm risks and formulating strategies to adapt alpine space mountain forests to changing climate in the foreseeable future.

3.1. Tree height

Among other factors, tree height is one of the most sensitive factors contributing to the damage probability. Both statistical models [9,10] and mechanistic models [11] concluded the same results that damage probability goes up along with increasing tree height. Schmidt et al (2010) simulated the change pattern of damage probability for Norway spruce with varying tree heights (see Figure 7). The model [13] was run by varying only tree height and DBH to maintain the fixed Height-DBH-ratio as 80 cm/cm, while keeping other predictors constant. This means the "standard" Norway spruce is assumed to be exposed to the same topographic exposition. The Numbers at the right side indicates 5 different locations in Baden-Württemberg, which stands for 5 different wind field characteristics. As a consequence, the overall trend of increasing damage probability with tree height going up is common in various geographical locations.

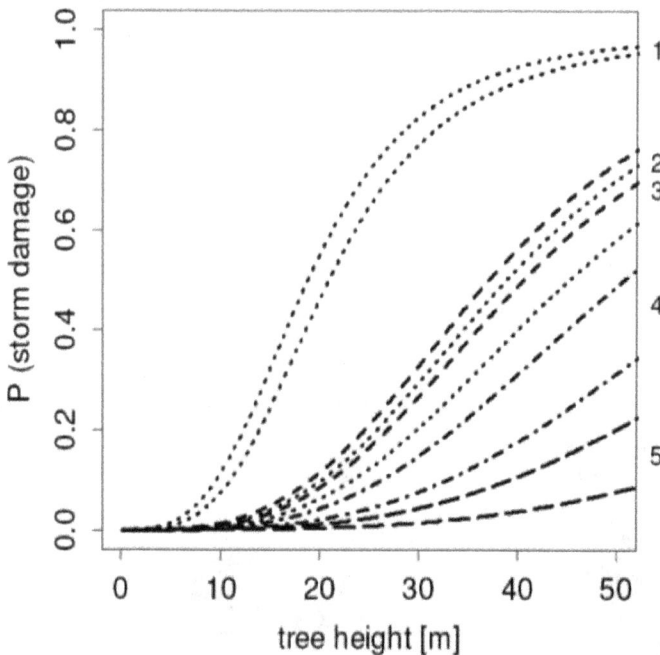

Figure 7. Impacts of tree height on storm damage probability (Retrieved from Schmidt et al.2010)

3.2. Tree species

In general, conifers are more vulnerable than deciduous trees in winter storm, whilst conifers might be less vulnerable than deciduous trees in summer thunder storms. The historical storm data is also in favor of this rule (see Figure 8). It has shown that there is a significant positive correlation (p= 0, Pδ = 0.381) between the volume of damage trees and the percentage of coniferous forest in NUTS 3 regions. The higher the percentage of coniferous forest is, the larger the volume of damaged trees is. In contrary with the increasing of area with broad leaved forests (p= 0, Pδ = - 0.381), which are apparently much more resistant to the damages produced by storms, the amount of damaged wood decreases. Some other authors [12,13] also came to the similar results who also discuss that coniferous species are considered more vulnerable to wind damage than broadleaved species, with Norway spruce (Picea abies (L.) Karst) being regarded as particularly vulnerable.

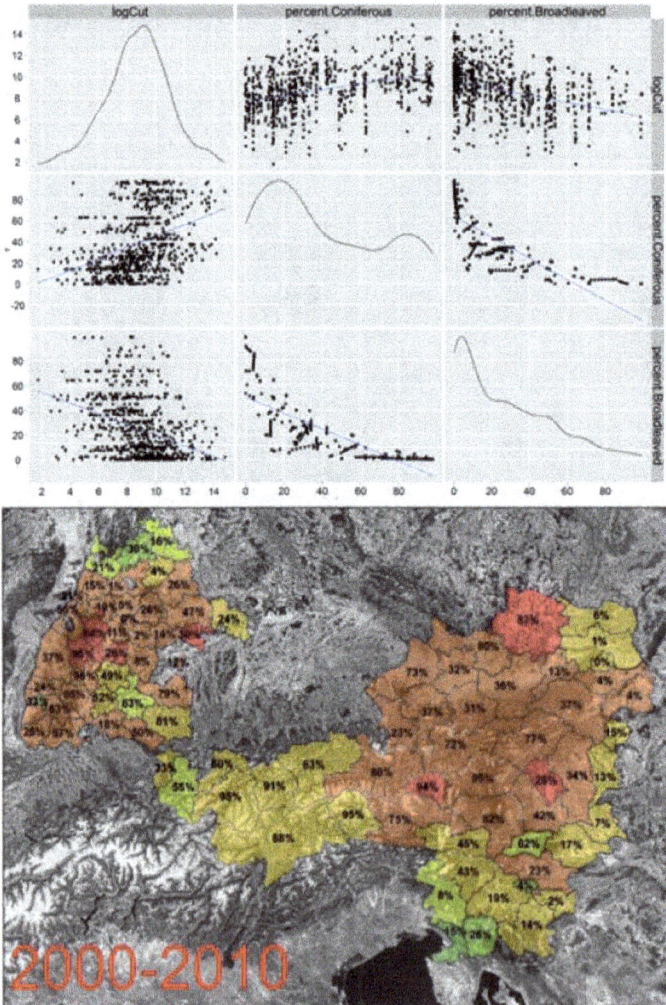

Figure 8. Percent of coniferous forest in 2006 (EEA,2006)and the NUTS 3 regions with maximum volumes of cut trees [m³/100 ha of forest] because of storm damage between years 2000 and 2010.

In terms of species' susceptibility to storms, some statistical studies showed that spruce and silver fir are the most and least vulnerable conifers, and poplar and oak for the broadleaves, respectively [2]. According to a robust statistical model developed by Schmidt et al [13], the predicted damage probability under *ceteris paribus* condition indicated that the ranking from highest to lowest damage probability is: Norway spruce, silver fir/douglas fir, scots pine/larch, beech/oak, and then other broadleaves. When the interaction between height and species are taken into consideration at the same time, the two species groups, e.g. Silver fir/ Douglas fir and Scots pine/larches showed higher damage probability than Norway spruce at lower tree heights, but lower damage probability at higher tree height [13]. Many other experts have also studied on the resistance of different species to windthrows and obtained somehow contra-dictory results [14,15,16,17]. This on the other hand reflects the fact that trees' vulnerability to windthrows depends on various factors, including, but are not limited to, stands structure, forest management, site condition, soil condition, and meteorological differences and so on. Nevertheless, even-green conifers are in general more sensitive to windthrows than deciduous trees, especially during winter storms.

3.3. Stand structure

According the report of European Commission, stand structure has no clear straightforward correlation with storm damage [2]. Although it is commonly expected that the interpenetration of the crowns, complex social structure of trees with stands and vertical stratification of rooting system could enhance the ability of stands to resist against storm, there is no clear conclusion on the effect of stand mixture on stability of stands [2]. Whether a stand with mixed tree species can have positive influence on stand stability is also controversial. Slodicak 1995 [18]and Schütz et al.2006 [16] found out species mixture has positive effects, while others did not draw such conclusion on the beneficial impacts of mixture [19,20].

3.4. Soil condition

Different soil conditions have largely differing impacts on storm damages. For example, Silin et al 2000 argued that forest soils that are deeply frozen in winter in North Europe are less vulnerable to windthrows, while moist soils with water saturation are more sensitive to wind [10]. Soil types and profiles determine its ability to resist wind damage too. For instance, Bock et al. [21] found out large damaged beech stands on superficial calcareous soils in Lorraine in Northeast France during Lothar 1999. As a matter of fact, forest stands that have deep rooting system and better root anchorage on frozen, not-saturated soils can reduce significantly the level of damage.

3.5. Topography

Schmidt et al 2010 developed a sound statistical model to explore the influence of local topographic exposure to air flow on storm damage probability at individual tree level based on the National Forest Inventory 2002 data, which was the best database to record the Lothar 1999 damages. The importance of each topographical exposure (TOPEX) index at various directions depends largely on major wind direction and aspects and exposure of forest stands to wind direction. For example, the Lothar crossed the state of Baden-Württemberg mainly from southwest to northeast. Therefore, the model [13] that was developed only using the Lothar damage data also show higher coefficient of the TOPEX index with this same direction as Lothar's track than other TOPEX index. This is to say, forest stands locating on the top of northwest to southeast running ridges would have higher damage probability than other topographical locations, given the same tree-related information and the same wind field variables.

3.6. Wind speed

Wind speed is one of the most dominating meteorological factors that is positively correlated to the potential storm damage magnitude. The levels of storm damages was classified into several categories based on the maximal gust wind speed: no appreciated damage, moderate levels of damage, high levels of damage and severe levels of damage which is defined by the percentage of damaged woods in the national growing stock [2]. On the other hand, the time duration of storm over a certain area also plays a key role in the total damage. Equally important is that the local natural and climatic condition can influence considerably the levels of damage, since trees can acclimate themselves to windy characteristics by adapting their rooting system to improve the root anchorage to resist against the windthrows [22]. As a consequence, more severe damages could possibly be found in areas where strong winds are common climate phenomenon than those areas with quite few strong winds, when these areas are exposed to the same storm event. High wind speed penetrate more into the inland countries in alpine space, not only the coastal countries, like Lothar, the gust wind speed peak reached up to 51 m/s and 64 m/m in Baden-Württemberg and in Switzerland, respectively[1]. This might put eastern alpine space member states under highly potential storm damage risks in the future.

4. Storm damage under changing climate

Firstly, the storm intensity is increasing. Leckebush et al. [23] developed an index to predict the storm severity for the past decades (1960-2000) and for the future under 2 climate change scenarios (SRES Scenarios A1B and A2). Both estimates showed the trend of increasing storm intensity. Not only had the projection indicated this tendency, but also the measured data. For example, in Switzerland, the measurement of wind gust speeds has also increased strongly since the beginning of records in 1933 [24]. Another tendency is that the hurricanes originated from Atlantic oceans tend to travel far further into the mainland of Europe with wider path. Although the 1999 lothar did not induced serious damage to southern and southeastern regions of alpine mountain, the alpine ridge might not be able to serve as a shield in the near future. This is to say, central Europe will be possibly exposed to larger storm damages as is different from the past. Regarding the return period of

---

1 http://en.wikipedia.org/wiki/Cyclone_Lothar

devastating storms, Della-Marta and Pinto [25] also stated that it will be reduced as a consequence of increasing storm intensity.

Secondly, climatic characteristic is also a crucial factor leading to the storm damage variations. Temperature and precipitation are the most influential climatic factors. As explained in the previous section, unfrozen soil imposes potentially higher risk to forest stands during windthrow events. With the increasing temperature due to global warm- ing, the unfrozen period in winter especially in northern Alpine space at high latitude will likely to be extended in the future, which makes the forests even less resistant to wind- throws. Heavy rains are normally accompanied to big storm events either in the preced- ing days of or during hurricanes. This evidence was found in many unforgettable storm events in Europe, like Vivian & Wiebke 1990, Lothar 1999. Under the tendency of climate change, precipitation in winter is expected to be increasing, which result in moister soil. As a consequence, the water-logged soil will further weaken the ability of forest soil to stand against storm damage. Therefore, the damage could be expected to be more serious in winter in the future than it is now.

However, it is also important to point out that a meteorologically strong storm event does not necessarily induce catastrophic forest damages. It is highly dependent on the actual forest conditions, e.g. the growing stock at the time of storm. The most devastating ever storm Lothar has triggered enormous damages in Europe, which was partly due to the well forest growth and increment in standing volume in the past decades before the storm event. It can be foreseen that the damage magnitude would be the same or higher if the forests continue to grow with less cuts, leading the net increase of growing stock. Otherwise, the serious damage could be mitigated if appropriate forest management regimes are to be implemented. On the other hand, trees can acclimate to windy condition by strengthen- ing their anchorage [24], so a storm might cause more serious damages in an area where high wind speed is rare than those areas with high wind speed as a common phenom- ena. Taking into consideration the different resistance capacity of different species to storm damage, silviculture strategies is also being adapted in many countries. For example, the damage inventory showed that 85% of damaged woods by Lothar in the German state of Baden-Württemberg were conifer species. After Lothar, more storm proof deciduous tress or mixed stands are planned for the regeneration in order to prevent storm damages. The goal of the share of species composition will be half deciduous and half conifer, with the aim to minimize the potential loss by future strong storm events.

Forest storm damage changes over time can be projected using different forest growth model and climate change scenarios. Schelhaas et al. [4] made a great effort to estimate the future storm damages for the whole EU27 plus Norway and Switzerland by multiplying the predicted growing stock by the observed percentage of annual storm damage between 1950 and 2010, as well as the percentage of maximum damage during 1950-2010. Two forest management scenarios were applied, e.g. business as usual and high demand. The estimation was made under the assumption of no changes in wind climate and no changes in vulnerability due to changes in age class distribution. The results showed that the annual average damage will increase from 17.5 Mil.m3 in 2000 to 31.0 Mil.m3 or to 24.1 Mil.m3 by 2050 under these two management scenarios, respectively. This is equal to 78% and 38% increase of annual damage by 2050 [4]. Even though these estimates were made for the whole Europe forests, the Alpine space forests as a core component of the European forests are expected to experience the similar damage changes.

In accordance with the trend of higher temperature, more precipitation, more intensive storm activities, the storm damage in forests can be expected to be substantially increasing by the end of this century. Gardinar et al [2] stated that it is expected to be at least double under these climatic trends together with even-growing and ageing forest standing volumes. However, projections are always accompanied by uncertainties. Various forest growth models produce varying results. In General, under warmer climate in the future particularly in the north Alpine regions, higher temperate provides good conditions for trees to grow faster, which might aggravate the damage levels under the same annual harvest amounts. The forest harvest practice can be altered by forest policies and planning or even policies beyond forest sectors. For example, if more agricultural land is need for some reasons, wooded land might be cleared, or if bio-energy is becoming more demanding, requiring more woods for power generation, then the annual harvest will go up sharply, leading to slow net increment in growing stock. This in turn reduces the potential storm damage risk. In other words, apart from the impact of natural climatic changes on storm damages, anthropogenic interventions is also of great influence on future storm damages.

Drawing on the Lothar model (Schmidt et al. 2010), two exemplary storm risk maps are made for the second case study region in the context of MANFRED project, which is a joint trans-national area consisting of Ravensburg (Baden-Württemberg, Germany), Allgäu (Bavaria, Germany) and Vorallberg (Austria). The risk maps illustrate the storm damage (either stem broken or uprooted) probability at single tree level. The color shifting from green to red indicates the damage probability increases. In order to compare the impact of two climatic characteristics, e.g. wind speed, the risk maps were made for pre-defined model trees, e.g. Norway spruce with height 40 meters, DBH 53 cm. The upper map (Figure 9) shows the potential risk of the model Norway spruce when it is exposed to high wind speed at each specific real topographic features. The lower map (Figure 10) deliver the same message under moderate wind speed scenario at the same geographical location. Figure 9 tells us that Norway spruce would be under very high risk to be damaged if a future storm event with similar wind flow characteristics as Lothar would hit this case study region.

When storm damages to other sectors are also taken into account, global and regional climatic models showed the consistent trend of increasing damages. For the perspective of insurance industry, a joint study by Swiss Re and the Swiss Federal Institute of Technology (ETH) Zürich found out that winter storms in Europe will cause a 16%-68% increase in average annual loss over the period 1975-2085 in constant currency depending on models applied [26]. In monetary term, the annual insured loss from winter storms would be 3.5 billion Euros in 2085. The effect of climate change will also result in greatly differing storm damages in various countries. For instance, Germany is expected to suffer insured loss almost three times the corresponding European value [28].

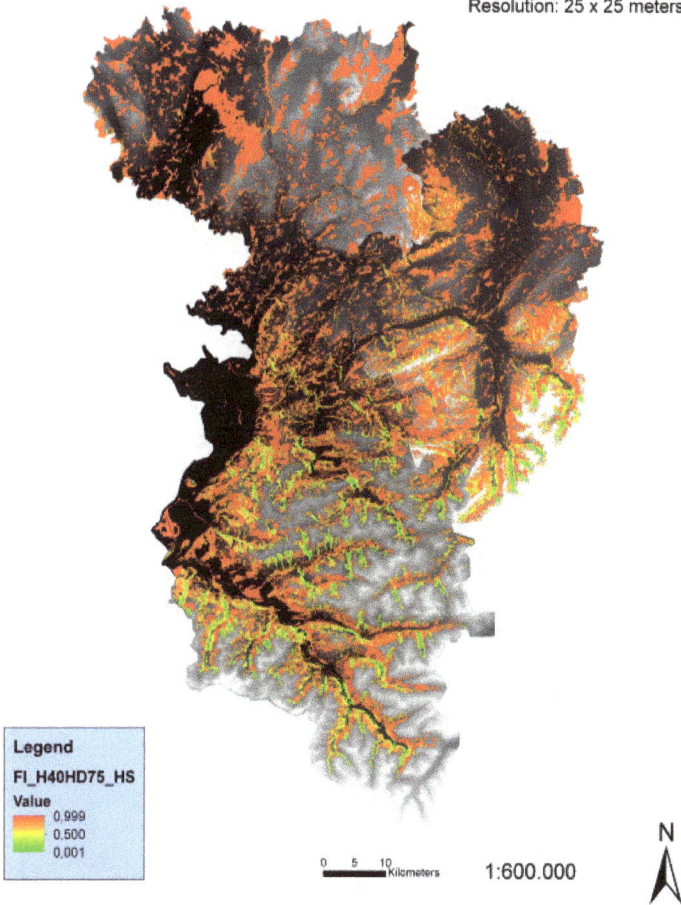

Figure 9. Storm damage probability under high speed scenario

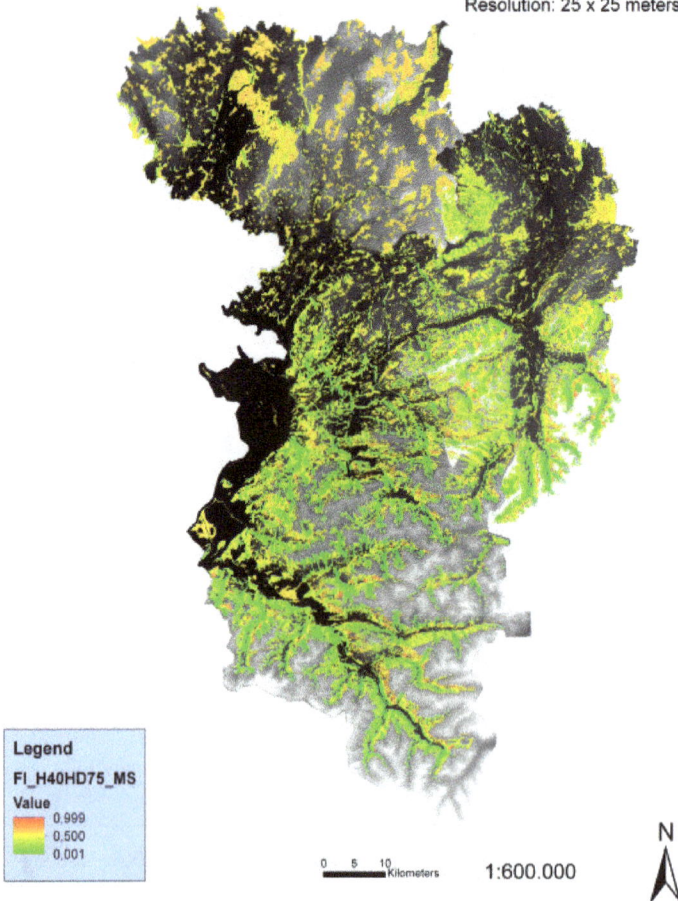

Figure 10. Storm damage probability under moderate wind speed scenario

5. Management strategies and policy recommendations to cope with storm damages

Over the period 1950 – 2000 storms were responsible for 53 % of total damages in European forests [9]. Because storms have a high potential to cause severe economic and ecological consequences in forests[15] it is important to minimize the damages as much as possible. In one of previous chapters it is discussed that many factors (climate, wind speed, topography, soil condition, forest structure) are important triggers for storm damages in forests. On most of those factors humans have no direct influence, but with appropriate silvicultural and forest management decisions it is possible for humans to control the structure of forest.

To minimize the forest damages produced by storms it is important to discuss two questions, first how to improve the resistance of forests to strong winds (PREVENTION) and second how to proceed if severe storm damage in forest happens (POST-STORM MANAGE STRATEGIES).

Review of most important actions to prevent storm damages in forests:

- Natural tree species composition is more resistant to storm damages than exchanged tree species composition [27].

- Conifers are more vulnerable than deciduous or mixed forests. Spruce and silver fir are the most and least vulnerable coniferous and poplar and oak for the broadleaves [2].

- Appropriate mixture of tree species reduces storm vulnerability of forest. Even a mixture of 10 % of broadleaved trees or wind-firm conifers (silver fir or larch) is very effective [16].

- Vertical forest composition plays an important role in stability of forest against strong winds. Unevenly-aged forests and selection forests are more resistant to storm damages than evenly-aged forests [28,29, 29].

- With increasing tree height also damage probability increases [13].

- With decreasing tapering or increasing height/DBH ratio increase storm damage probability [30].

- Susceptibility of a stand to wind damage increases with large mean diameter and with higher stand age [19] or overmature stands [31]. All of this indicates that leaving stands past the normal rotation age involves greater risk.

- Thinning is also an effective measure for improving stability of the stand, but it must be done soon enough. The key to effective thinning is to enhance stability factors (stem, form or crown) without loosening stand cohesion [16].

Despite appropriate use and implementation of reviewed preventive measures, it is not always possible to avoid storm damages in the forests. In cases of disasters it is therefore necessary to ensure rapid and coordinated sanitation of the damaged forests. Successful post-storm management requires participation of different disciplines from federal authorities to forestry actors. From this point of view good preparations on those events are essential and to achieve this regional or country risk management systems should be established. Improvements in the following areas should be done [32]:

- Optimal management of the crisis

- Organization of the labour force and equipment

- Selection of appropriate techniques and methods

- Implementation of logistics

- Determination of the financial requirements

Until time of writing of this text it is already possible to find some guides which aims to help both public institutes and wood-chain professionals to take decisions, prepare their future actions and improve cooperation in the event of hurricanes [33,34]. In the future all EU countries should establish effective post-storm management systems

Author details

Bin You[1*] and Mitja Skudnik[2]

*Address all correspondence to: bin.you@forst.bwl.de

1 Forest Research Institute of Baden-Württemberg, FVA, Freiburg, Germany

2 Slovenian Forestry Institute, Slovenia

References

[1] Eurostat(2009). Forestry statistics.2009 edition.Puplications Office of the European Union, Luxembourg.

[2] Gardiner, B, Blennow, K, Carnus, J. M, et al. (2010). Destructive Storms in European Forests: Past and Forthcoming Impacts, Final report to European Commission-DG Enviroment

[3] Schelhaas, M. J. (2008a). Impacts of natural disturbances on the development of European forest resources: application of model approaches from tree and stand levels to large-scale scenarios. DissertationesForestales 56, Alterra Scientific Contributions 23.

[4] Schelhaas, M, Hengeveld, J, Moriondo, G, Reinds, M, Kundzewicz, G. -J, Maat, Z. W, & Bindi, H. t. M. (2010). Assessing risk and adaptation options to fires and windstorms in European forestry. Mitigation and Adaptation Strategies for Global Change, , 15, 681-701.

[5] EuroGeographicsAdministrative or Statistical units ((2010). Eurostat.

[6] EEACORINE Land Cover ((2000). Commission of the European Communities.

[7] EEACORINE Land Cover ((2006). Commission of the European Communities.

[8] Nilsson, C, Stjernquist, I, Bärring, L, Schlyter, P, Jönsson, A. M, & Samuelsson, H. Re-corded storm damage in Swedish forests Forest Ecology and Management ((2004)., 1901-2000.

[9] Schelhaas, M-J, Nabuurs, G-J, & Schuck, A. Natural disturbances in the European forests in the 19th and 20th centuries. Global Change Biology ((2003).

[10] Usbeck, T, Wohlgemuth, T, Dobbertin, M, Pfister, C, Bürgi, A, & Rebetez, M. (2010a). Increasing storm damage to forests in Swizerland from 1858 to 2007. Agri. Forest Meteo., 150, 47-55.

[11] Dorland, C. Tol RSJ, Palutikof JP. Vulnerability of the Netherlands and Northwest Europe to Storm Damage under Climate Change. Climatic Change ((1999).

[12] Kellomaki, S, & Peltola, H. (1998). Silvicultural strategies for predicting damage to forests from wind, fire and snow. Research notes 73, University of Joensuu, Faculty of Forestry, Finland: 151

[13] Schmidt, M, Hanewinkel, M, Kändler, G, Kublin, E, & Kohnle, U. (2010). An inventory-based approach for modeling single tree storm damage- experiences with the winter storm of 1999 in sourtheastern Germany. Canadian Journal of Forest Resources , 40, 1636-1652.

[14] Peltola, H, Kellomäki, S, Väisänen, H, & Ikonen, V. A mechanistic modelfor assessing the risk of wind and snow damage to single trees and stands of Scots pine, Norway spruce, and birch. Can. J. For. Res. 29 (6): 647-661. doi:10.1139/cjfr-29-6-647, 1999.

[15] Klaus, M, Holsten, A, Hostert, P, & Kropp, J. P. Integrated methodology to assess windthrow impacts on forest stands under climate change. Forest Ecology and Management ((2011).

[16] Schütz, J. P, Götz, M, Schmid, W, & Mandallaz, D. (2006). Vulnerability of spruce (Piceaabies) and beech (Fagussylvatica) forest stands to storms and consequences for silviculture. European Journal of Forest Research , 125, 291-302.

[17] Bazzhiger, G, & Schmid, P. (1969). Sturmschaden und Fäule. Schweiz. Z. Forstwes. 120,521.-535.

[18] Grayson, J. (1989). The 1987 Storm: Impacts and Responses Forestry Commission Bulletin 87, 46pp.

[19] Jalkanen, A, & Mattila, U. U. (2000). Logistic regression models for wind and snow damage in northern Finland based on the National Forest Inventory data Forest Ecology and Management 135 1-3: 315-330.

[20] Kohnle, U, & Gauckler, S. (2003). Vulnerability of forests to storm damage in a forest district of southwestern Germany situated in the periphery of the 1999 Storm (Lothar). In: Ruck B., Kottmeier C., Mattheck C., Quine C., Wilhelm G. Eds, Proceedings of the International Conference Wind Effects on Trees. University of Karlsruhe. Germany September Laboratory for Buildingand Environmental Aerodynamics, Institut for Hydromechanics, Karlsruhe 151-155., 16-18.

[21] Slodicák(1995). Thinning regime in stands of Norway spruce subjected to snow and wind damage . In Wind and Trees . M.P. Couts and J. Grace (eds). Cambridge University Press, Cambridge

[22] Colin, F, Brunet, Y, Vinckler, I, & Dhote, J. F. (2008). Résistance aux vents forts des peuplementsforestiers, etnotamment des mélanges d.'espèces. Revue forestièrefrangise, LV (2) 191-205.

[23] Colin, F, Vinkler, I, Rou-nivert, P, Renaud, J, Hervé, P, Bock, J-C, & Piton, J. B., (2009). Facteurs de risques de chablisdans les peuplementsforestiers : les lecnstirées des tempêtes de 1999. In: Birot Y., Landmann G., Bonhême I. eds., La forêt face aux tempêtes. Editions Quae, , 177-228.

[24] Bock, J, Vinkler, I, Duplat, P, Renaud, J, Badeau, P, & Dupouey, V. J.-L., (2005). Stabilité au vent des hêtraies : les enseignements de la tempête de 1999. Revue forestièrefrancise, , 57(2), 143-158.

[25] Nicoll, B. C, Gardiner, B. A, & Peace, A. J. (2008). Improvements in anchorage provided by the acclimation of forest trees to wind stress. Forestry , 81, 389-39.

[26] Leckebusch, G. C, Renggli, D, & Ulbrich, U. and application of an objective storm severity measure for the Northeast Atlantic region. MeteorologischeZeitschrift , 17, 575-587.

[27] Usbeck, T, Wohlgemuth, T, Pfister, C, Volz, R, Beniston, M, & Dobbertin, M. (2010b). Wind speed measurements and forest damage in Canton Zurich (Central Europe) from 1891 to winter 2007. International Journal of Climatology , 30, 347-358.

[28] Della-marta, P. M, & Pinto, J. G. (2009). Statistical uncertainty of changes in winter storms over the North Atlantic and Europe in an ensemble of transient climate simulations. Geophysical Research Letters 36: L14703.

[29] Swiss Re(2006). Swiss Re Focus Report: the effects of climate change: Storm damages in Europe on the rise

[30] Intermühle, M, Raetz, P, & Volz, R. LOTHAR UrsächlicheZusammenhänge und Risikoentwicklung. Synthese des Teilprogramms 6. In: Umwelt-Materilien ((2005). Bern: BundesamtfürUmwelt, Wald und Landschaft. 145

[31] Dvorák, L, Bachmann, P, & Mandallaz, D. Sturmschäden in ungleichförmigenBeständen I Storm damage in irregular stands. SchweizerischeZeitschrift fur Forstwesen ((2001).

[32] Mason, W. L. Are irregular stands more windfirm? Forestry ((2002).

[33] Gardiner, B. A, Stacey, G. R, Belcher, R. E, & Wood, C. J. Field and wind tunnel as-
      sessments of the implications of respacing and thinning for tree stability. Forestry
      ((1997).

[34] Ruel, J-C. Factors influencing windthrow in balsam fir forests: from landscape stud-
      ies to individual tree studies. Forest Ecology and Management ((2000).

[35] STODAFORTechnical Guide on Harvesting and Conservation of Storm Damaged
      Timber--Pischedda D, ed. ((2004). FVA. 115.

[36] JuttaOdenthal-Kahabka (2003). Hurricane "Lothar" and the forest of Baden-Würt-
      temberg   (Germany)-   damages,   impacts   and   effects,   http://
      www.centreacer.qc.ca/PDF/Publications/Colloques/2003/Jutta%20english.pdf

[37] STATFORStorm damaged forests- efficient and safe harvesting and log conservation
      methods (2004) http://www.ctba.fr/stodafor/index.htmaccessed 2.6.2012, (2012). Type
      of Medium).

Chapter 7

Abiotic Stressors – Fire Hazard

Bruna Comini, Giampaolo Cocca, Elena Gagliazzi,
Paolo Nastasio, Enrico Calvo, Roberto Colombo,
B. Di Mauro, Lorenzo Busetto, Mitja Skudnik,
Tomaz Sturm and Andrej Breznikar

Additional information is available at the end of the chapter

1. Introduction

Forest fires are worldwide recognized as one of the main factors affecting the forest ecosystem equilibrium, leading to direct and indirect impacts on the functions provided by forests (production, protection, wildlife, tourism, etc..). Forest fire ignition and propagation are closely linked to site-specific conditions: fuel characteristics, forest structure and composition, weather and topography. Within the MANFRED project, ERSAF was aimed to identifying potential evolution scenarios of forest fires danger due to climate change in the Alpine Space.

2. Forest fires in the Alpine space

Some statistics and considerations about the state of the art of forest fires in the Alpine Space, can be derived from the forest fire database, collected through a joint action between the project MANFRED and ALPFFIRS (Alpine Space Programme). The database contains information on forest fires for all the Alpine regions of Austria, France, Germany, Italy, Slovenia and Switzerland. All the events occurred in the period 2000-2009, with the exception of Germany (2005 -2009), were collected and analyzed.

Looking at the geographical pattern (Figure 1), two different situations can be distinguished. The 90% of the fire events in Alpine Space occur in France and Italy. In the countries located in the in the north of the Alps (Austria and Germany) the occurrences showed an overall lower incidence (5% of the events). At yearly level, in the analyzed decade, 2003 emerged as the year

with the highest fires frequency (Figure 2). It was characterized by very high summer temperatures and a prolonged drought period. At seasonal level, it can be observed that, unlike
the European countries of the Mediterranean area (F: Méditerranée – I: Liguria) showing a
prevalence of forest fire events during the summer, the rest of the Alpine Space is characterized
by a winter-spring regime of fires. At monthly level, the highest frequency of forest fires was
recorded between March and April, with a secondary peak, in July and August (Figure 3). The
only exception is France where the maximum frequency is commonly reached during the
summer. This result is influenced by the events occurred in the Southern region of France
where Mediterranean climate is responsible for this kind of fire regime (Wastl et al., 2013).

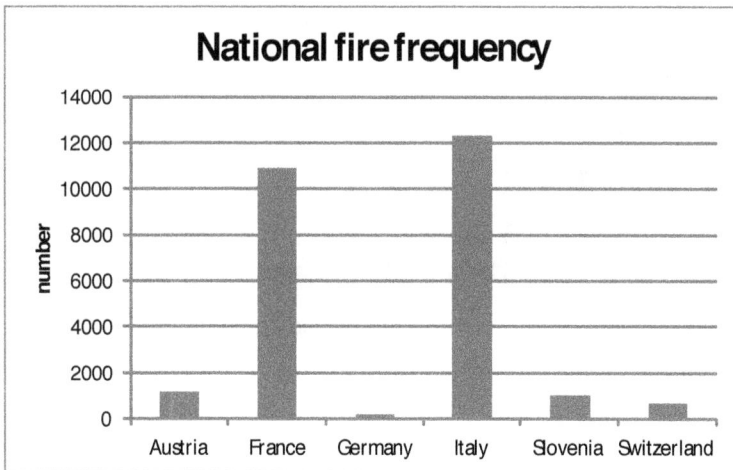

Figure 1. National fire frequency in Alpine Space (2000-2009)

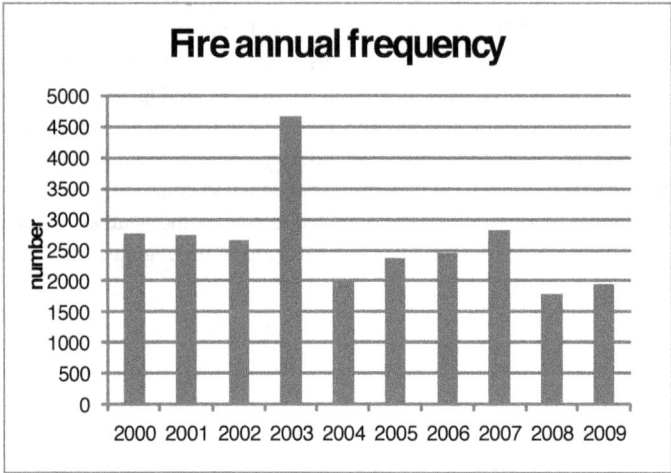

Figure 2. Fire annual frequency in Alpine Space (2000-2009)

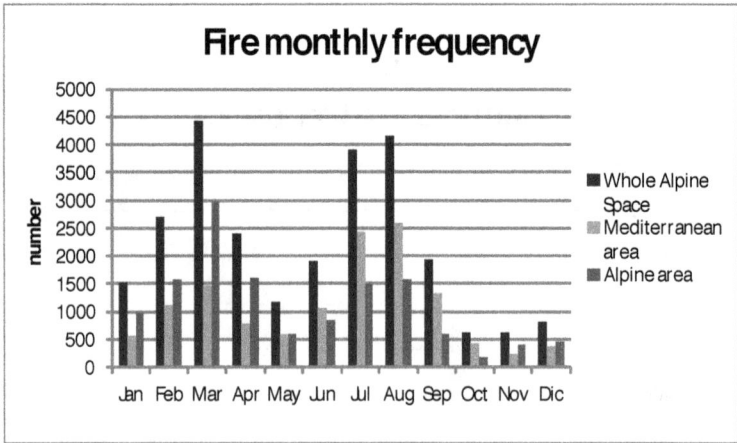

Figure 3. Fire monthly frequency in Alpine Space (2000-2009)

3. 4FI.R.E. (Forest Fire Risk Evaluator) — Tool for forest fire risk calculation

In the framework of MANFRED, ERSAF developed a tool for forest fire hazard and vulnerability evaluation. **4 FI.R.E.** is a stand-alone application composed by two separate modules

4FI.R.E – Hazard, capable of generating, from a set of input spatial layers, a series of forest fire hazard maps, and **4FI.R.E – Vulnerability**, capable of generating, from a set of input spatial layers, a series of forest fire vulnerability maps.

The tool was tested at the Alpine Space and regional/country level (Lombardy Region / Slovenia), while the vulnerability model has been applied only at pilot area level (Valle Camonica, ERSAF pilot area).

3.1. 4FI.R.E — Hazard

The 4FI.R.E – Hazard module is dedicated to forest hazard mapping. This module is almost mature and functional, and a screenshot of the main user interface in visible in Figure 4.

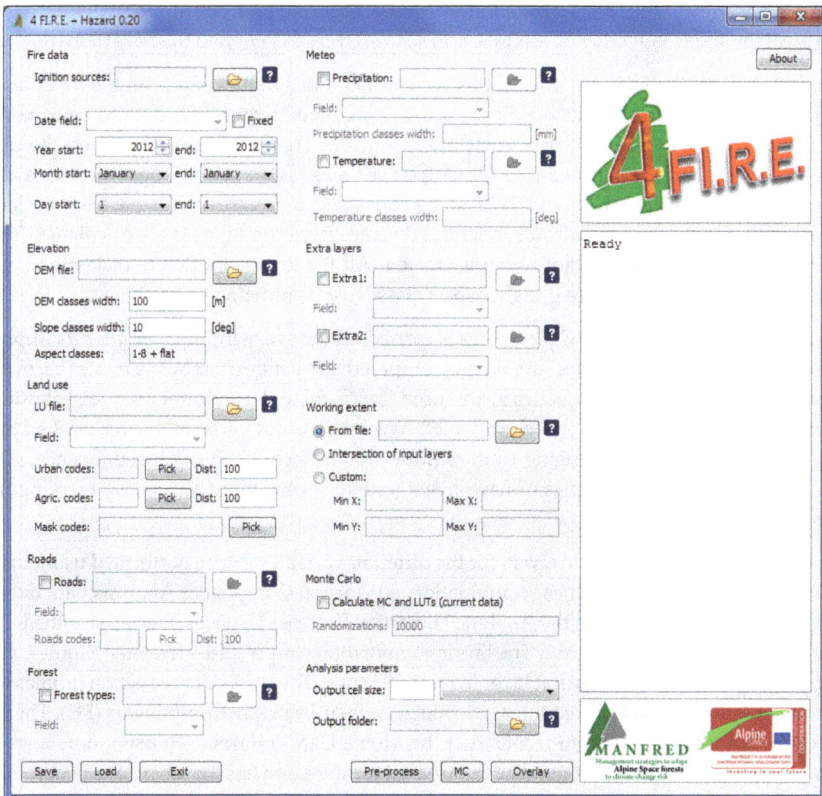

Figure 4. R.E Hazard main user interface

The 4FI.R.E software automates a series of procedures for data import and consolidation, for statistical analysis of spatial layers and for output creation. The creation of the hazard maps involves the statistical analysis of the distribution of historical fire events to reclassify input layers in terms of their correlation to fire events. See Conedera et al. (2009) for details on methodology.

The Hazard module requires a set of input spatial layers, some compulsory, others optional. The compulsory layers are the historical fire data, the Digital Elevation Model (DEM), and the land use map. The optional layers include the roads map, the forest map, the rainfall and temperature maps. The module also allows for the selection of two additional layers to fit specific user requirements. The input layers can be of raster of vector type. In the former case, the ESRI ASCII GRID raster format is supported; in the latter the ESRI Shapefile format is supported.

Conceptually, the hazard mapping is made of three steps:

1. Import of input layers; 2. Calculation of Monte Carlo maps from input layers based on historical fire data; 3.Overlay of Monte Carlo scores layers to generate the hazard maps.

The data import step includes a set of processing operations that depends on the input layer characteristics. Nevertheless, since the two following steps require the data to be in raster format, all the input layers, if necessary, are rasterized. Then, they are clipped on the user specified extent, and resampled into the output cell size. More specific operations are also performed on input data: from the DEM, slope and aspect maps are calculated; from the land use map, distance from agricultural areas and distance from urban areas are calculated; from the roads map the roads distance map is created. All the layers are finally reclassified into discrete classes, pre-defined (e.g. eight aspect classes) or user defined.

The calculation of the Monte Carlo scores is carried on for every input layer after its import. The distribution of the historical fire data is compared against every input variable (i.e. every input layer). Fire event frequencies are used to calculated random-based distributions (typically 10000 simulations for each variable). As a result, input variables are reclassified into seven significance classes, ranging from -3 (little significance) to +3 (high significance). The output of this process is a series of new layers (one for every input variables), carrying the Monte Carlo ranks.

The overlay of the Monte Carlo layers for the different variables can be performed using three different methods: a simple linear combination, a Principal Component Analysis data reduction, and an overlay based on the Analytic Hierarchy Process. The user interface for the layer combination is visible in Figure 2. The linear combination simply adds up all the Monte Carlo maps and then rescales the output layer in the 1-10 range. The PCA data reduction implies the calculation, from all the input Monte Carlo maps, of their Principal Components (PC). Then, a backward transform is used to reconstruct the Monte Carlo information using only a small number of PCs (typically just one). The PCA-based combination has the advantage of eliminating, if present, data redundancies of input layers. The AHP combination requires a series of ancillary input data, where one or more experts evaluate the relative importance (in terms of fire hazard) of every input variable when compared to every other variable. The tool is capable of importing AHP data from txt files, to calculate metrics, such as the Consistency Ratio (CR), to evaluate the validity of AHP data. A sub-module is also present to input AHP data.

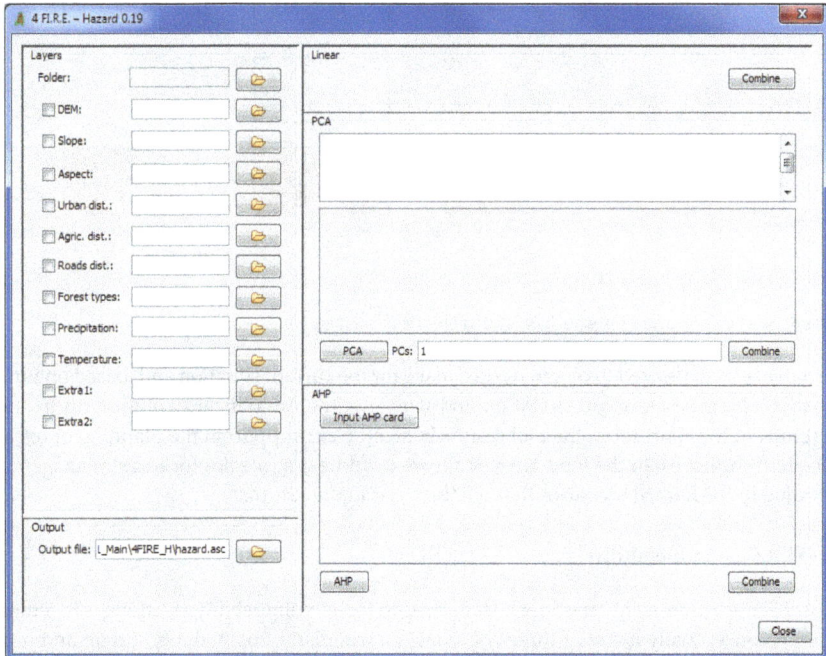

Figure 5. Layer combination user interface

The procedure described so far applies to fire hazard analysis performed on actual data. The Hazard module provides also a means to create hazard maps for future scenarios. The procedure to analyse future data is the following:

1. Import of input layers for actual scenario; 2. Calculation of actual Monte Carlo maps for input layers based on historical fire data; 3. Generation of Lookup Tables (LUTs) to link every actual variable class to its Monte Carlo score; 4. Application of the LUTs to future data to derive future Monte Carlo maps; 5. Overlay of future Monte Carlo maps to generate the hazard maps for future scenario.

This procedure is based on the assumption that the relationship between every variable and its contribution to fire hazard is constant in time. The two different data flows for the current and future data processing are depicted in Figure 3.

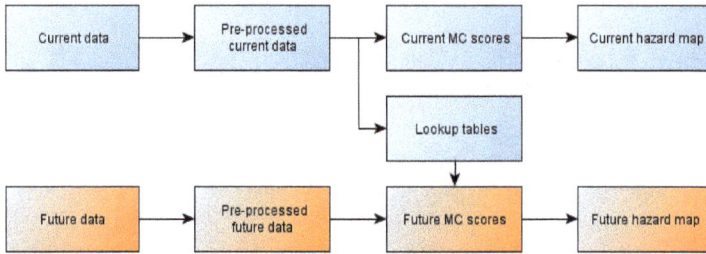

Figure 6. Data flows for actual scenario (blue) and future scenario (orange).

The tool has been utilized to obtain hazard maps for the current situation and, based on future scenarios of temperature, precipitation and land use, the fire danger evolution up to 2080. Therefore, in the short-term, the tool results in an efficient support in the planning of hazard mitigation strategies. In the long-term, it allows to address strategies for forest management according to the hazard scenarios that will be faced in the future.

3.2. 4FI.R.E — Vulnerability

The 4FI.R.E – Vulnerability module is dedicated to forest vulnerability mapping. The vulner-ability is conceptually made of three components: woodland vulnerability, urban and infra-structure vulnerability, and population vulnerability. These three components are included in the Vulnerability module as three sets of input layers. The input layers are imported, rasterized (if needed), resampled and linearly combined to yield the final vulnerability map.

The woodland vulnerability component is made of several sub-components: vegetation resistance, vegetation resilience, vegetation protective function, vegetation productive function, vegetation naturalistic function, vegetation touristic function, and vegetation carbon stock function. Typically the input layers are classified in three classes: low, medium, high. The naturalistic function is given by the combination of the protected areas present. If the protective function map or the productive function map is not available, they can be replaced by proxy data. The protective function map can be replaced by the slope map. The productive function map can be replaced by two maps: the biomass map, and an accessibility map. The latter map is generated by the Vulnerability module from an input DEM combined with the roads map. A screenshot of the main user interface in visible in Figure 4.

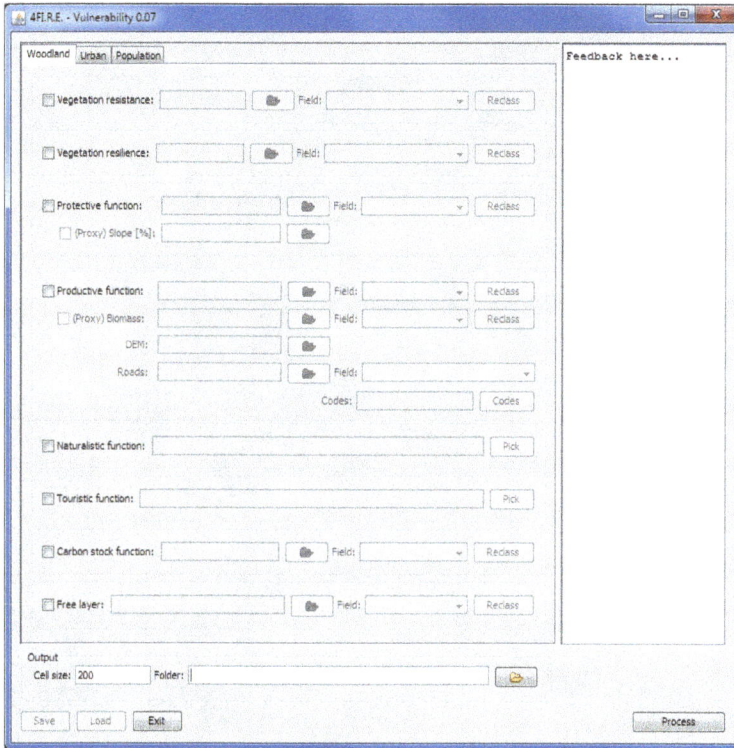

Figure 7. R.E Woodland vulnerability main user interface.

The urban and infrastructure vulnerability map is generated from a combination of a series of urban areas and features map given in input. The distance map from urban areas and features is calculated, and then reclassified into three classes.

The population vulnerability map is derived from the resident population map, reclassified in three classes.

The vulnerability map is given by the linear combination and normalization of all the pre-processed and reclassified input layers. The relative weights of the single input layers can be specified by the user.

3.3. 4FI.R.E — Risk

The definition of risk embedded in the tool refers to the likelihood of having a disaster or outcome, combining the probability of the hazard event with the expected consequences of the hazard (Allen, 2003; Brooks et al., 2005). Risk is defined by the following formula:
Risk = Hazard × Vulnerability

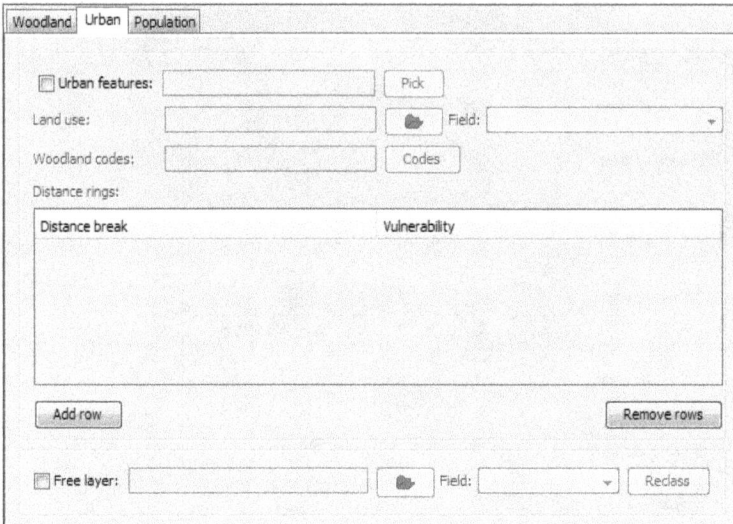

Figure 8. R.E Urban vulnerability main user interface.

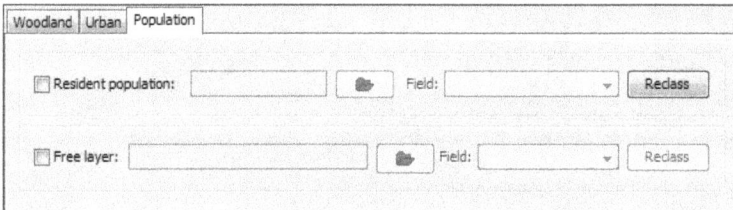

Figure 9. R.E Population vulnerability main user interface.

The final risk map builds on the integration of the hazard and vulnerability maps, where for a defined region, the level of risk is related to hazard potential, vulnerability or both.

4. Fire risk in the Alps: Today and future scenarios

4.1. Alpine space scale

The aim of this analysis was to evaluate forest fire hazard within the Alpine Space focusing the attention on current situation (year 2011) and a future scenario (year 2080). The probability maps of forest fires were produced by means of the tool 4FI.R.E (see previous section). The following data sources were used to derive maps of fire hazard: forest fire ignition points, Digital Elevation Model (elevation, slope, aspect), Land use (Urban and Agricultural areas -

buffers), Vegetation Map, Seasonal Precipitation and Seasonal maximum length of dry episodes (30-year mean; Winter: Dec./Jan./Feb.; Spring: Mar./Apr./May; Summer: Jun/Jul/Aug; Autumn: Sept./Oct./Nov.). Data on precipitation and maximum length of dry episodes for current and future scenarios were provided by AIT - Austrian Institute of Technology, whereas data on land use for both scenarios were made available by WSL - Swiss Federal Institute for Forest, Snow and Landscape Research. In detail, the hazard calculation was performed using the A1b scenario data (Intergovernmental Panel on Climate Change - IPCC, 2000). The ignition point coordinates were available only for a subset of the data (i.e. Italy, Austria, Slovenia, Switzerland), as shown in Figure 10. The seasonal hazard maps for the current and future hazard scenarios in the Alpine Space were produced by using the available ignition point coordinates, required as input for the hazard processing.

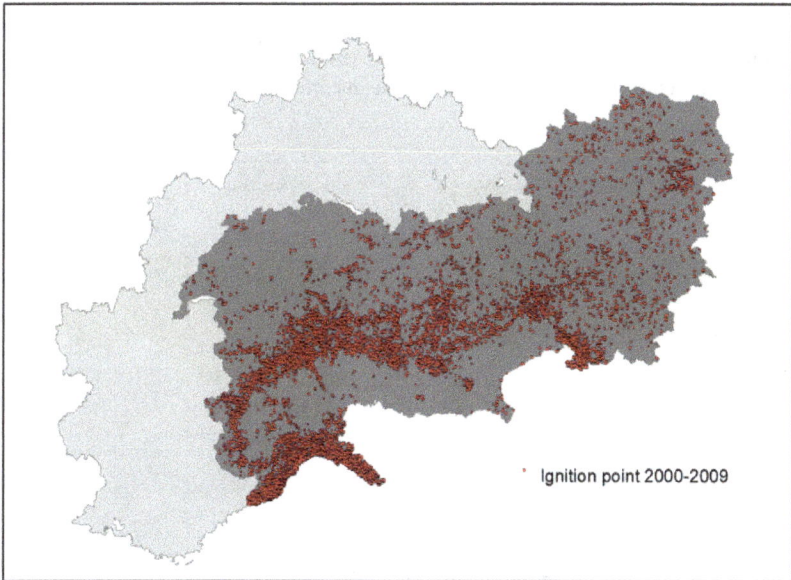

Ignition point 2000-2009

Figure 10. Alpine Space, ignition points 2000-2009

The comparison of the seasonal hazard maps (2011 vs 2080,) showed some significant changes between summer and winter seasons (Figure 11 and 12). According to the future scenario at 2080, it's possible to state that larger part of the analyzed area will be affected by an increase of the likelihood of forest fires events.

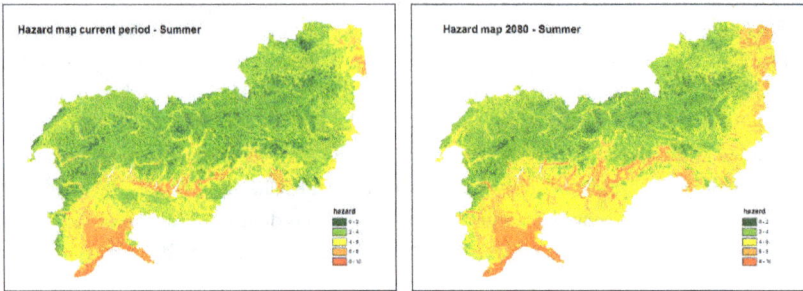

Figure 11. Alpine Space, maps of summer hazard (2011 - 2080)

It is likely that the adopted future climate scenarios showed an overall reduction in precipitation and increase in maximum length of dry episodes both in winter and summer seasons. This fact is probably related to the concentration of precipitation events in short time periods. The trend outlined for the hazard fire depicts a lower hazard in spring and autumn driven by the increase of rainfall and a slight increase in the maximum length of dry episodes.

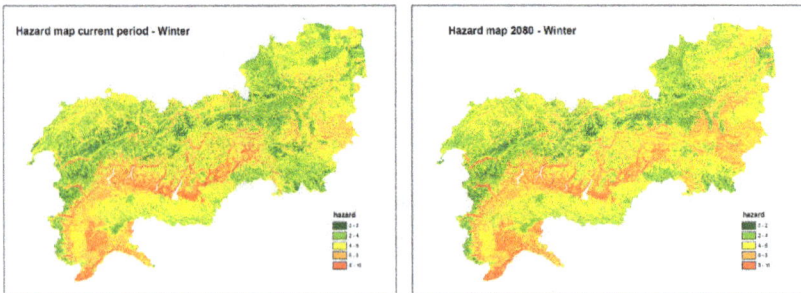

Figure 12. Alpine Space, maps of winter hazard (2011 - 2080)

4.2. Regional scale: Lombardia region and Slovenia

4.2.1. Lombardia region

As in the previous analysis carried out at Alpine scale, the objective was to have a picture about forest fire hazard of the current (2011) and future (2080) scenarios at regional level. Again, the hazard maps of scenario were produced by means of 4FI.R.E tool. using the following data: forest fire ignition points, Digital Elevation Model (elevation, slope, aspect), Land use (Urban and Agricultural areas - buffers), Vegetation Map, Seasonal Precipitation and Seasonal temperatures (30-year mean; Winter: Dec./Jan./Feb.; Spring: Mar./Apr./May; Summer: Jun/Jul/

Aug; Autumn: Sept./Oct./Nov.). Precipitation, Temperature and land use data under current period and future scenarios were made available by WSL - Swiss Federal Institute for Forest, Snow and Landscape Research. In detail, the hazard calculation was performed using A1b scenario data (IPCC, 2000).

The hazard maps comparison (2011 *Vs* 2080) showed for the future scenario a general increase in the hazard, lead mainly by temperatures increase. During summer and winter seasons Lombardia region is likely to face a the most important change in fire hazard with a shift from low to medium hazard classes. In autumn the future trend depicts an increase in the higher hazard classes. The spring scenario shows a reduction of risk closely related to the future increase in rainfall events. Following a clear spatial pattern, the eastern part of the region will be most likely to face higher hazard levels than the western part. Graph 1 shows the number of past fires events (2000-2009) divided per season and danger class. The number of past fire occurrences increases linearly with the hazard values. Only the class 9-10 shows a decrease in the number of fires events in respect to the previous classes.

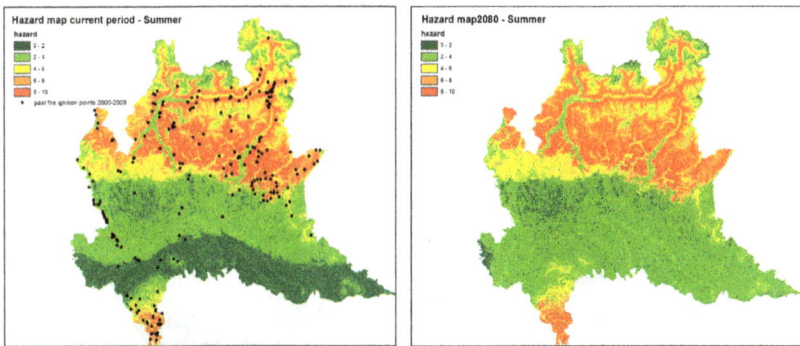

Figure 13. Lombardia Region, maps of summer hazard (2011 - 2080)

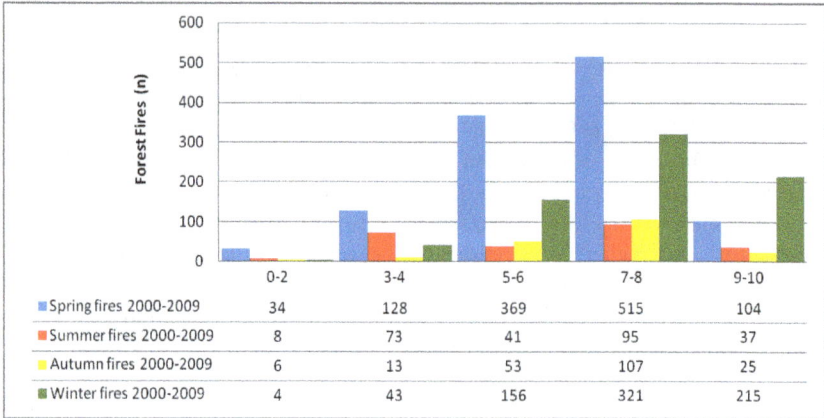

Forest Fires (n)	0-2	3-4	5-6	7-8	9-10
■ Spring fires 2000-2009	34	128	369	515	104
■ Summer fires 2000-2009	8	73	41	95	37
■ Autumn fires 2000-2009	6	13	53	107	25
■ Winter fires 2000-2009	4	43	156	321	215

Figure 14. Intersection of past fire occurrences (2000-2009) and 2011 hazard map

Here we have to take into account that the hazard maps were produced considering, as relevant factors, morphological features (elevation, aspect) and meteorological conditions (precipitation, temperature).

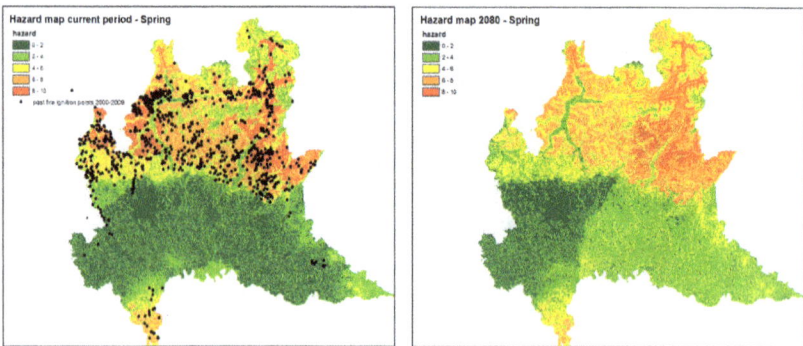

Figure 15. Lombardia Region, maps of spring hazard (2011 - 2080)

Some other factors, mainly linked to human interference were only partially considered due to lack of information.

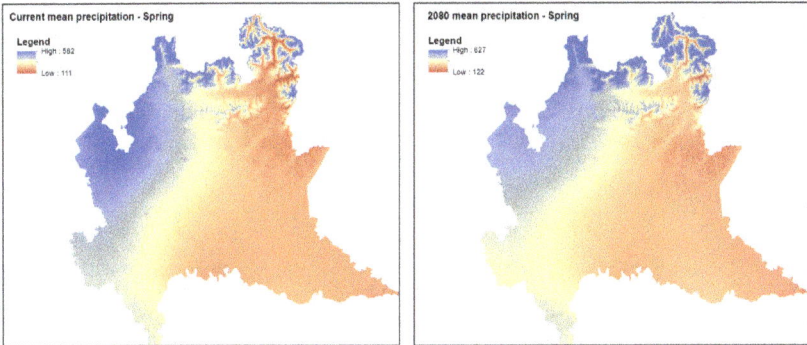

Figure 16. Lombardia Region, maps of spring precipitation (2011-2080)

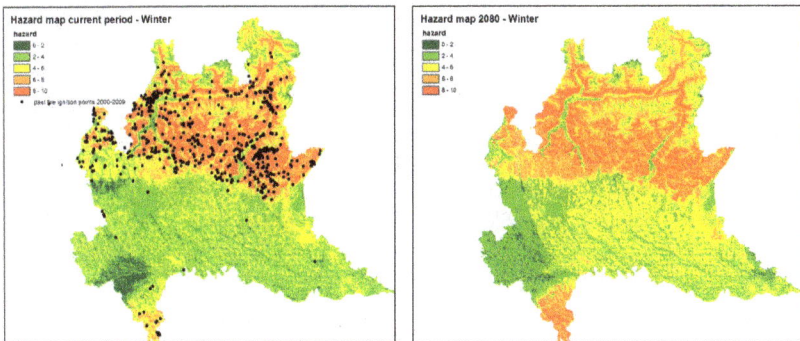

Figure 17. Lombardia Region, maps of winter hazard (2011-2080)

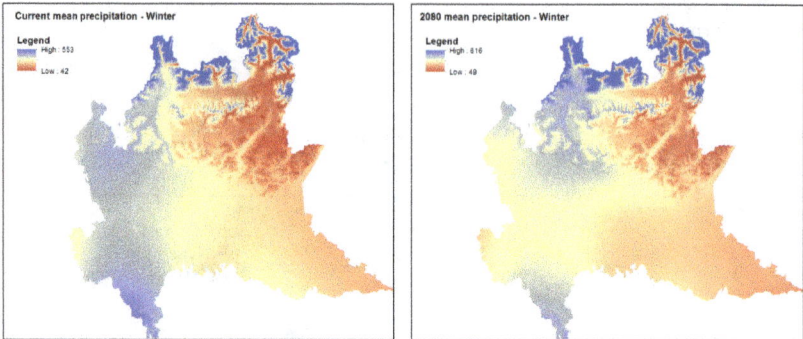

Figure 18. Lombardia Region, maps of winter precipitation (2011-2080)

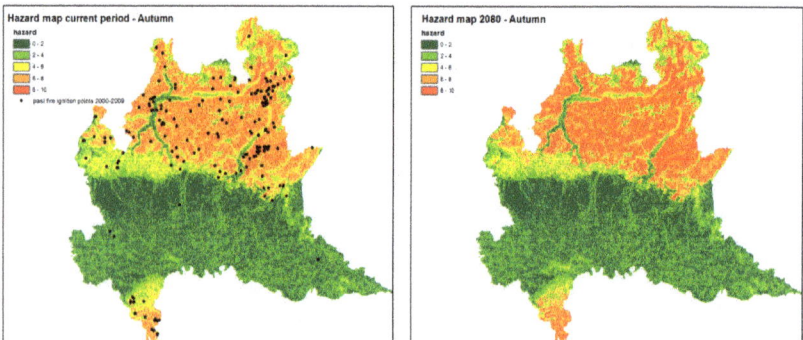

Figure 19. Lombardia Region, maps of autumn hazard (2011 - 2080)

4.2.2. Slovenia

Objective of the following analyses is to obtain some additional information about present and future fire hazard risk scenarios in Slovenia. For this purpose hazard tool 4FI.R.E was used.

GIS layers that were used in the model were:

location of past fire ignition sources (points)[1] (from years 1995 until the end of 2009);

digital elevation map[2]–cell size 100x100 m;

1 Dataowner: SlovenianForestryService, 2012

2 Dataowner: The Surveying and Mapping Authority of the Republic of Slovenia, 2005

map of land use[3] (Urban and Agricultural areas)– cell size 100x100 m;

1. mean precipitation for months April - September (summer) and October- March (winter) for the present time (year 2011) and the future (year 2080),

2. mean monthly temperature averaged for months April - September (summer) and October - March (winter) for the present time (year 2011) and the future (year 2080).

Precipitation and temperature data for present and future scenarios were obtained from WSL - Swiss Federal Institute for Forest, Snow and Landscape Research. The hazard calculation was performed based on the data from A1b scenario (IPCC, 2000), model name "mpi_clm_echam5_ar4_wccru". Land use scenarios maps provided by WSL were not used in this model due to their coarser precision (cell size 250x250 m), so national land use maps were used which are in vector format. We assumed that the change of land use will not be so obvious in the near future comparing the resolution differences of both maps.

Comparison of modeled hazard maps created for present time (summer and winter) with the locations of past fire events in Slovenia shows good matching results for the Karst area (figure 17 - left). This part of the country was in the past also the most endangered part concerning the number and size of fire events. In addition, we have to stress that created hazard maps have some deviances, especially in central (south of the Ljubljana) and in south-eastern par tof Slovenia. In these two areas the hazard maps show high fire risk, but as could be inferred from past fire ignition data (1995 - 2009) there were few fire occurrences in the past. The north-western (only for summer time) and north-eastern part were endangered in the past with fire events while the hazard maps predicted low fire risk.

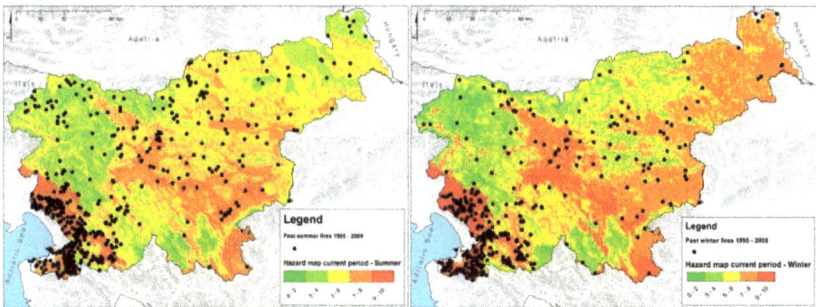

Figure 20. Slovenia summer hazard maps (left) and winter hazard map (right) with locations of past fire ignition points

Intersection of past fire occurrences with modelled hazard maps created for present time (Table and Graph 2) show proper agreement with hazard values 0 to 8, e.g. the number of past fire occurrences increases linearly with the higher hazard values, but there are some deviances in class 9-10.

---

3 Dataowner: Ministry of Agriculture and the Environment, 2012

Value of hazard	Summer fires	Winter fires
0-2	8	2
3-4	105	33
5-6	260	139
7-8	343	259
9-10	32	94
Total number of past fires	764	541

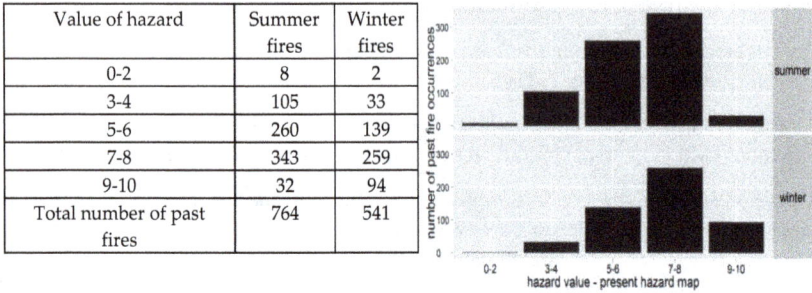

Figure 21. Intersection of past fire occurrences (1995 – 2009) with the present hazard maps

Those areas should be the most endangered but the number of past fires events is very low. Nevertheless we have to take here into the consideration that the hazard maps show the potential hazard area which is based on elevation, aspect, precipitation etc. but not on some other factors, e.g. anthropogenic interference.

The comparison of hazard maps (summer period) for 2011 and 2080 shows that Karst area will remain the most endangered part of Slovenia. Fire ignitions are predicted to increase in the eastern and especially in the north-eastern part of Slovenia. The area with hazard classes 9-10 will decrease from 7 700 ha to 4 100 ha, but the sum of area of classes 7-10 will increase from 268 600 ha to 272 000 ha in next 70 years. One of the key factors affecting the increase of the fire risk in eastern parts of Slovenia could be the predicted reduction of precipitation (figure 18) during the summer. On the contrary, the decreased hazard in central Slovenia could be a result of predicted higher summer precipitations.

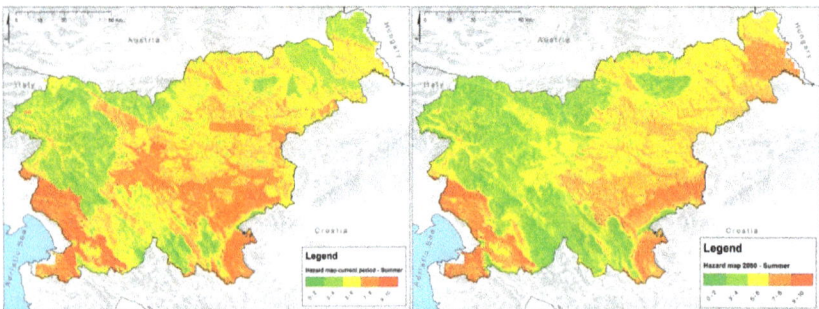

Figure 22. Slovenia summer hazard maps (left: current hazard situation; right: future hazard situation 2080)

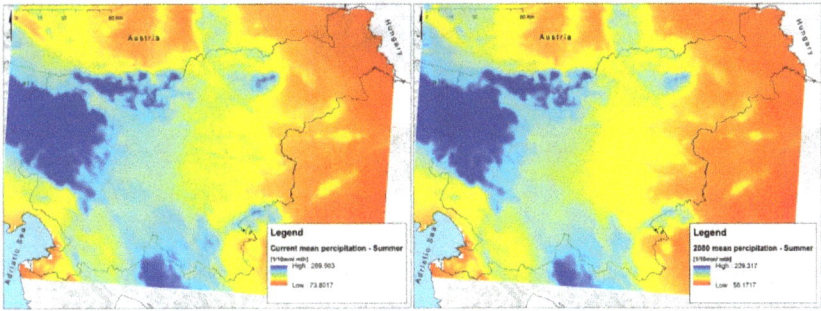

Figure 23. Summer precipitation maps for Slovenia for current period and prediction for the year 2080 (source: WSL)

The comparison of hazard maps (winter period) for 2011 and 2080 show that in the future we can expect the decrease of hazard at the Inner Carniola-Karst region, small part of the Southeast Slovenia and in the northern part of Central Slovenia. The highest increase of hazard is supposed in the northern part of Slovenia. The cross tabulation of both maps show that at this time 38 120 hectares of Slovenia is under high hazard risk (class 9-10) and this area will increase to 40 294 hectares in the year 2080.

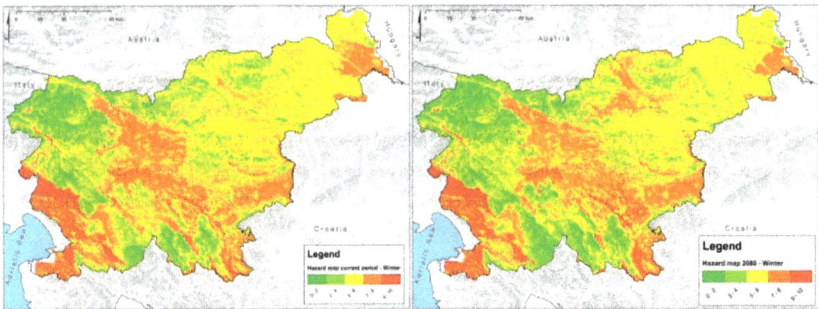

Figure 24. Slovenia winter hazard maps (left: current hazard situation; right: future hazard situation 2080)

From created fire hazard maps for year 2011 and 2080 we can speculate that the number of summer and winter fires will increase in the eastern part and decrease in the central part of Slovenia. One of the main reasons for these changes could be found in predicted changes of precipitation regimes. The Slovenian Karst region will remain a high risk area also in the future.

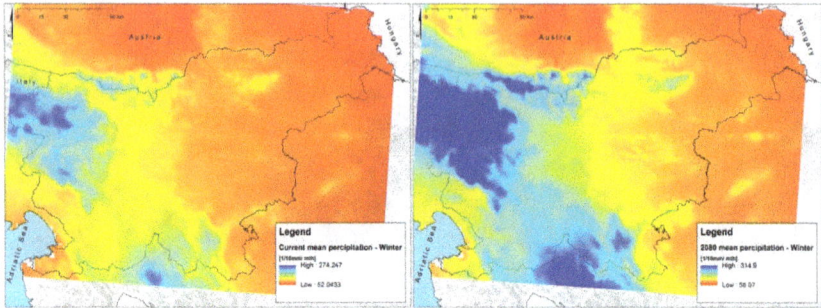

Figure 25. Winter precipitation maps for Slovenia for current period and prediction for the year 2080 (source: WSL)

5. Introduction and objectives

It is well known that wildfires create profound changes in the ecosystem, and in terms of remote sensing signal these changes provoke variations in surface reflectance, moisture and temperature. The use of spectral vegetation indices is a common tool in many studies regarding forest regeneration after fire disturbances. For example, the NDVI (Normalized Difference Vegetation Index), is associated to vegetation greenness and provides a means for monitoring density and vigor of green vegetation, being the most extensively used to assess and monitor post wildfire processes. Another index, sensitive to canopy greenness and canopy structure is the EVI (Enhanced Vegetation Index) that is becoming common in studies related to burned area mapping and assessment. Satellite remote sensing has been successfully employed to evaluate post-fire dynamics and, depending on the eco-regions, a post-wildfire recovery period has been found vary from 5 to 9 years (boreal forest of Canada) and more than 13 years in Siberian forests. However, no study has been carried out to evaluate the performance of satellite vegetation indices to analyze forests after a wildfire disturbance in Alpine areas.

This study investigates the dynamics of postwildfire in Alpine forests exploiting time series of NDVI and EVI indices derived from Moderate Resolution Imaging Spectroradiometer (MODIS) on board the Terra satellite. Time series from 2000 to 2010 were used to analyze characteristic temporal patterns and post-wildfire dynamics in the Lombardia Region (Italy) in order to evaluate different behavior of post-wildfire recovery period.

5.1. Methodology

Study area and forest burned area database

The study area is located in Lombardia region, Northern Italy and it is composed by a mosaic of deciduous, evergreen forest and pasture, where wildfires mainly occurs in winter period.

The forest burned area map used in this study was produced by ERSAF Lombardia and it was used to select the samples for this investigation. Overall, dataset consists of 3385 polygon occurred from 2000 to 2009. For this study, based on MODIS data, only large fires were considered (area> 40 ha) and a subset of 84 events was considered. Post-wildfire dynamics were investigated by analyzing the selected burned areas in comparison with paired adjacent unburned areas (control plots). Particular attention was paid to extract burned and unburned surfaces. Pixel of interest were identified by intersecting the MODIS grid (250x250 m) with the map of fires (polygons), retaining only the pixels which showed at least 80% of burned surface area.

Development of satellite time series - MODIS 16 days composite NDVI and EVI data with 250 m spatial resolution acquired from Terra platform in the 2000-20010 period were used to evaluate the post-fire dynamics. Maximum NDVI and EVI values were automatically extracted and the start and end of season dates were then computed as the day of the year (DOYs) corresponding to the first and to the last zeroes of the third derivative of the fitted curve. The pairs of burned and unburned areas were grouped according to the year of the fire. The difference of the vegetation indices was then calculated for each pair to analyze the response of NDVI and EVI to disturbance over time, accounting for the influence of interannual variability and other environmental factors captured in the unburned areas. In this case the zero on the time scale of the time series represents the year of burn.

5.2. Results and discussion

The EVI time series of burned and unburned area along 10 years of the Monte Argua Covallo winter fire event that involved 75 ha of a mixed deciduous forest type is shown in Figure 26. The EVI shows reduced values at the time of wildfire followed by a gradual period of recovery. A one-way ANOVA indicated that after 4-5 years after the wildfire, the differences between the burned and unburned areas were not significant at the 95% confidence level for the EVI.

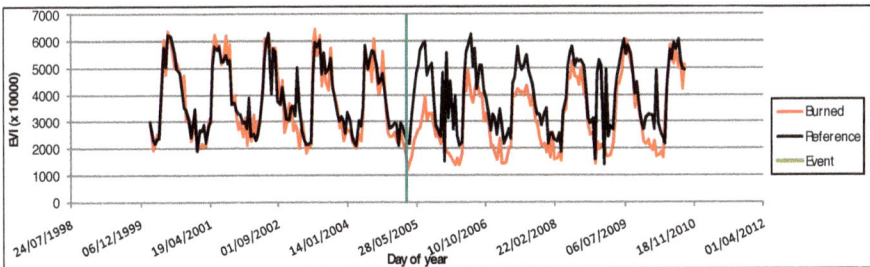

Figure 26. Example of the MODIS EVI time series of the of Monte Argua Covallo winter fire and for the reference surface.

Wildfire effects on EVI max and NDVI is clearly evident in Figure 27.

The drop in values at the time of the wildfire event corresponds to a massive or total loss of healthy, life vegetation combined with the reflectance signal of the ash covering the soil and burned materials. The subsequent increase in the first years following the fire is often caused by a rapid growth of the herbaceous layer and possibly surviving shrubs (Colombo et al., 2011). In the subsequent years, seed regeneration occurs. This gradually increases the amount

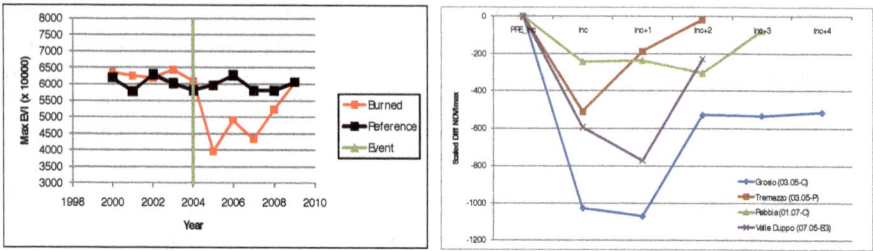

Figure 27. Example of the MODIS EVI maximum values time series for the of Monte Argua Covallo winter fire and for the reference surface (left) and the temporal variability of fires occurred on different land cover

of healthy green vegetation and re-introduces a vertical structure, causing a further increase in both NDVI and EVI.

The difference between the burnt area and the control sites can be considered a measure of the impact of wildfire on the NDVI/EVI. Although for the Monte Argua Covallo this impact is not very large, it continues until 4 years after the fire. After this, the forests follow a similar trend towards full recovery of the EVI signal. These observations are consistent with those found in previous study with similar recovery periods for different ecoregions.

The analysis of the length of the growing season shows that fire affects its temporal variability differently. It was found that some events are characterized by strong variations in the overall length of the growing season, while other recover quickly after fire events.

5.3. Conclusions

The post-wildfire regeneration of Alpine forests in Lombardia region was investigated using a 10-year observation period of remotely sensed MODIS NDVI and EVI time series. Satellite images allowed us to preliminarily explore the damage and vegetation recovery following wildfires. In this study, the terms regeneration or recovery of NDVI and EVI were related to the process of post-wildfire regeneration or recovery as measured by these indices and not the process occurring on the ground (e.g. the satellite signal may be the same even if different vegetation species are present after the fire). This means that, for example, the NDVI signal after a stand-replacing wildfire may return to the pre-wildfire levels but the vegetation on the ground is not in the same state (e.g. there are different species before and after the wildfire) as it was before the wildfire. By coupling data from adjacent burned and unburned control sites, we were able to separate interannual variations caused by the local climate from changes in the behavior due to a wildfire. Results indicate that spectral indices and phenological parameters can be useful to evaluate the response of the vegetation to fire events. Sometimes it takes more than 5 years for the NDVI and the EVI signal to fully recover after severe wildfire and the length of the growing season may decrease of about two weeks. In other cases, the effects are smaller, or not visible at all.

6. Forest management strategies

6.1. Risk mitigation: Key concepts

Prevention strategies are recognized as a key aspects in forest fire risk management and mitigation. On the basis of the adopted strategies, preventive measures against forest fire risk can be distinguished into indirect and direct (Figure 24).

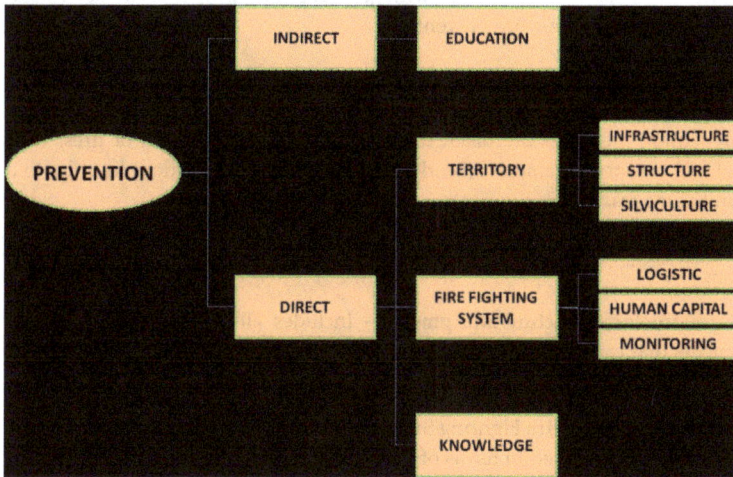

Figure 28. Summary scheme of preventive measures.

Indirect prevention includes strategies and activities aimed at controlling determinant causes of forest fires. It encompasses the set of measures designed for dissemination, training and education towards population. In detail, the activities are mainly focused on:

- reduction of forest fires due to human causes;

- protection of life and properties;

- increasing awareness and self protection abilities.

Direct prevention includes prevention activities acting on predisposing factors that may favor spread or control of fires. This is expressed through direct actions on the ground, on the fire alarm system (system organization, training and monitoring) and on knowledge system (research, planning).

6.1.1. Indirect prevention

Indirect prevention refers to activities against forest fires mainly oriented on determinant causes such as education and information on forest fires problem to population. In fact, at

Alpine level, the greatest part of forest fires are caused by human activities, often due to careless and unintentional actions linked to agricultural and forestry operations. Hence, targeted public education become a key issue in those areas showing an higher forest fire risk. Target groups need to be identified and adequately informed about fire risk by means of education and media (radio, television, press releases and news). In the process of target group definition it is always advisable to analyze the causes of forest fire. For instance, if fires are mainly caused by a carelessly tossed cigarette, then indirect prevention actions must be addressed to smokers, identifying the most appropriate channels to spread information. In general, it is important to increase awareness about forest fire risk with general public, properties owners, stakeholders and people involved in forest fire prevention.

6.1.2. Direct prevention

Direct prevention measures are taken to avoid outbreak and spread of fires, working on ecological and forestry parameters (i.e. density, structure, composition) in order to decrease danger and damages.

Three main categories can be considered in the analysis of direct prevention and forest fire risk mitigation:

• **Fire Infrastructure/Structure/Equipments** - Includes all the infrastructures, structures, equipments that can be used in forest fire fighting (roads, fixed tanks, water points, lakes and reservoirs for aircraft, helicopter landing sites, etc.).

• **Organization of Forest Fire Fighting System (FFFS)** - FFFS's has a central role in forest fires fighting. FFFS is expressed in terms of: number of teams, volunteers, vehicles, equipment (mobile tanks, modules, and blowers), support equipment (radio fixed, mobile and vehicular).

• **Planning and Monitoring (prevention)** - It includes action plans for forest fire fighting at ground level and monitoring measures carried out by means of aerial flights (air patrol) and/or fixed cameras positioned in strategic areas. This kind of prevention enables to seize peculiarities and problems of each territory, providing appropriate measures to achieve mitigation objectives.

6.1.2.1. Direct prevention — Territory

Infrastructure

The accessibility of the territory ensures the possibility of territorial monitoring and an early intervention in case of fire.

Road network. An efficient road network is important to allow access of forest fire fighting personnel, equipment and means, in areas prone to forest fires. Forest roads in particular are important to control fire spread, acting as physical barriers to fire movement and increasing accessibility for fire suppression activities. New forest roads should be planned according to forest fire risk plans in those regions more sensitive and prone to forest fire ignition and spread. Technical characteristics of agro-forestry road network, related to planning and management purposes, should facilitate forest fire fighting operations (e.g. setting of temporary water supply points). Forest road maintenance should be regularly carried out over time, planning interventions (e.g. cleaning of the channels), in order to keep the road surface in good condition, removing any obstacles that restrict the access to forest fire means and vehicles (i.e. logs or boulders). Accessibility is another key point to ensure early and appropriate interventions, avoiding that a fire develop into an uncontrolled event.

Structure

The presence of structures, functional for prevention and forest fire control, is an essential element in any efficient fire management system.

Water supply. Water supply (natural watercourses, dams, wells, ponds, underground water tanks, etc.) placed in strategic locations can be used in firefighting operations. Water reservoirs once implemented should be regularly checked for functionality and accessibility to enable their efficient use during fire suppression phases. Water supply can be distinguished in permanent water supply and mobile water supply. Permanent water supply are located and usually dimensioned in view of the mean burned area (derived from statistic data on forest fire) of a specific area. In order to allow a correct fire management, in the Alpine area the capacity level of permanent water supply have to reach at least 20000 liter per hour (l/h). The water has to get at least 1 m depth corresponding to the minimum level allowing a correct water supply of helicopters.

Permanent and temporary and helipads. Permanent and temporary helipads are settled up in those areas showing a higher probability of forest fire occurrence. Helipads need to be dimensioned to helicopter size (with at least a 30 meters side) and road network characteristics. It is also advisable that the location selection considers the minimum number of helicopter discharge during a forest fire event (15-20 per hour).

Silviculture

Preventive silviculture aims at preventing occurrence and spread of fires by decreasing the amount of flammable material and promoting ad hoc forest management practices to reduce forest susceptibility to fire. The main objective of silvicultural measures is to avoid that a ground fire transforms to a crown fire (difficult or almost impossible to extinguish). In this sense, one of the most effective ways to reduce the likelihood of having a crown fire is to reduce the amount of fuel. Silvicultural practices must be addressed to avoid situation of monospecificity, supporting at the same time conversion of coppice to high stand forests (highly recommended in chestnut forests) and presence and maintenance of less represented forest types. The most common silvicultural practices addressed to forest fire management and risk reduction are:

a. thinning, cutting and removal of small trees in presence of forests characterized by high tree densities or physiological stress (may include the removal of dead trees and shrubs);

b. selection cutting performed in adult and monospecific forests to improve structural complexity and to increase the presence of deciduous trees. Increasing the proportion of deciduous trees decrease the likelihood that a ground fire evolves into a crown one;

c. cultural practices adopted in intensively managed coppice, to decrease dead fuel availability;

d. conversions applied in degraded coppice located in areas with high danger of forest fires;

e. reafforestation programs to enhance the restoration of degraded forest habitats in presence of monospecific structure or over-managed areas;

f. environmental cleanup: post-fire treatments to remove dead vegetation;

Post-fire restoration generally refers to long-term efforts required to restore habitat quality, resilience, and productivity. According to the main forest function (timber production, disaster protection, recreation, environmental...) measures consist of:

1. cutting of burned trees: cutting of burned trees and release of subjects with the highest survival probability;

2. direct seeding: seeding of herbaceous and woody species to prevent surface run-off;

3. reafforestation in order to mitigate potential increases in runoff and erosion which can occur immediately after a wildfire and promote wide range of forest associations less prone to forest fires, more resilient, productive and with higher biodiversity values

4. construction of fire breaks (vegetated fire breaks, protective strips and fuel breaks):

5. vegetated breaks consist of strip of land (100-300 meters wide buffer strip) with presence of trees offering an higher degree of resistance to fire passage. The strip is suitable to contain the spread of forest fires increasing the probability that a crown fire downgrades into an easier to control fire

6. protective strip is a strip of land (20 to 30 meters wide buffer strip) covered by tree from which readily flammable material (twigs, shrubs, dry or deadwood) has been removed. The strip is created with the aim to decrease the flame front speed thus aiding fire suppression operations

7. fuel break is a strip of land (5 to 20 meters wide buffer strip) constructed and maintained along railways, motorways and highways and forest roads in those areas showing an high forest fire danger rate. In particular, fuel breaks along forest road network should be made on both sides of the roads

8. prescribed fire. Prescribed fire is the application of fire to a specific land area to achieve both prevention purposes (i.e. fire hazard reduction) and resource management objectives (e.g. plant community restoration, landscape and habitat conservation, etc). In forest fire prevention this technique is widely used to reduce the amount of ground fuel readily flammable. Fuel reduction is essential to prevent crown fires which burn at high intensity and are capable of causing irreversible damages. At Alpine level this fire management

tool has been recently introduced and fruitfully applied mainly for hazard reduction purposes.

6.1.2.2. Direct prevention — Fire fighting system

Logistic

The changing technology and increasingly demanding for safety led to the need of a regular equipment adaptation, means and resources for forest fire operators. Logistics includes PPE (Personal Protective Equipment), up to date means (aerial and terrestrial), office and field support for mobilizating firefighting resources during suppression operations (e.g personnel with GIS & GPS expertise).

Human capital

Human capital refers to the knowledge, skills, and abilities which people involved in forest fire fighting gathered over time trough education, training and culture. One of the key factors influencing incidence and spread of forest fires is the inadequate organizational system of fire prevention, prediction and active fight. This issue is mainly related to:

- poor capacity of cooperation;
- lack of knowledge about behaviour of the phenomenon;
- preparedness response and recovery

For that, long term training programs need to brought into play to provide (to qualified personnel) definite techniques and procedures to be adopted during suppression activities.

Monitoring

Remote sensing systems. Traditional and remote sensing systems for automatic monitoring and early detection of forest fires are of great importance in the context of prediction and prevention of fire danger and risk. The most used systems refer to ground or aero-satellite monitoring.

Ground monitoring stations. Ground monitoring stations are based on cameras located in strategic places. Terrestrial systems based on ground monitoring stations (video cameras, thermal imaging cameras, infrared spectrometers) allow an early detection of sources of heat with a low rate of false alarms.

Aero and satellite systems. Ground monitoring system is integrated by aerial monitoring. Planes are equipped with high spatial resolution cameras and a GPS system. GPS data allow for georeferencing and for orthorectification of the captured images in post-processing. Acquired images and data can also be send in real time to the operative centers of civil protection.This system allow to:

- provide information and images on forest fire behaviour;
- survey and mapping burned surface in real time;

- provide, during a forest fire, real time data useful for better coordination of suppression activities to local entities (municipalities, mountain communities, etc.).

Recently new remote sensing technologies like light detection and ranging (LiDar) and near infrared images (NIR) have produced significative advances in forest fire prevention.

Forest fire hazard indices. Forest fire hazard indices are a early warning tool designed for monitoring and predicting forest fire danger. Fire hazard indices are normally calculated on a daily basis starting from meteorological data as input (obtained by weather stations or radars). Fire hazard indices are recognized as a valuable tool allowing alert, preparedness and readiness of fire fighting services and civil protection. The most widespread index used to assess fire danger is the Canadian Forest Weather Indices (FWI).

6.1.2.3. Direct prevention — Knowledge

Knowledge management refers to the achievement and combination of technical and scientific advances, to accomplish resource and fire management goals and objectives. These information are directly transferred into prevention programs (i.e. fire management plans) and implemented on a broad legal and institutional framework. Fire management plans are the key element to identify and analyze risks caused by forest fire hazards providing long-term and sustainable solutions to address prevention, preparedness, readiness and recovery of communities, infrastructures, biodiversity and ecosystem.

Acknowledgements

This work was conducted under the European Project MANFRED, funded by the European Regional Development fund of the Alpine Space Program, reference number 15-2-3-D (MANFRED). We acknowledge the Italian National Forestry Corps and Valle Camonica Mountain Community for their support in this work.

Author details

Bruna Comini[1], Giampaolo Cocca[1], Elena Gagliazzi[1], Paolo Nastasio[1], Enrico Calvo[1], Roberto Colombo[2], B. Di Mauro[3], Lorenzo Busetto[2], Mitja Skudnik[3], Tomaz Sturm[3] and Andrej Breznikar[3]

1 Regional Agency for Development in Agriculture and Forestry (ERSAF) of Lombardia Region, Italy

2 Remote Sensing of Environmental Dynamics Lab., University of Milano-Bicocca, Italy

3 Slovenian Forestry Institute - Department of Forest and Landscape Planning and Monitoring, Slovenia

References

[1] Allen, K. (2003). Vulnerability reduction and the community-based approach, in Pelling (ed.), Natural Disasters and Development in a Globalising World. , 170-184.

[2] Brooks, N, Adger, W. N, & Kelly, P. M. (2005). The determinants of vulnerability and adaptive capacity at the national level and the implications for adaptation. Global Environmental Change. , 15, 151-163.

[3] Colombo, R, Busetto, L, & Fava, F. Di Mauro, B., Migliavacca, M., Cremonese, E.,Galvagno, M., Rossini, M., Meroni, M., Cogliati, S., Panigada, C., Siniscalco, C., Morra di Cella, U., (2011). Phenological monitoring of grassland and larch in the Alps from Terra and Aqua MODIS images. Italian Journal of Remote Sensing. , 43(3), 83-96.

[4] Conedera, M, Torriani, D, Neff, C, Ricotta, C, Bajocco, S, & Pezzatti, G. B. (2011). Using Monte Carlo simulations to estimate relative fire ignition danger in a low-to-medium fire-prone region. Forest Ecology and Management. , 261, 2179-2187.

[5] IPCC(2000). Nebojsa Nakicenovic and Rob Swart (Eds.) Cambridge University Press, UK. , 570.

[6] Wastl, C, Schunk, C, Lüpke, M, Cocca, G, Conedera, M, Valese, E, & Menzel, A. (2013). Large-scale weather types, forest fire danger, and wildfire occurrence in the Alps. Agricultural and Forest Meteorolology , 168, 15-25.

Chapter 8

Ozone Fluxes to a Larch Forest Ecosystem at the Timberline in the Italian Alps

Giacomo Gerosa, Angelo Finco, Antonio Negri,
Riccardo Marzuoli and Gerhard Wieser

Additional information is available at the end of the chapter

1. Introduction

Ozone is a phytotoxical pollutant causing negative effects on vegetation at biochemical, physiological, individual and ecosystem level [1]. Alpine forests, could experience high level of ozone concentrations, in particular in the southern part of the Alps. The reason of this phenomena is explained more in detail in the chapter 9, as well as the ozone effects on vegetation.

At the moment, from a regulative point of view, AOT40 (Accumulated Ozone over a Threshold of 40ppb) is the only instrument to evaluate the ozone hazard, even if its scientific soundness is widely discussed [2,3]. In fact, the AOT40 is an exposure index which does not take into account the physiology of the vegetation which is exposed to that ozone concentration. Since the greatest damages to vegetation are produced by the ozone entering through the plant stomata and plants can regulate stomatal opening as a response to the environmental parameters, a new approach based on stomatal ozone fluxes as been proposed by UN/ECE. In fact, the magnitude of the negative effects of ozone on vegetation is related to the real amount of this pollutant taken up through stomata [4], i.e. the dose or stomatal flux, and high environmental ozone concentrations in the air do not necessarily lead to high ozone doses, representing only a potential risk (more correctly an hazard).

The effective ozone dose, based on the flux of ozone into the leaves through the stomatal pores, represents the most appropriate approach for setting future ozone critical levels for forest trees. However, uncertainties in the development and application of flux-based approaches to setting critical levels for forest trees are at present too large to justify their application as a standard risk assessment method at a European scale [5].

The scientific community is hence moving toward an evaluation of the ozone risk based on stomatal ozone fluxes. This can be realized by means of measures or by models. Measurements allow to estimate only local ozone risk but they are necessary to parameterize and validate the models, which, once all the input data are available, allow to estimate ozone risk at regional, national or continental level.

Measurements of ozone fluxes are hence the first fundamental step for a proper evaluation of the ozone risk. The most used technique for the measure of ozone fluxes is called eddy covariance. This technique is based on the atmospheric turbulence and it requires particular instrumentation, which must be able to measure at least ten times per second the three wind components, the air temperature and the ozone and other gases concentrations. This kind of instrumentation must be mounted above the studied ecosystem. Additional meteorological measurements are useful for a better comprehension of the exchange process.

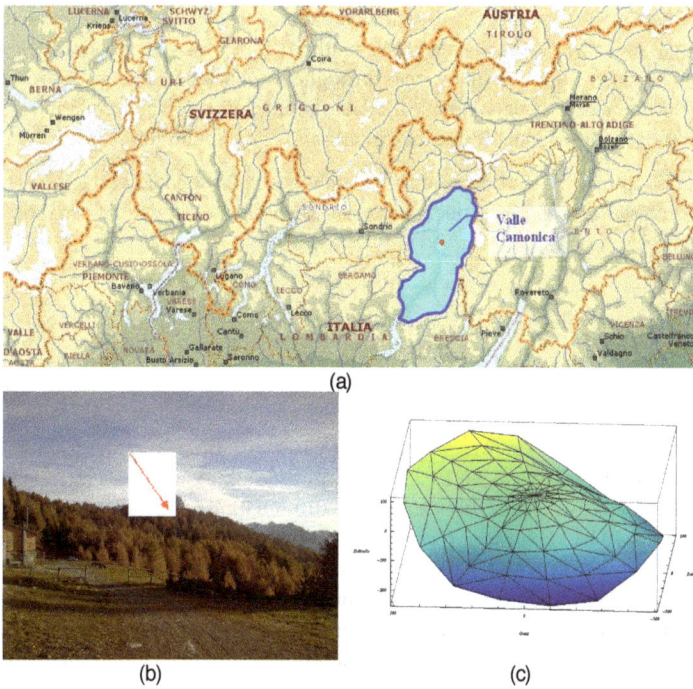

Figure 1. a) Geographical location of the Case Study 4: Valle Camonica, Northern Italy. The red dot indicates the position of the eddy covariance tower.; b) the studied larch forest, the photo was taken from a nearby refuge; c) local orography of the measurement site, the centre of the figure corresponds to tower localization

The eddy covariance technique allow to measure the total ozone deposition to the whole ecosystem, without distinguishing between how much ozone enters through plant stomata (the most harmful pathway) and how much is destroyed on the plant surfaces or on the terrain. In order to estimate the ozone stomatal fluxes, water is used as a tracer. In fact, it is assumed that ozone can enter into plant stomata when they are open during carbon uptake and plant transpiration: for this reason concomitant water fluxes measurements allow for the partition the total ozone fluxes between the stomatal and the non-stomatal component [6].

Once the stomatal flux is known it is possible to calculate the ozone dose (which is simply the sum of the stomatal ozone fluxes) and the Phytotoxical Ozone Dose (POD_1, which is the cumulated dose over the threshold of 1 nmol O_3 m^{-2} s^{-1}) [4]. The POD_1 have been introduced in the UN/ECE scientific community because it takes into account the internal capability of the vegetation to detoxify part of the ozone entering through the stomata. Moreover, many experiments have showed that the POD_1 is better correlated with the biomass reduction than the simple ozone dose, allowing thus to estimate the harmful effects of ozone on vegetation.

In this chapter an example of micrometeorological measurements of ozone fluxes over a high elevation larch forest, performed within the framework of the MANFRED project, are presented and discussed. The aims of this field campaign were to quantify the ozone deposition to a forest ecosystem at the timberline in the southern Alps (the more exposed to ozone) and to assess the actual ozone taken up by trees through stomata; a further aim was to gather data for the development of a stomatal uptake model to be employed for the simulation of the ozone deposition to timberline forests in future climatic and ozone pollution scenarios.

Till now ozone uptake by larch at treeline has only been investigated at the twig level [7], so the Eddy Covariance measuring technique has been chosen to get information at whole ecosystem level.

2. Methodology

2.1. Measurements

A micrometeorological tower was run for two years in Paspardo, Valle Camonica, in the Brescia district in Italy (46°2′40″N, 10°23′04″) (Figure 1a). Measurements interested only the summer season of the two years: July-October in 2010 and June-September in 2011. The studied ecosystem was a 10 ha larch forest (*Larix decidua*, Mill) with a mixed grass understorey used as a cattle pasture at 1750 m a.s.l.. Trees were between 100 and 350 years old and their average height was about 26 m. The trees coverage of the population had a LAI equal to 0.9 and a SAI equal to 0.7, while the understorey grass and shrubs had a LAI equal to 1.3 and a SAI equal to 0.2. The ecosystem lays on a westward gentle slope, with an inclination ranging between 3° and 13° (Figure 1b and Figure 1c).

The selected micrometeorological technique was the eddy covariance which requires fast response instrumentation for wind, temperature, water and ozone. For this sake an ultrasonic three axial anemometer-thermometer (USA-1, Metek, D) was installed at 29 m, on the top of the tower, just above the canopy. At the same height an open path krypton-light hygrometer (KH2O, Campbell, USA) was set. Ozone concentrations were sampled from air drawn near the sonic anemometer with two instruments: a fast response one (COFA, Ecometrics, I) and a standard UV photometer (1308, SIR, E) used as a reference since COFA uses cumarine targets whose sensitivity decays in 5 days and have to be changed. At four different levels of the tower additional thermo-hygrometrical (HD9000, Deltaohm,I), radiation (LPAR01, Deltaohm, I) were installed in order to obtain vertical profiles of each parameter. Moreover three reflectometers (CS616, Campbell, USA) and three thermopiles (SHFP, Hukseflux, NL) were deepen into the ground to measure soil water content and soil heat fluxes. In order to assess the energy balance closure a net-radiometer (NR-LITE, Kipp&Zonen, NL) as well as a pyranometer (LI 200 SZ, LI-COR, USA) were installed on the top of the tower. Finally a rain gauge (mod. 52202,Young, USA) was set in a little nearby clearance and a leaf wetness sensor (mod.237, Campbell, USA) was used too. Fast sensor data were sampled 20 times per second and collected by a personal computer through a customized program, saving data in a new file every half an hour. Slow sensor data were sampled by a datalogger (CR10x and AM416, Campbell, USA) every 15 seconds and the 30 minutes averages of each parameter were stored.

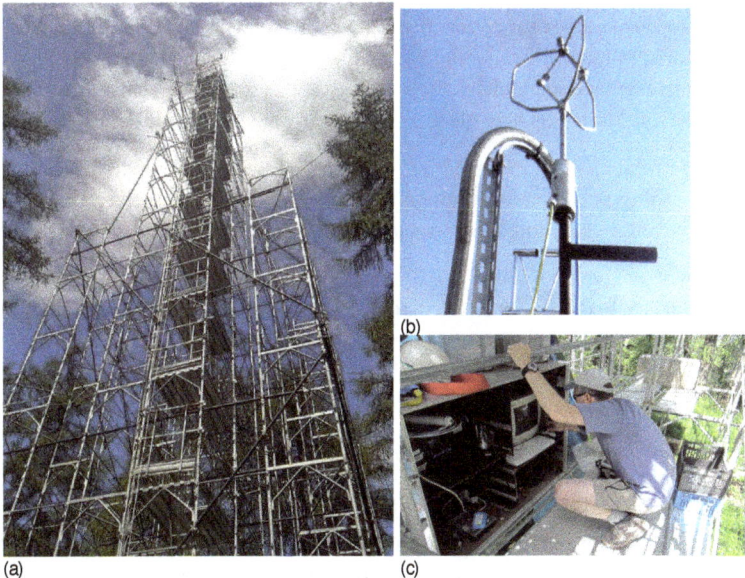

Figure 2. a) The flux tower in Paspardo; b) photo of the instrumentation installed on the top of the tower: the ultrasonic anemometer, the fast hygrometer and the inlet of the fast ozone analyzer; c) box at the bottom of the tower, inside the box there was the data acquisition system and the ozone reference analyzer

2.2. Data processing

Vertical fluxes of sensible (H) and latent (LE) heat and ozone (FO₃) were calculated from the 30 minutes data files of the fast instrumentation as the covariance between the vertical component of the wind and the corresponding scalar quantity, following the eddy covariance theory [8]. In particular these fluxes are obtained with the following equations:

$$H = \rho c_p \overline{w'T'} \tag{1}$$

$$LE = \lambda \rho \overline{w'q'} \tag{2}$$

$$FO_3 = \overline{w'O_3'} \tag{3}$$

where ρ is the air density, c_p is the specific heat at constant pressure, w is the vertical component of the wind intensity, T is the air temperature, λ is the latent heat of vaporization, q is the specific humidity and O_3 is ozone concentration; primed variables mean the fluctuations around their 30 minutes averages which are represented as overscript bars [7].

The covariances in the above mentioned equations are usually obtained from the rotated covariance matrix, following the methodology proposed by McMillen [9]; the rotations of the covariance matrix remove small tilts of the sonic anemometer from the verticality. In this case, since the forest was located on a gentle slope an *ad hoc* rotation procedure was developed.

The coordinate rotations align the "tower" vertical axis to the perpendicular of the wind streamlines. In this way the mean w component is zeroed and the advective flux component is removed. The angle of the rotation varies in agreement to the slope in the upwind direction (Figure 1c).

Once the total ozone fluxes have been calculated, a flux partition procedure has been employed in order to separate the two main deposition pathways: stomatal uptake by leaves and non-stomatal ozone disruption by both chemical sinks and ozone removal from non-living surfaces.

The partition procedure is based on an electrical analogy which is typical of the SVAT (Soil Vegetation Atmosphere Transfer) models where the flux corresponds to an electrical current flowing through a resistances network. Every resistance describes a part of the whole deposition process whose driving force is represented by the ozone concentration differences between the measuring height z and the substomatal leaf cavity at the standard height $d+z_0$ (d 2/3 of canopy height, $z_0$1/10 of canopy height, for further details see, for instance, [2], where ozone concentration is assumed to be zero [10].

The total resistance, which is obtained directly from measurements as the ratio between the ozone concentration at the measuring height and the total ozone fluxes, is equal to three series resistances:

$$R_{tot}(z) = R_a(d + z_0, z) + R_b + R_c \tag{4}$$

where R_a is the aerodynamic resistance that ozone faces during the turbulent transport from the height z to the height $d+z_0$ (momentum sink), R_b is the resistance faced by ozone while crossing the thin layer of still air surrounding leaves (diffusive transport) and R_c is the integrated resistance of the exchanging surface (leaves, stems, soil).

R_a was calculated following the formulation proposed by Dyer [11] while R_b was calculated using Hicks et al. [12] equation. R_c is finally obtained as a residual from equation (Eq. 4) all the others variables being known.

R_c is hence considered equivalent to two parallel resistances:

$$R_c^{-1} = R_{ST}^{-1} + R_{NS}^{-1} \tag{5}$$

where R_{ST} is the stomatal resistance and R_{NS} is the non-stomatal resistance.

R_{ST} could then be deduced from the Penman-Monteith equation [13] which describes the water loss process between a wet surface and the atmosphere. This equation is based on the energy balance closure at the evaporating surface so, when the net incoming energy (net radiation) as well as all the other energy losses (soil, sensible and latent heat fluxes) are known -as in our case-, R_{ST} is the only unknown term which can be derived by the inversion of the equation.

Finally the stomatal flux is obtained, using the Ohm's:

$$F_{ST} = \frac{R_c}{(R_a + R_b + R_c)R_{ST}} C_m \tag{6}$$

where C_m is the ozone concentration at the measuring height.

The stomatal dose, D, is simply given by the integral of F_{ST} over the measuring period (from t_a to t_b).

$$D = \int_{t_a}^{t_b} F_{ST}(t)dt \tag{7}$$

The UNE/ECE POD$_1$ is the dose which exceeds the instantaneous threshold of 1 nmol m^{-2} s^{-1}.

$$POD_1 = \int_{t_a}^{t_b} \left[F_{ST}(t) - 1 \right] dt \quad \forall F_{ST} \rangle 1 \; nmol O_3 m^{-2} s^{-1} \tag{8}$$

For comparison purposes, the currently set AOT40 ozone exposure index was calculated too, as it follows:

$$AOT40 := \sum_{\substack{[O_3]_i > 40\,ppb \\ RadGlob > 50W/m^2}} ([O_3]_i - 40) \cdot \Delta t \tag{9}$$

2.3. Estimation of the ozone uptake by Larch trees

The dose calculated in the previous paragraph takes into account the ozone fluxes entering all the stomata of the whole ecosystem, i.e. the stomata of grass and trees. In order to estimate the uptake of the larch needles only, a two layers resistive model was developed (Figure 2). Differently from the SVAT model of the previous paragraph, this model is a prognostic model which tries to predict the stomatal ozone flux from meteorological and atmospheric turbulence data. This model simulates separately the stomatal behaviour of the trees (1st layer) and of the understorey grass (2nd layer) as well as the ozone removal by chemical reactions with terpenes in the trunk space and the ozone deposition to the underlying soil and the external non transpiring surfaces (cuticles, branches, stems). The stomatal processes were modelled using the Jarvisian approach adopted also by UN/ECE *Manual on the Methodologies and Criteria for Modelling and Mapping Critical Loads and Levels and Air Pollution Effects, Risks and Trends*, hereafter simply called Mapping Manual [5]. A maximum stomatal conductance of 125 mmol m^{-2} s^{-1} was chosen for larch, following the findings of Sandford and Jarvis [14] and Wieser et al.[15], while the other parameterizations were taken from the generic continental conifers in the UN/ECE Mapping Manual, since for larch nothing else was available. The second layer was parameterized as the generic grass in UN/ECE Mapping Manual, which prescribes a maximum stomatal conductance of 270 mmol m^{-2} s^{-1}.

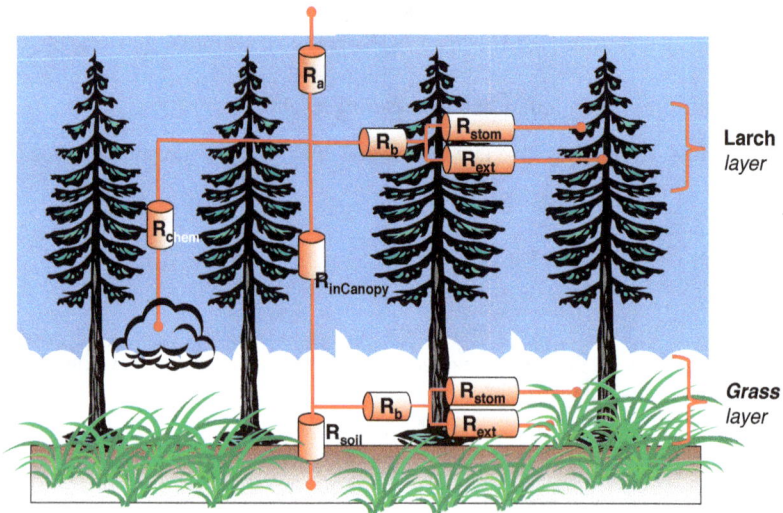

Figure 3. Deposition scheme of the prognostic multi-layer model: going from the top to the bottom the first resistance is the aerodynamic one (R_a). Each vegetation layer (i.e. larch one and grass one) is composed by a sub-laminar resistance (R_b) followed by two parallel resistances: a stomatal one and a non stomatal one. The larch and the grass layer are connected passing by an in-canopy resistance. In parallel to this one there is also the chemical resistance (R_{chem}). Below the lower layer (the grass one) there is the soil resistance (R_{soil}).

3. Results

The meteorological conditions of the two measuring periods are reported in Table 1. Comparing the common trimester (July-September) of the two years (2010 and 2011), it is evident that 2011 summer was hotter and drier than 2010. The average temperature ranged between 9.4 °C and 15.4 °C in 2010 and between 13.6 °C and 18.1°C in 2011. In the same period the rain received by the forest ecosystem was 806 mm in 2010 and 431 mm in 2011.

	Unit	July 2010	August 2010	September 2010	October 2010	June 2011	July 2011	August 2011	September 2011
$T_{Average}$	[°C]	15.4	13.3	9.4	6.2	12.7	13.8	18.1	13.6
T_{Max}	[°C]	20.3	18.2	12.8	10.9	19.5	19.2	23.7	16.5
T_{Min}	[°C]	11.0	8.2	3.6	0.7	8.7	9.4	12.02	3.9
Rainfalls	[mm]	89	291	426	55	90	67	156	208
$RH_{Average}$	[%]	70.4	77.0	78.9	85.7	79.5	74.2	72.9	74.2
RH_{Min}	[%]	43.9	38.5	43.5	66.7	57.7	49.4	47.1	51.8
PAR_{Max}	[μmol $m^{-2} s^{-1}$]	637	596	469	304	638	616	529	403
$SWC_{Average}$	[%]	23	28	29	30	31	23	25	26

Table 1. Meteorological and soil measurements; $T_{average}$ is the monthly average of the air temperature, T_{Max} is the monthly maximum value of the daily averages of air temperature, T_{Min} is the monthly minimum value of the daily averages of air temperature, Rainfalls are the cumulated rainfalls in the month, $RH_{average}$ is the monthly average of the relative humidity, RH_{min} is the minimum of daily averages of the relative humidity, PAR_{Max} is the monthly maximum of the daily averages of PAR, SWC average is the monthly average of soil water content. Air temperature, air relative humidity and PAR measurements are referred to top canopy measurements.

Most of the energy received by the ecosystem as solar radiation (R_n) was dissipated backward to the atmosphere as sensible heat (H) which reached on average a daily maximum value around 300 W m^{-2} at 3.00 PM local time (GMT +2) (Figure 3). A minor part of the energy was employed to drive the evapo-transpirative processes (LE) with a daily maximum of 150 W m^{-2} and an almost constant minimum of 20 W m^{-2} in the dark hours. The residual energy, a very little part with respect to the other two components, heated the ground (G) with a maximum average daily value of 12 W m^{-2} in the late afternoon. The heat stored in the soil during the daylight hours was completely returned to the ecosystem.

Figure 4 shows a detailed example of the energy fluxes (Figure 4a) as well as of the ozone concentrations and fluxes (Figure 4.b) measured at the top of the tower. Each point in Figure 4 is a semi-hourly averaged sample from 36000 rapid measurements collected 20 times per second.

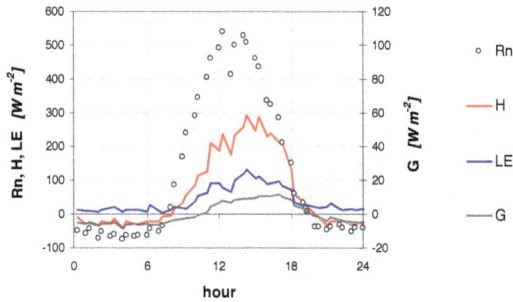

Figure 4. Energy fluxes at the Paspardo flux tower between July and September 2010. Mean daily courses of Net radiation (Rn), sensible heat flux (H), latent heat (water) flux (LE), soil heat flux (G). The scale of G curve has been enhanced five times for a better reading.

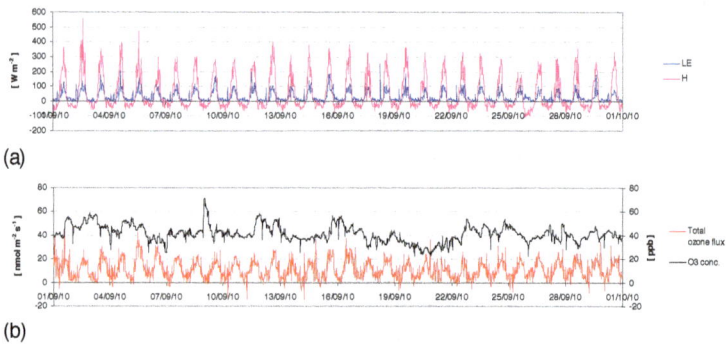

(a)

(b)

Figure 5. Example of semi-hourly measurements of a) sensible (H) and latent heat (LE) fluxes, and b) ozone concentrations (right axis) and total ozone fluxes (left axis) in a selected month (September 2010) at the Paspardo flux tower.

The ozone fluxes showed in Figure 4b are the net ozone amount received by the ecosystem from the atmosphere. In the convention adopted here a positive ozone flux value means a flux directed from atmosphere toward the ecosystem, while a negative value means the opposite direction. The studied ecosystem behaved as a net ozone sink with very rare episodes of ozone effluxes in the night-time. Despite ozone concentrations varied during the presented month ranging from 23 to 71 ppb the behaviour of the total ozone fluxes was almost the same. This fact highlights that the ozone deposition is not only driven by the atmospheric chemistry but also by the atmospheric turbulence and the plant physiology, which in turn responds to the soil water content.

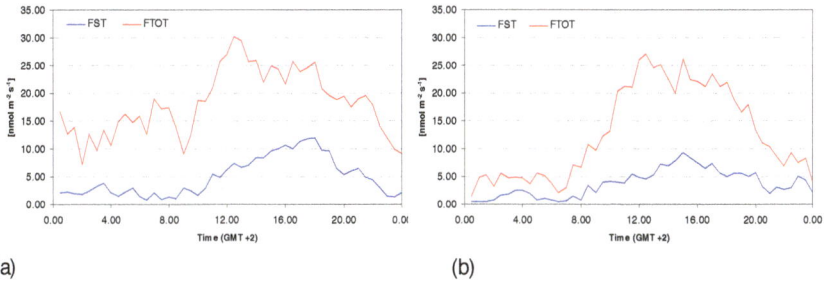

Figure 6. Ozone flux partition a) in 2010 and b) in 2011.

The importance of the plant physiology is further confirmed by the results of the flux partition (Figure 5), where it is evident that a significant part of the ozone deposition was removed by the stomatal activity. This process accounted for a maximum flux of about 10 nmol m^{-2} s^{-1} which was nearly 50 % of the ozone deposition in the afternoon of both years, even if this daily stomatal peak was located in the late afternoon in 2010 and earlier in 2011. During the night a residual stomatal uptake around 2 nmol m^{-2} s^{-1} was observed, revealing an incomplete stomatal closure as already reported by Matyssek and Innes [16].

On average the daily peak of the total ozone deposition (30 nmol m^{-2} s^{-1} in 2010 and 27 nmol m^{-2} s^{-1} in 2011) was measured at noon when turbulence is usually high. At the same time, the non-stomatal deposition, i.e. the ozone disruption on the external non-living surfaces, was maximum as can be inferred by the difference between the two curves in Figure 5 which is just the non-stomatal deposition.

The diurnal behaviour of stomatal and total fluxes, of course, showed a clear dependence on the seasons as it can be observed in Figure 6, which presents as an example the average daily course of every month in the 2010 measuring period. The total flux (Figure 6b) increased until August, when it reached its maximum values, and then decreased until October when it became four times lower. On the contrary, the stomatal flux maximum was observed in July and, in the following months, it decreased gradually reaching values three times lower in October.

The stomatal fraction of the total ozone deposition, during the central hours of the day, was 40%, 24%, 23% and 37% respectively in July, August, September and October. The non-stomatal deposition hence was always the main ozone sink for this measuring site.

Ozone concentrations (Figure 6a) reached their maximum values in August and after that slowly decreased. It is worth noticing the reversed bell shape of the ozone concentration daily courses which is typical of high elevation sites [17]. The concentration minimum occurred when the total ozone deposition was maximum. This fact highlights the sink role of the forest in the ozone removal from the atmosphere, a role which decreased as the end of the growing season was approaching.

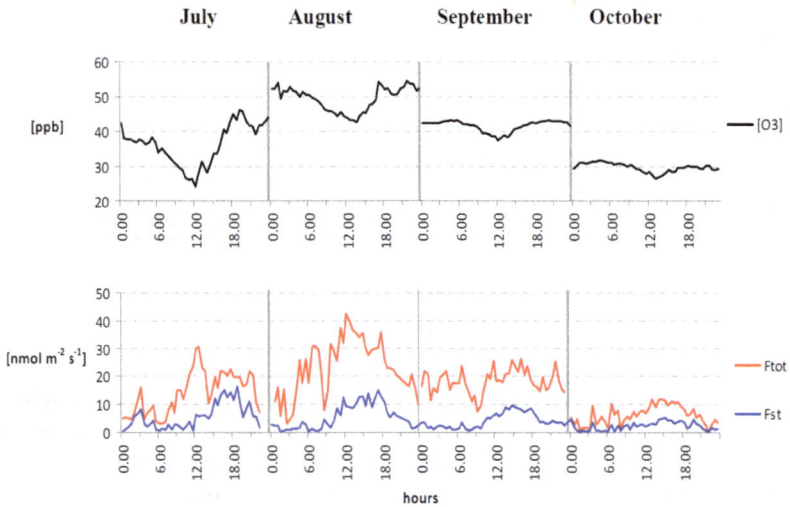

Figure 7. Mean daily courses of ozone concentration (a) and ozone fluxes (b), both total and stomatal fluxes, registered in the four months July-October 2010 at the Paspardo flux tower.

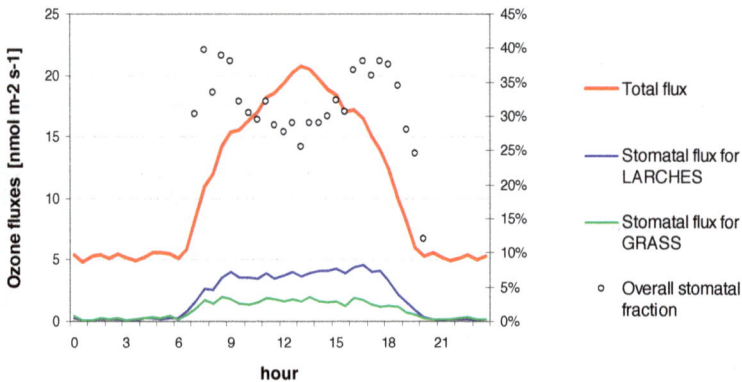

Figure 8. Model sub-partition of stomatal flux into a larch component and a grass component (2010 data)

In order to assess the ozone risk for the larch only, it was necessary to separate the stomatal flux taken up by the whole ecosystem into a stomatal component for the larch forest and a stomatal component for the understorey grass. The results of the sub-partition modelling exercise on 2010 data are showed in Figure 7. The model results were in satisfactory agreement with the measured fluxes, taking into account the correction by Moelder et al. [18] for the measurements taken in the roughness sub-layer, as in our case.

In the central hours of the day, the ozone fraction taken up by the larch needles through stomata was around 70% of the bulk stomatal flux calculated for the whole ecosystem. The remaining part (around 30%) was due to the understorey grass uptake. On average, the larch stomatal flux was about 25% of the total ozone deposition to the whole ecosystem.

Taking into account the uptake of both larch and grass, their overall stomatal uptake was on average 32% of the total flux, thus confirming the results of the flux partition obtained from the measurements. It is worth noticing that most of the ozone was destroyed by non-stomatal processes (68%) which include deposition on soil, leaf cuticles and stems.

Using the outputs of this model in order to estimate the phytotoxical ozone dose received by the larch trees during the whole measuring period, a POD1 value of 17.9 mmol m^{-2} was found, which is more than twice the critical flux level set by UN/ECE to protect Norway spruce, the species which was chosen as representative for all conifers.

By applying the flux-effect relationship proposed by UN/ECE for Norway spruce, a biomass reduction around 4 % can be predicted.

This reduction should be intended as the missing plant growth in the measuring period. Even if this value seems low, it is not negligible considering a pluriennial time scale. In fact it could be responsible of a slowing of the forest development as well as a minor plant capability to cope with other biotic and abiotic stressors.

It is important to highlight that the UN/ECE flux effect relationship had been developed on epigean biomass data only. However some studies showed a more significant effect of ozone uptake on root development [19, 20]. On a long time scale this effect could lead to a general impairment of the plant capability to stabilize mountain slopes and to prevent hydro-geological instability.

In any case it should be important to remark that the quantification of the effects is based on Norway spruce, the only species for which data are currently available in literature. The effects on larch could be lower or even worse but this is the best that could be done with the present knowledge. Ozone will obviously affect even the understorey vegetation and the other ecosystem components. So the estimated negative effects are likely higher than those predicted for larch only.

From a regulative point of view, the larch forest experienced an ozone exposure of 5.1 ppm h in the measuring period. This value was calculated as AOT40, the exposure index currently in use, following the EU and Italian legislation. Using the relationship between epigean biomass reduction and AOT40 for Norway spruce [5], in this case a biomass reduction of 0.8 % can be expected, a value that is 5 times lower than the corresponding biomass reduction estimated with the POD1 approach. This great difference underlines once more the criticism addressed toward AOT40, which does not take into account the real interaction between ozone and the plant, but only mimes it by the exclusion of the night-time measurements.

In this case, although this situation is typical of the mountain sites and particularly of the most elevated ones, the relatively low AOT40 value is the consequence of the ozone concentrations minima experienced in the midday hours of the daily concentration profile (Figure 6a). On the contrary the highest concentrations were always measured at night, when the AOT40 index is not calculated. It is worth noticing that the highest stomatal fluxes are experienced just when the ozone concentrations are lower, thus causing this risk assessment discrepancy. For this reason, besides the fact that the plant physiological processes are taken into account, the flux-based approach is considered more scientifically sound and advanced.

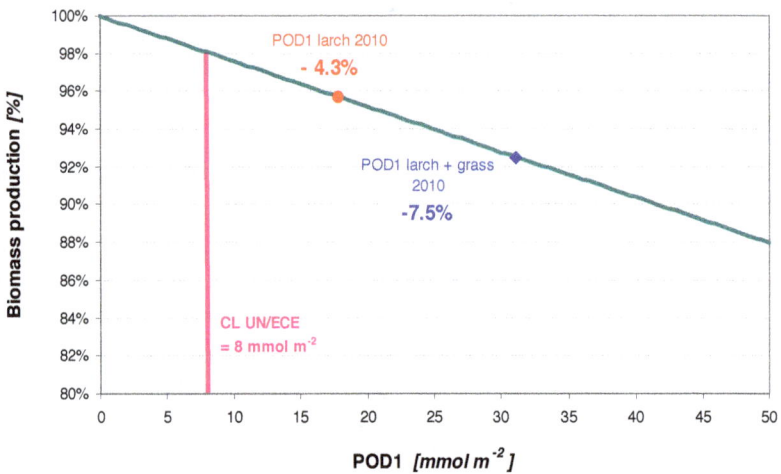

Figure 9. Estimated reduction of biomass production in Larch due to the ozone negative effects. See text for explanation.

The flux-based approach allows further considerations on the forest ecosystem services and in particular on the forest capability to remove the air pollutants, an aspect which is difficult to quantify, and it is not often treated.

In order to assess this aspect it necessary to consider also that part of ozone deposition which is called non-stomatal flux. This process is responsible for the removal of 38 kg O_3 ha[-1](Figure 9), which are destroyed without harming the vegetation, while the total amount of ozone removed by the forest ecosystem is equal to 53 kilograms of ozone per hectare in three months, an amount that is remarkable. In fact, considering an air volume with a basis of 1 ha and a thickness of three meters, containing ozone at a concentration of 90 ppb (i.e. nearly 180 μg m^{-3}, the attention threshold for the Italian regulation) the amount of ozone is only 5 grams.

This significant ecosystem service has a counterpart of forest growth reduction of about 4%.

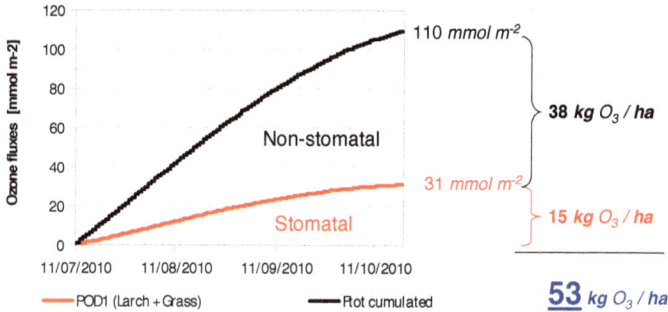

Figure 10. Evolution of stomatal and total fluxes during the measuring period in 2010.

4. Conclusions

The MANFRED project allowed to run a micrometeorological tower to measure directly ozone fluxes over a larch forest ecosystem for the first time in the Alpine region, with the eddy covariance technique.

Results showed that ozone removed by the forest was significant (53 kg ha^{-1}), but the quantity absorbed by leaves was the minority (between 23 to 40%), the most being non-stomatal.

A double layer (4-sinks) SVAT model was developed to study the sub-partition of the stomatal fluxes between the larch trees and the understorey grass, as well as to give an insight to the non-stomatal processes. The model is now suitable for simulations of ozone uptake in climate change scenarios.

Larch needles uptake 25% of the total ozone flux, around 70% of the bulk stomatal flux. The remaining 30% of the bulk stomatal flux was absorbed by the understorey grass.

Even taking into account the detoxifying capability of the plants (instantaneous flux threshold) the phytotoxically active ozone dose (POD1) appears above the critical level for fluxes provisionally set by the UN/ECE to 8 mmol m^{-2} for Norway spruce, the species chosen as a reference for conifers. At such ozone flux levels, the calculated POD1 for larch suggests a possible trees growth reduction of 4 – 5% in a growing season. This missing growth is counterbalanced by the offer of an ecosystem service, that is the removal of 53 kg ha^{-1} of ozone from the atmosphere.

Acknowledgements

This publication was funded by the Catholic University's program for promotion and divulgation of scientific research. The authors thank the Adamello Park and the mountain community of valle Camonica, themunicipality of Paspardo and in particular dr G.B. Sangalli, dr A. Ducoli and dr D. Orsignola for their support in realizing this field campaign. The authors would like also to thank Stefano Oliveri and Luca Francesco Garibaldo for their important help during the installation. A sincere thank to Michela Scalvenzi for her support in 2011.

Author details

Giacomo Gerosa[1], Angelo Finco[1], Antonio Negri[1], Riccardo Marzuoli[1] and Gerhard Wieser[2]

*Address all correspondence to: giacomo.gerosa@unicatt.it

1 Mathemathics and physics department, Catholic University of the Sacred Heart, Italy

2 Department of Alpine Timberline Ecophysiology, Federal Research and Training Centre for Forests, Natural Hazards and Landscape, Innsbruck, Austria

References

[1] Fuhrer, J, Skarby, L, & Ashmore, M. R. Critical levels for ozone effects on vegetation in Europe. Environmental Pollution(1997).

[2] Gerosa, G, Finco, A, Mereu, S, Vitale, M, & Manes, F. and Ballarin Denti, A.: Comparison of seasonal variations of ozone exposure and fluxes in a Mediterranean Holm oak forest between the exceptionally dry 2003 and the following year, Environmental Pollution (2009). , 157, 1737-1744.

[3] Sofiev, M, & Tuovinen, J. P. Factors determining the robustness of AOT40 and other ozone exposure indices. Atmospheric Environment (2001). , 35, 3521-3528.

[4] Mills, G, Pleijel, H, Braun, S, Büker, P, Bermejo, V, Calvo, E, Danielsson, H, & Emberson, L. González Fernández I., Grünhage L, Harmens H., Hayes F., Karlsson P.-E., Simpson D.. New stomatal flux-based critical levels for ozone effects on vegetation. Atmospheric Environment (2011). , 45(28), 5064-5068.

[5] Ece, U. N. Mapping Manual Revision, (2004). UNECE convention on long-range transboundary air pollution. Manual on the Methodologies and Criteria for Modelling and Mapping Critical Loads and Levels and Air Pollution Effects, Risks and Trends. <www.icpmapping.org>.

[6] Gerosa, G, Vitale, M, Finco, A, Manes, F, Ballarin-denti, A, & Cieslik, S. Ozone uptake by an evergreen Mediterranean forest (Quercus ilex) in Italy. Part I: Micrometeorological flux measurements and flux partitioning. Atmospheric Environment (2005). , 39, 3255-3266.

[7] Wieser, G, & Havranek, W. M. Environmental control of ozone uptake in Larix decidua Mill.: a comparison between different altitudes. Tree Physiology (1995). , 15, 253-258.

[8] Gerosa, G, Marzuoli, R, Cieslik, S, & Ballarin-denti, A. Micrometeorological determination of time-integrated stomatal ozone fluxes over wheat: a case study in Northern Italy. Atmospheric Environment (2004).

[9] Mcmillen, R. T. An eddy correlation technique with extended applicability to nonsimple terrain. Boundary-layer Meteorology (1988). , 43, 231-245.

[10] Laisk, A, Kull, O, & Moldau, H. Ozone concentration in leaf intercellular air spaces is close to zero. Plant Physiology (1989). , 90, 1163-1167.

[11] Dyer, A. J. A review of flux-profile relationships. Boundary-Layer Meteorology (1974). , 7, 363-372.

[12] Hicks, B. B, Baldocchi, D. D, Meyers, T. P, Hosker, R. P, & Matt, D. R. A Preliminary multiple resistance routine for deriving dry deposition velocities from measured quantities. Water, Air and Soil Pollution (1987). , 36, 311-330.

[13] Monteith, J. L. Evaporation and surface temperature. Quarterly Journal of the Royal Meteorological Society (1981). , 107, 1-27.

[14] Sandford, A. P, & Jarvis, P. G. Stomatal responses to humidity in selected conifers. Tree Physiology (1986). , 2, 89-103.

[15] Wieser, G, Häsler, R, Goetz, B, Koch, W, & Havranek, W. M. Role of climate, crown position, tree age and altitude in calculated ozone flux into needles of Piceaabies and Pinuscembra: a synthesis. Environmental Pollution (2000). , 109, 415-422.

[16] Matyssek, R, & Innes, J. L. Ozone- A risk factor for trees and forests in Europe? Water Air and Soil Pollution (1999). , 116, 199-226.

[17] Loibl, W, Winiwarter, W, Kopcsa, A, Zuger, J, & Baumann, R. Estimating the spatial distribution of ozone concentrations in complex terrain using a function of elevation and day time and Kriging techniques. Atmospheric Environment(1994). , 28(16), 2557-2566.

[18] Moelder, M, Grelle, A, Lindroth, A, & Halldin, S. Flux-profile relationships over a boreal forestroughness sublayer corrections, Agriculture andForest Meteorology, (1999).

[19] Grulke, N. E, Andersen, C. P, Fenn, M. E, & Miller, P. R. Ozone exposure and nitrogen deposition lowers root biomass of ponderosa pine in the San Bernardino Mountains, California, Environmental Pollution (1998).

[20] Mccrady, J. K, & Andersen, C. P. The effect of ozone on below-ground carbon allocation in wheat, Environmental Pollution (2000). , 107(3), 465-472.

Assessing Present and Future Ozone Hazards to Natural Forests in the Alpine Area — Comparison of a Wide Scale Mapping Technique with Local Passive Sampler Measurements

Angelo Finco, Stefano Oliveri, Giacomo Gerosa,
Wilfried Winiwarter, Johann Züger and
Ernst Gebetsroither

Additional information is available at the end of the chapter

1. Introduction

1.1. Ozone and vegetation

Ozone is considered as one of the most phytotoxic pollutant for vegetation [1]. Ozone penetrates the leaves through the stomata and it dissolves in the water film enveloping the cellular walls of the substomatal cavity. The reaction of ozone and water releases oxidative radicals like hydrogen peroxide, superoxide and similar compounds, which are known as ROS (Reactive Oxygen Species). The ROS are responsible for oxidative stress causing biological injuries. The injuries begin with the membranes disruption through the peroxidation of the membrane lipids, and go on with the oxidation of the reduced groups of biomolecules, particularly enzymes with a subsequent change of the biochemical pathways. Once ROS enter the chloroplasts, the chlorophylls of the photosystem reaction centers are damaged leading to a general reduction of the plant growth and productivity [2, 3, 4].

The microscopic effects can appear as visible injuries at leaf level, such as chlorosis, bronzing and necrosis, as increased transparency at crown level, as a reduced root growth, and as a general increased susceptivity to both biotical and abiotical stresses [5]. Nevertheless in many cases ozone damages may not show any visible symptoms but only physiological changes which can however lead to a productivity decrease [6,7].

At ecosystem level this stressor can cause slow structural changes to the plant community composition, favoring more resistant species[8]; in fact, different level of plant sensitivity to ozone were observed. Functional changes at ecosystem level can affect ecosystem services reducing carbon storage capability, increasing water loss and reducing the forest capability of stabilizing mountain slopes.

No direct ozone sources, both anthropogenic and natural, exist, ozone in fact is a secondary pollutant which is produced by photochemical reactions between nitrogen oxides (NO_x) and volatile organic compounds (VOC). These reactions are burst by UV radiation and hence the ozone production is obviously greater in summer and at higher elevation, provided not limiting air concentrations of the precursors (i.e. NO_x and VOC). NO_x mostly come from high temperature combustions both from vehicles and energy production plants, while VOC can be emitted also by vegetation itself (isoprene and terpenes) as well as by the same anthropogenic sources of NO_x when the combustion is not perfectly balanced. In mountain regions the breeze cycle generates pollution transport from valley bottom or planes, where typically precursor sources are located, to higher elevation forested areas. As a consequence mountain forests can experience higher ozone concentrations than the valley floor. In the southern side of the Alps these processes are enhanced because of the higher solar radiation and temperature, the higher elevation and the great number of precursor sources in the Po valley; furthermore the enclosed morphological configuration of the Po basin can contribute to trap the ozone precursors and to push them towards the mountain tops.

1.2. Ozone hazard assessment an EU legislation

Ozone hazard was widely studied during last decades. The United Nations Economic Commission for Europe (UN/ECE) gave birth in 1979 to the Convention on Long-range Transboundary Air Pollution (LRTAP). The Convention is composed by more 50 states from Europe, Asia and North-America and several protocols were signed to identify specific measures to limit their emissions of pollutants and gradually reduce pollution levels and also long-range transboundary air pollution. In particular, the Gothenburg protocol (1999) fixed the critical levels of ozone for vegetation with target values for the following years. A following EU directive (2002/03) acknowledged the Gothenburg protocol, which already explained how to assess the ozone hazard, introducing the AOT40 (Accumulated Ozone Exposure over a threshold of 40 ppb) index. The AOT40 is defined as:

$$AOT40 := \sum_{\substack{[O_3]_i > 40ppb \\ RadGlob > 50W/m^2}} ([O_3]_i - 40) \cdot \Delta t \tag{1}$$

that is the AOT40 is calculated as the sum of the concentrations of ozone (when they are above 40 ppb) minus the threshold of 40 ppb multiplied by the measuring time which is usually one hour. This sum is calculated only during daylight hours in order simulate the stomatal behavior of vegetation. Even though this index reflects the concept of exposure, the choice of calculating it only during daylight hours was an attempt to mime the concept of dose, that is the amount of ozone taken up by plants through stomata, whose opening is regulated by sun light (but also by other environmental factors).

The critical levels set by EU for vegetation is 18000 µg m⁻³ h over a period of six months, from 1st April to 30th September. This critical level can be expressed also in terms of ppb h, and its value is 9000 ppb h. UN/ECE, after the signing of the Gothenburg protocol, enhanced and promoted studies about ozone effects on vegetation and suggested a new critical level for forest (5000 ppb h during the growing season).

The choice of this new critical level was based on some experiments where seedlings of beech and birch (two sensitive species) where exposed to different levels of ozone concentrations, and hence different levels of AOT40 were measured; after the growing season a biomass reduction of the plants exposed to ozone was observed and plotted versus the corresponding values of the AOT40 (Figure 1). The critical level of 5000 ppb h corresponds to a potential biomass reduction of 5% (intercept of the two dashed lines). Knowing the AOT40 to which an ecosystem was exposed (for example 20.000 ppb h, that is equal to 20 ppm h), the potential biomass reduction can be estimated following the blue arrow first and then the red one, which leads to a 16% potential biomass reduction.

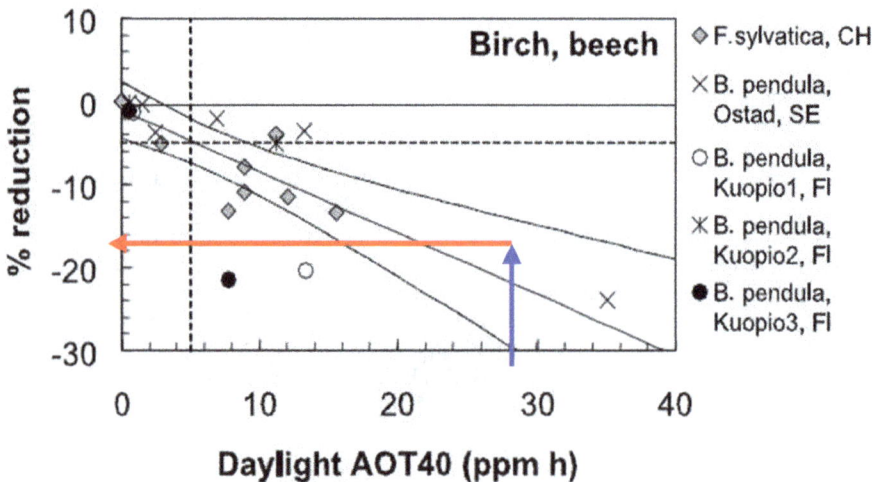

Figure 1. Modified from [9].The relationship between percentage reduction in biomass and AOT40, on an annual basis, for the deciduous, sensitive tree species category, represented by beech and birch. The blue and red line are used to estimate the biomass reduction as a function of the AOT40 value.

The EU directive gives recommendations to the member state about how to measure ozone and how to calculate AOT40, but, since the monitoring networks are mostly dedicated to estimate the risk for the population rather than vegetation, ozone data in mountain regions are not so often available, leading to a not proper evaluation of the ozone hazard for forests. Three main options are available to avoid this lack: mobile laboratory, passive samplers and modeling predictions.

Mobile laboratories are vans which are equipped with a small meteorological station and automatic instrumentation to measure continuously the pollutant concentrations (Figure 2). However, due to their high electric power consumption, mobile labs are seldom suitable for monitoring in forest remote areas.

Figure 2. The mobile laboratory employed in the MANFRED project

Passive samplers, on the contrary, do not require electricity since they are tubes which are filled with a substance which reacts to ozone (Figure 3). Passive samplers must be exposed to ambient air, typically for one week, and then collected and sent to chemical lab to obtain the average ozone concentration of the exposure period. Passive samplers are a valuable option for measurements in remote area even tough the time resolution of these devices is their main disadvantage.

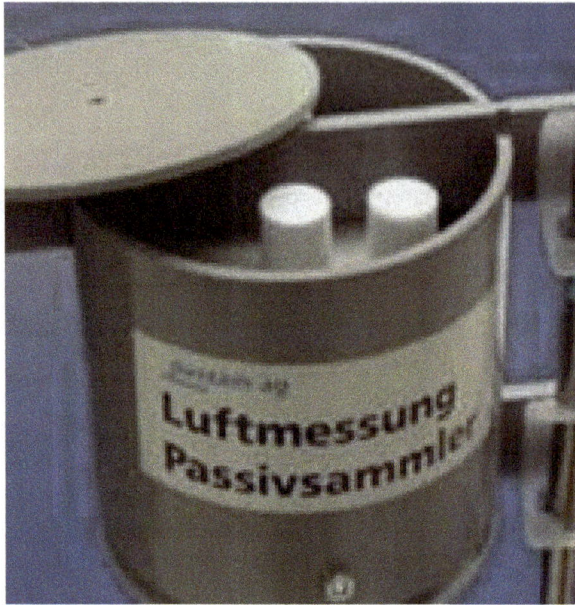

Figure 3. Ozone passive samplers

Ozone concentrations modeling is another valuable option which allows to map ozone hazard at wide scale, in some way avoiding the technical, economical and logistical constraints involved in the use of the instrumentation. However the goodness of the modeling outputs is strictly correlated to the quality of the input data supplied. This could be an important limitation as well as the fact that the model predictions should be validated with direct measurements in the studied area.

In this chapter the modeling of ozone hazard and the related AOT40 maps at Alpine scale will be showed and compared with direct measurements performed at a finer scale in a selected case study area. Finally an attempt of estimating the future ozone hazard in the Alpine area will be showed too.

2. The present ozone risk in the Alps and a case study in depth analysis

2.1. Mapping AOT40 at Alpine scale level

2.1.1. Model description

During the years 1990 - 1994 the Foresight & Policy Development Department of AIT developed a model to generate hourly ozone concentration maps using ozone monitoring data,

Assessing Present and Future Ozone Hazards to Natural Forests in the Alpine Area — Comparison
of a Wide Scale Mapping Technique with Local Passive Sampler Measurements

127

considering the influence of complex terrain on diurnal ozone variation within mountain areas. The model was designed and tested within the Geographic Information System (GIS) ARC/ INFO. Validity tests showed that the spatio-temporal dynamics of ground level ozone concentrations were calculated with high accuracy and spatial resolution. In 1994 the model was rewritten and included into the ozone monitoring network visualization interface at the Federal Environment Agency Austria (www.ubavie.gv.at).

The idea of the interpolation approach is to make use of the well known elevation dependence of ozone to calculate the overall spatial trend and to include monitoring data to consider seasonal and local influences. The approach needs a digital elevation model of the study area and a monitoring network.

The model is divided into several steps:

1. The core of the interpolation model is based on a function that reflects the dependence of ozone concentration on relative elevation and daytime in mountain regions.

2. The function generates daytime-specific standard ozone/elevation curves showing the increase of ozone concentration with increasing elevation.

3. Hourly ozone concentration measurements are used to include seasonal and day-specific influence to the average ozone concentration by fitting the standard ozone/elevation curves.

4. The day-specific curves and a digital elevation model are used to generate day-specific ozone concentration surfaces per daytime.

5. The deviations of the observed from the calculated ozone concentrations reflect the local influences at the monitoring stations. The interpolation of those deviations leads to a deviation surface.

6. The combination of the (hourly) day-specific ozone concentration surface with the hourly deviation surface leads to actual daytime- and day-specific ozone distribution maps of the study area considering the local influences.

The relative elevation is the vertical distance between the valley bottom and any location within the mountains. It is calculate out of a digital elevation model (DEM) where the valley bottom was assumed as the lowest absolute elevation within a distance of 5km around every DEM - grid point [10, 11].

The diurnal variation of ozone concentration depending on daytime and elevation is described by an analytical function. It was approximated by analyzing scatter plots of hourly ozone concentrations against relative elevation of the monitoring sites [10, 11].

Eq. 2 shows the basic function to model standard ozone concentrations:

$$O_{std}(h_{rel}, t) = (a_1 + a_2 \cdot e^{-(t-a_3)^2 a_4}) \cdot \ln\left(\frac{b_5 \cdot h_{rel}}{100} + b_1 + b_2 \cdot e^{-(t-b_3)^2 b_4}\right) - \frac{c_1}{1-(t-c_2)} \cdot \frac{h_{rel}}{c_3} \tag{2}$$

with

O_{std} = standard-ozone-concentration,

h_{rel} = relative height, t = time of day,

a_x, b_x, c_x = parameters of the function

This function generates a 2-dimensional trend surface of ozone concentration reflecting the dependencies of ozone on elevation and daytime (Figure 4). Slicing the surface parallel to the time-(x)-axis one gets the diurnal variation for any elevation level, slicing the surface parallel to the elevation-(y)-axis one gets ozone/elevation curves for any time of day. These standard ozone/elevation curves show the increase of ozone concentration depending on increasing elevation at a specific time of the day and are the basis to generate daytime-specific ozone concentration.

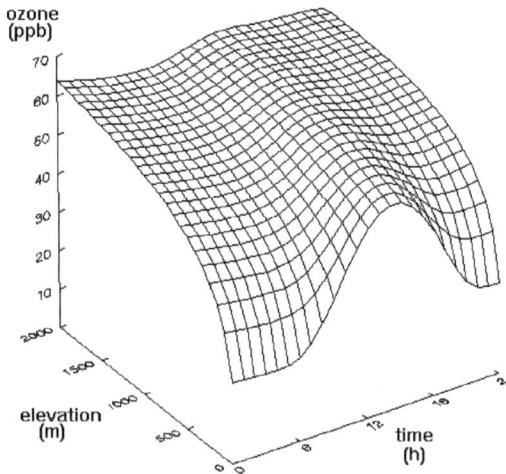

Figure 4. Trend surface of the standard ozone/daytime/elevation function

2.1.2. Ozone measurement data

The Ozone monitoring data have been taken from AirBase[1]. This is a public air quality database containing hourly air monitoring information for 35 countries throughout Europe. In total there are more than 2500 surface ozone measurement stations available. Approximately 330 AirBase ozone measurement sites can be found within the Greater Alpine Region, the study Area of the MANFRED project. The data availability for the countries surrounding the Alps (A, CH, D, F, I, FL, SLO) is shown in Figure 5.

_____

1 http://acm.eionet.europa.eu/databases/airbase/

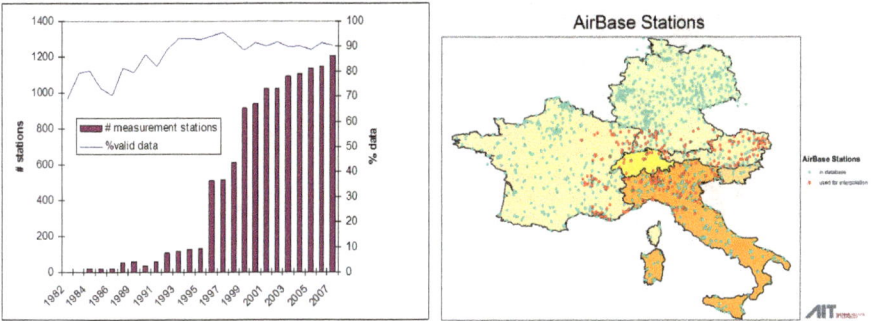

Figure 5. Number of measurement stations and amount of valid data for countries surrounding the Alps (left) and the Geographical coverage in 2007 (right)

2.1.3. Results for the Greater Alpine Region (GAR)

Ozone interpolation was done from 2003 to 2007 with 1 hour time and 1 km spatial resolution resulting in 43824 maps with 1018x668 grid points and app. 400 GB of data. In addition AOT40 maps with 1km spatial resolution were calculated for the years 2003 to 2007, which can be seen in Figure 6.

Figure 6 shows that the interannual variability of the AOT40 values is significant (2003 was an extreme year). The 5 yr mean (Figure 6, lower right map) gives an overview about the spatial distribution of the AOT40 values. In general a higher AOT40 value can be expected in the south of the alpine main ridge.

2.2. Measurements at local scale with passive samplers: The Valle Camonica case study

Valle Camonica was chosen as a case study area for ozone since this pollutant is a well known problem for the Lombardy region where it is located [12]. Even tough Valle Camonica is one of the biggest forested valley in Lombardy, few information on ozone hazard for forest were available for this valley and some ozone-like foliar symptoms were observed on beech leaves.

In summer 2010 a six month field campaign was performed with passive samplers located in 11 sites throughout all the valley: 10 of them were placed in remote forest areas while the remaining one was placed in Darfo, near the only automatic monitoring station available in Valle Camonica (Regional environmental agency, ARPA). The sites elevation ranged between 300 m and 1800 m a.s.l. spatially covering all the forest areas of the valley.

The employed passive samplers were polypropylene tubes with a glass fiber filter dipped in a solution of DPE (1,2-di(4-pyridyl)-ethylene) in acetic acid at the closed end (Passam ag, CH). The ozone diffusing in the tubes reacts stoichiometrically with DPE forming an ozonide which undergoes cleavage and yields an aldehyde. The amount of aldehyde is then determined spectrophotometrically in lab by the MBTH (3-methyl-2-benzothiazolinone hydraze ethylene)

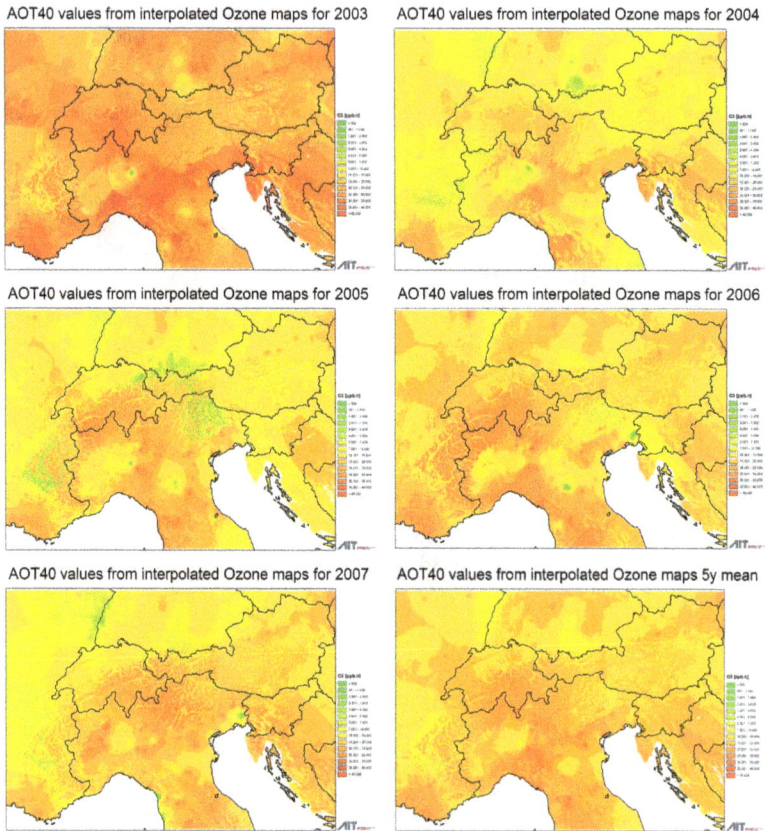

Figure 6. AOT40 maps for 2003 to 2007 and 5 yr mean, derived from interpolated 1-hourly ozone maps

method at 442 nm, giving directly the mass of the captured ozone. The ozone concentration in the air (μg m^{-3}) is calculated using the Fick's law, knowing the diffusion coefficient of ozone in the air and the tube geometry.

Two polypropylene tubes were exposed for one week in each site, inside a protective shelter preventing the advective ozone capture due to the wind (Figure 3). The first exposure began at 6th April 2010 and the last one (the 26th one) ended at 6th October 2010.

In order to calculate the AOT40, the collected ozone air concentrations were converted into volumetric ratios (ppb) taking into account the elevation of each site, following the methodology described in [13]. For this purpose temperature and pressure data from meteorological stations in the valley were used.

In order to calculate the AOT40 it is necessary to estimate hourly ozone concentrations starting from the weekly averages, the only data available from passive sampler measurements. For this sake the Loibl function (Eq. 2) was used. This function models the ozone concentration hour by hour given the relative height of the site, i.e. the difference between the elevation of the site and the elevation of the lowest point within a range of 5 km. The Loibl function has been shifted so that its average was set equal to the passive sampler value (weekly average) and it was also assumed that the daily course of all the days of the exposure period was the same. Further details on this methodology can be found in [14]. The Loibl function capability of predicting the ozone daily course in our case study has been tested by setting up 8 intensive monitoring campaigns in different sites of the valley with a mobile lab equipped with continuous analyzers.

AOT40 was then calculated for each measurement site (Eq. 1) using the hourly ozone concentration obtained from the shifted and replicated Loibl function for all the days of each exposure week. Weekly values of AOT40 were mapped on the Valle Camonica domain using a geostatistical technique known as ordinary kriging [10, 11], which requires a model of the spatial data variability estimated from the semivariogram plots.

Further details on this mapping technique can be found in [14]. Then the 26 weekly AOT40 maps were summed to obtain the ozone hazard map for forests in Valle Camonica (6th April – 6th October).

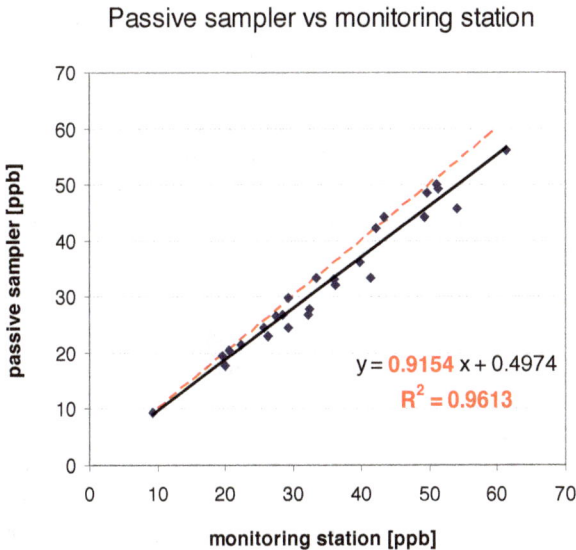

Figure 7. Passive sampler measurements compared with a nearby continuous analyzer

The performance of the passive sampler was very good as it can be verified from Figure 7 where the passive sampler measurements were compared with the calibrated continuous analyzer ones (R^2= 0.96, slope= 0.91).

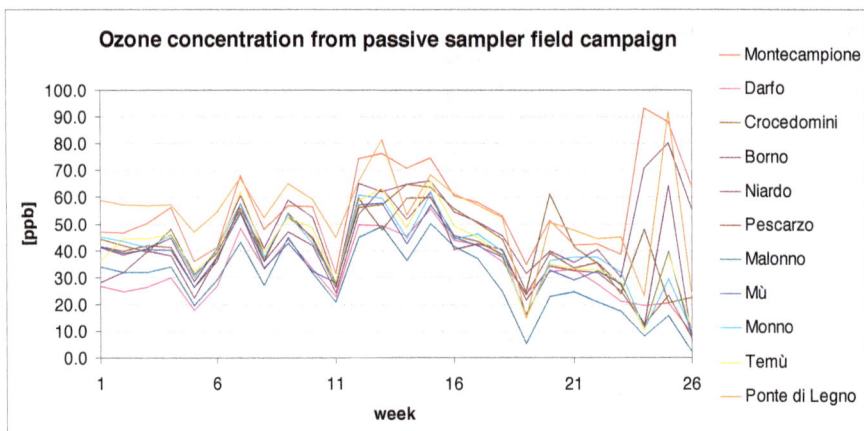

Figure 8. Ozone concentration measurements of all the sites in the measuring period

Figure 8 reports all the measurements of all the sites in the whole measuring period. There is a general increase of the ozone concentrations with increasing elevation. It is also evident a clear variability between one week and the other, which highlights the influence of local meteorology on the ozone production. Finally ozone levels show a seasonal trend which follows the intensity of the solar radiation, with a slight increase until the end of June, a maximum in July and a decrease from August. September values appears anomalous with respect to this general trend and they are likely due to ozone intrusion from the upper troposphere after the autumnal repositioning of the tropopause. In fact these anomalous values were measured mainly in high elevation sites.

Figure 9 shows monthly average maps of ozone concentrations in the studied domain. It can be clearly seen that ozone concentration is low in valley floor and it increases regularly with the elevation reaching the maximum values in the Adamello massif, at the eastern edge of the valley. High concentrations were observed also in the southern part of the valley, namely in the Borno upland and at the Montecampione relief, both above at an elevation of 1000 m a.s.l.. In June and in July a general increase of the ozone background level was measured with a marked increase in south-eastern part of the domain. In August average concentrations lowered in most of the valley with the exception of Borno and Montecampione areas.

The final map assessing the ozone exposure for forests, i.e. the ozone hazard expressed as AOT40, is showed in Figure 10. With the only exception of the valley floor nearby Malonno,

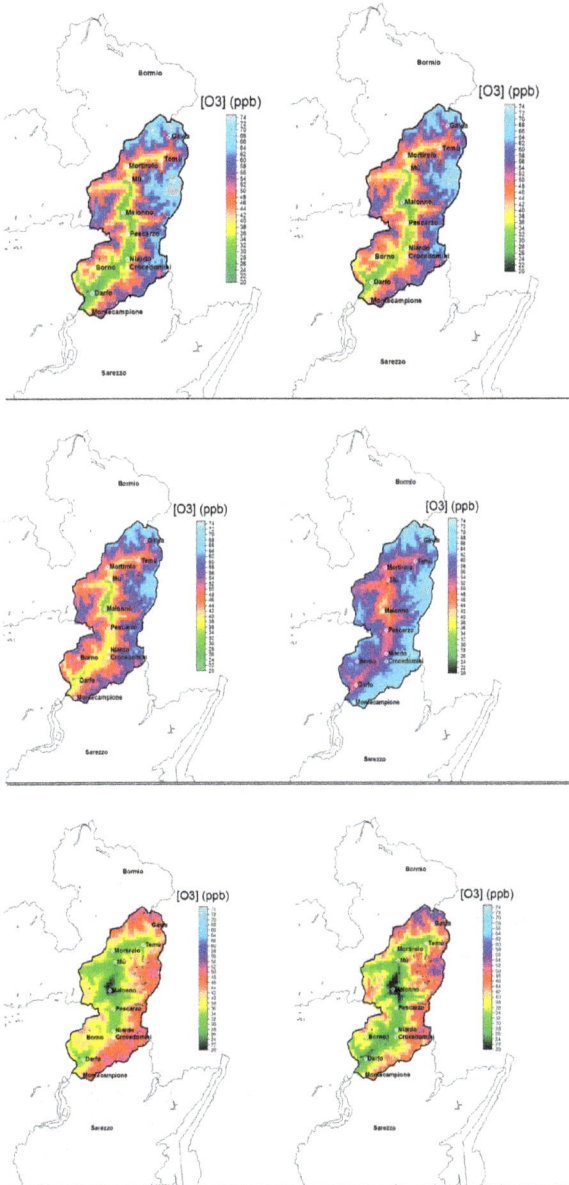

Figure 9. Monthly average maps of ozone concentrations in the Case Study 4

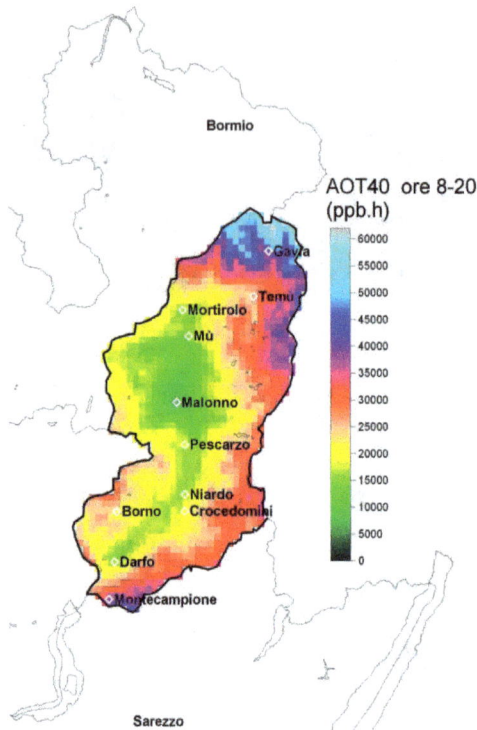

Figure 10. AOT40 map obtained from passive samplers measurements.

all the valley is above the EU critical level for forest (9000 ppb h), with exceedances up to 6 times the critical level in the most elevated areas. If the UN/ECE critical level is considered the situation is even worse. This map confirm a potential ozone risk for vegetation in this area.

For a deeper analysis of the risk for the forested area in Valle Camonica refer to the chapter about this case study area.

2.3. Comparison with the wider scale estimation (the case of Valle Camonica)

To compare interpolation results, as described in chapter 2.1, with the measurements (chapter 2.2) AIT started to run the ozone interpolation model for Valle Camonica on a very local scale. UNICATT provided ozone data for the period April, 1st 2010 to September, 30th 2010 from 19 measurement stations in the vicinity of Valle Camonica. In a first step we tried to use the interpolation routine in the same way as we did for the entire alpine region. With that we faced some troubles, mainly depending on the low number of stations. Originally the interpolation routine was developed to work with numerous stations (more than 300 for the Greater Alpine Area) and an acceptable station density. Especially the fact that a maximum of two stations at higher altitudes were available was unsatisfactory. As a result of this the algorithm was unable to fit the dependency curve in the top layer and therefore the whole interpolation did not work correctly. We removed the whole day specific fitting part from our routines and did the interpolation just based on the standard elevation/daytime ozone dependency curve. With these modifications the model worked satisfactorily. The following Figure 11 shows the location of measurement stations and AOT40 values from 2010 calculated with the modified interpolation method using 19 local ozone measurement stations. Additionally a comparison of different methods to estimate AOT40 values was done. Because the continuous measurement of ozone in forest areas is difficult often passive samplers are used. The disadvantage of this method is the coarse time resolution, which does not allow calculating AOT40 values directly from measurements. As we found a reasonable correlation between monthly mean ozone and AOT40 values we tested some variants of calculation. The top row in Figure 12 shows results for AOT40 values derived from interpolated 1-hourly ozone maps, calculated with monthly mean/AOT40 correlation function derived from 2010 Valle Camonica data and a function derived from 5-year data of all measurement stations used for the interpolation for the Greater Alpine Region.

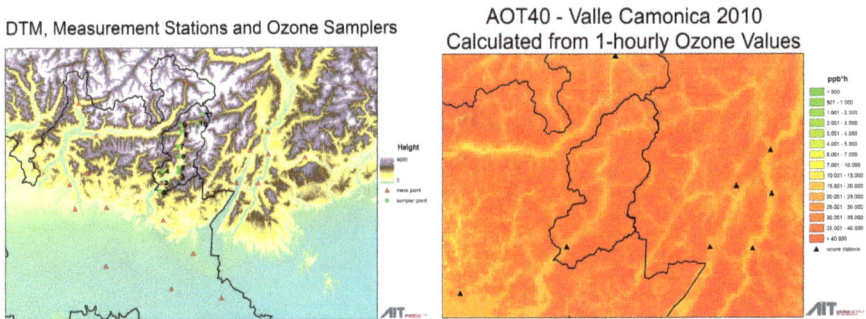

Figure 11. Location of passive samplers and ozone measurement stations (left) and AOT40 map for 2010 derived from 1-hourly Ozone maps produced by local interpolation for Valle Camonica (right)

3. AOT40 estimations in the Alpine area for the future

As for the future no hourly based estimations of Ozone concentration are available, we investigated the possibility to calculate the AOT40 values out of monthly means. For all years 2003 to 2007 the 1-hourly maps were aggregated to monthly means and the values at the locations of measurement stations were extracted. As shown in Figure 13 this estimations show a significant correlation between the monthly means and AOT40 values. So we used this approximation for the calculation of the future AOT40 values according to different ozone concentration scenarios.

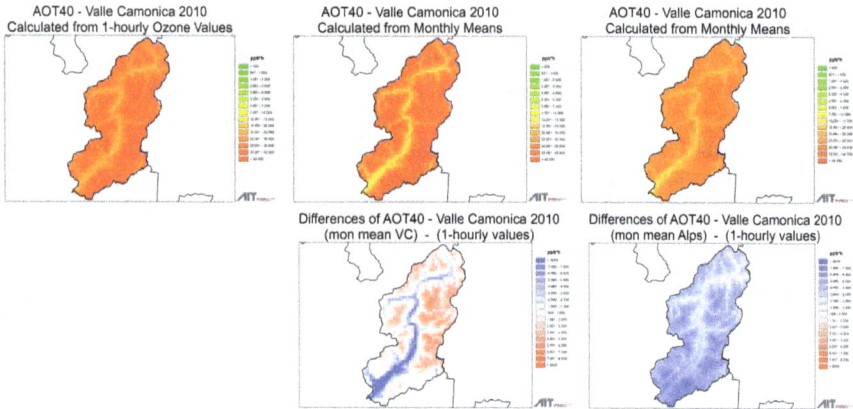

Figure 12. AOT40 maps for Valle Camonica 2010 (top row) derived from 1-hourly Ozone maps (left), correlation monthly mean/AOT40 using 2010 Valle Camonica data (center), using 2003-2007 measurement data from the alpine area (right) and differences compared to AOT40

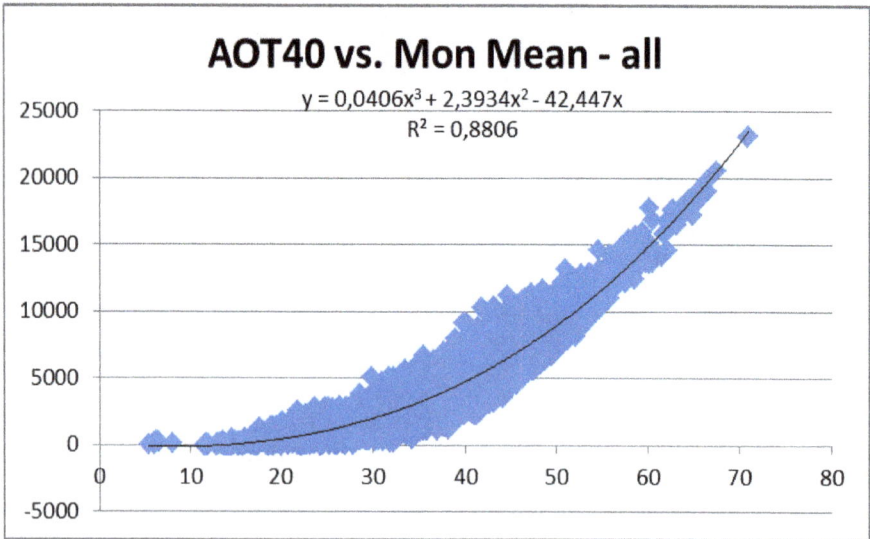

Figure 13. Correlation between monthly mean ozone and AOT40 values for all available stations in the GAR

Two scenarios[2], originally developed to research future climate conditions and providing ozone concentrations globally until the final decade of this century, were used to estimate the ozone hazard to forests. While application towards air pollution exceeds the scope of climate scenarios, at least some guidance on possible future situations may be derived. Scenarios chosen are "RCP2.6", developed as a low climate forcing scenario that minimizes anthropogenic climate impacts, and the "business as usual" scenario "RCP8.5" which extends current trends. They represent very different views of how the world may look in 2100.

The climate change influence on ozone concentrations from the first to the last decade of the 21st century has been obtained from the World Climate Research Program[3] (Figure 14) [15]. This data represent 10-year means of monthly mean ozone concentrations on a global scale with a spatial resolution of 5°. Within MANFRED this data were used to estimate changes of the background concentrations towards the end of this century, upon which we added the fine resolution data as obtained by the interpolation method.

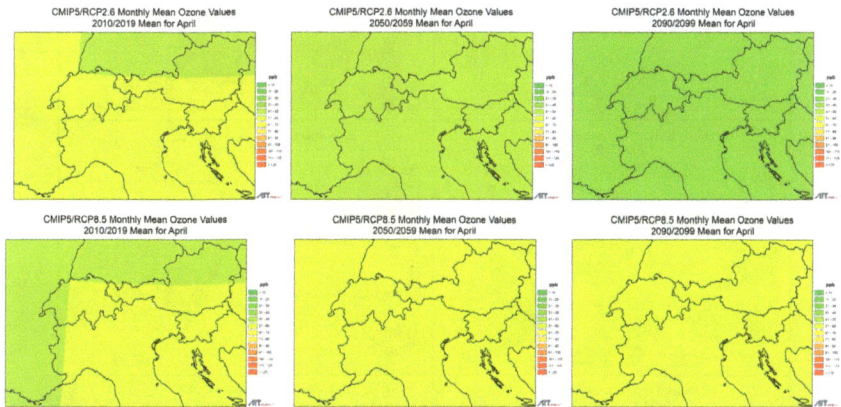

Figure 14. Estimation of future monthly mean ozone concentration – April 10-year means for scenario RCP2.6 (top row) and scenario RCP8.5 (bottom row) for decades 2010/2019, 2050/2059 and 2090/2099 (from left to right)

The change signals found in this dataset (Figure 15 second row) were added to the 5-year means (2003-2007) of monthly mean ozone concentration derived from the 1-hourly interpolated ozone concentration maps (Figure 15 first row). For future scenarios AOT40 maps have been estimated by using the dependencies between monthly mean ozone concentrations and monthly parts of the AOT40 (Figure 15third row). Possible changes of AOT40 values towards the end of the century are shown in the fourth row of Figure 15.

---

2 The Representative Concentration Pathway or RCP scenarios describe future emission situations at different climate forcing (2.6 W/m² and 8.5 W/m²) in the year 2100, but encompass air pollutant emissions also.

3 (WCRP's) Coupled Model Intercomparison Project phase 3 (CMIP3) multi-model datasetis made available publicly by the modelling groups, following the RCP scenarios. This data represent monthly mean ozone concentrations on a global scale with a spatial resolution of 5°. The Program for Climate Model Diagnosis and Intercomparison (PCMDI) collects and archives this data, and the WCRP's Working Group on Coupled Modelling (WGCM) organized the model data analysis activity. The WCRP CMIP3 multi-model dataset is supported by the Office of Science, U.S. Department of Energy.

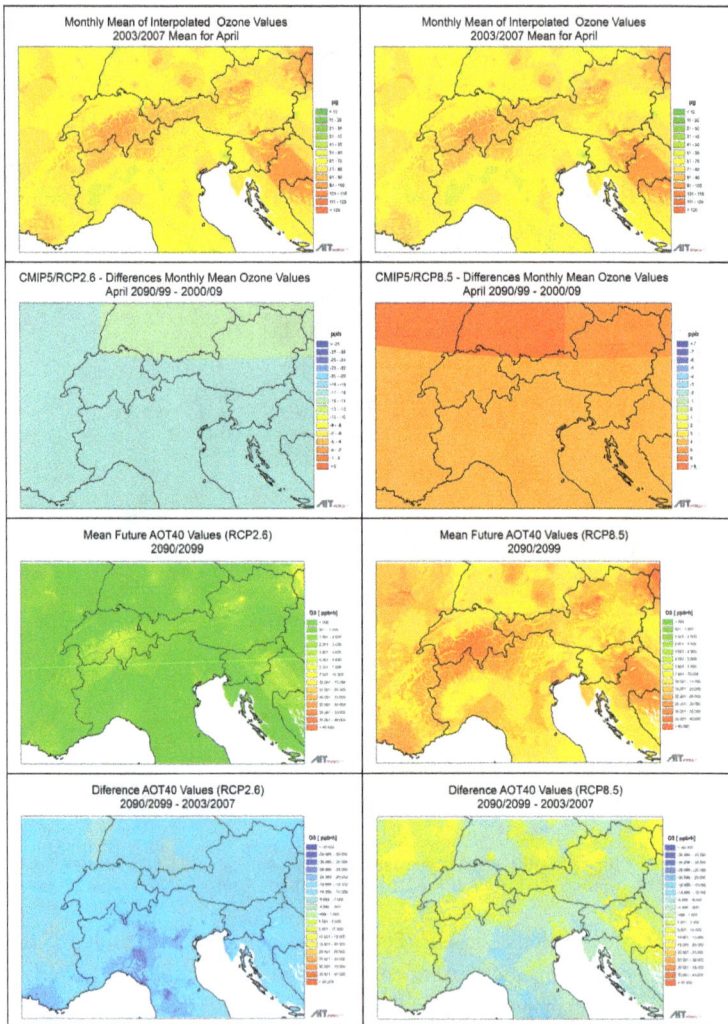

Figure 15. Monthly mean ozone concentrations for April, potential changes in the future, expected values towards the end of 21st century and differences to the period 2003-2007 following scenario RCP2.6 (left) and RCP8.5 (right).

The two RCP scenarios has lead to two significantly different forecasts for ozone risk assessment, mostly due to the characteristics of the scenarios. In fact, The RCP8.5 (the results on the right in Figure 15) can be understood as a scenario that comprises economic growth as well as increased energy demand including associated emission trends. Some technological improvements on abatement technologies are overcompensated by growth [16]. At least in mountainous regions this scenario reflects an increase compared to current levels of ozone accumulation which result in an increased potential ozone risk for forest. There is an overall increase in a large part of the Alpine region but in particular in the southeast of the Swiss and in the central and west part of Austria. The situation for the RCP2.6 Scenario [17] is in plain contrast. As cobenefits to the reduction in energy consumption needed to achieve the low climate forcing, also pollution is reduced drastically leading to ozone levels clearly below the current adverse conditions. The strong reduction of the pollution levels might lead also the AOT40 to values which might be under current critical level of the AOT40 index.

It is worth noticing that AOT40 represents only a potential risk [9] and that this index does not take into account the plant physiology, so even in a positive situation like RCP2.6, the real situation on field could be different, in fact other climatic results of the project highlighted an increase in summer drought and in autumn and winter precipitation. The combined effect of these two results lead to a decrease of the ozone uptake during the summer, as a consequence of a reduced stomatal activity but also to a possible increase of the ozone uptake in winter. This period is usually not considered in the evaluation of the ozone risk assessment, mainly because the vegetation activity and the ozone concentrations are low, but taking also into account the increase in temperature of all climatic scenarios, the ozone risk during the winter months can significantly increase. Furthermore it should be remarked that the two chosen scenarios represent in one case a very good situation (strong reduction of pollution) and in the other a balanced situation (economic growth taking care of the atmospheric emissions) and no pejorative scenario was considered.

4. Conclusions

The ozone risk assessment was evaluated for the first time at the whole Alps scale, realizing AOT40 maps for 5 consecutive years (2003-2007) and averaging them as suggested in [9]. A great interannual variability was observed and a potential higher ozone risk was identified for the southern part of the Alps.

The experimental field campaign run in Valle Camonica and the following analyses lead to develop local scale maps of the ozone risk which were in a substantial good agreement with the one obtained from monitoring stations only. Some differences were observed, especially at high elevations, but this was mostly due to the denser grid of passive sampler measuring point in the valley.

This intercomparison allowed to develop a semi-empirical relationship between monthly AOT40 values and the monthly average concentration. This relationship was used to estimate future AOT40 values, in fact the monthly average concentrations are the only output of chemical composition models. In this way it was possible assess the ozone risk in the period 2090/2099: in the more optimistic scenario, with strong pollution reduction, a substantial decrease of the AOT40 values was observed, which might result lower than the critical level;

in the other scenario, where economic growth is balanced by more attention to the emissions, a significant increase of the AOT40 in the central Alps is forecast.

Acknowledgements

This publication was funded by the Catholic University's program for promotion and divulgation of scientific research. The authors thank the Adamello Park and the mountain community of valle Camonica and in particular dr G.B. Sangalli, dr A. Ducoli and for their support in realizing this field campaign. A sincere thank to dr E. Soncina and Federico Fausti for their support in 2010.

Author details

Angelo Finco[1], Stefano Oliveri[1], Giacomo Gerosa[1], Wilfried Winiwarter[2], Johann Züger[2] and Ernst Gebetsroither[2]

1 Catholic University of the Sacred Heart, Mathemathics and Physics Department, Catholic University of the Sacred Heart, Brescia, Italy

2 Austrian Institute of Technology GmbH, Vienna, Austria

References

[1] Fuhrer, J., Skarby, L., Ashmore, M.R., 1997. Critical levels for ozone effects on vegetation in Europe. Environmental Pollution 97 (1–2), 91–106.

[2] Wustman, B.A.,Oksanen, E., Karnosky, D.F., Noormets, A., Isebrands, J.G., Pregitzer, K.S., Hendrey, G.R., Sober, J., Podila, G.K.. Effects of elevated CO2 and O3 on aspen clones varying in O3 sensitivity: can CO2 ameliorate the harmful effects of O3? Environmental pollution 2001; 115, 473-481.

[3] Fredericksen, T.S., Joyce, B.J., Skelly, J.M., Steiner, K.C., Kolb, T.E., Kouterick, K.B., Savage, J.E., Snyder, K.R.. Physiology, morphology, and ozone uptake of leaves of black cherry seedlings, saplings, and canopy trees. Environmental Pollution 1995; 89, 273–283.

[4] Lee, J.C., Skelly, J.M., Steiner, K.C., Zhang, J.W., Savage, J.E.. Foliar response of black cherry (Prunusserotina) clones to ambient ozone exposure in central Pennsylvania. Environmental Pollution1999; 105, 325–331.

[5] Bermejo, V., Gimeno, B. S., Sanz, M. J., De La Torre, D., and Gil, J. M.: Assessment of
     the ozone sensitivity of 22 native plant species from Mediterranean annual pastures
     based on visible injury, Atmospheric Environment 2003; 37, 4667–4677.

[6] King, J. S., Kubiske, M. E., Pregitzer, K. S., Hendrey, G. R., Mc Donald, E. P., Giardi-
     na, C. P., Quinn, V. S., and Karnosky, D. F.: Tropospheric O3 compromises net pri-
     mary production in young stands of trembling aspen, paper birch and sugar maple
     in response to elevated atmospheric CO2. New Phytologist2005; 168, 623–636.

[7] Felzer, B., Kicklighter, D., Melillo, J., Wang, C., Zhuang, Q., and Prinn, R.: Effects of
     ozone on net primary production and carbon sequestration in the conterminous
     United States using a biogeochemistry model, Tellus B 2004; 56, 230–248.

[8] Kolb T.E., Matyssek R.. Limitations and perspectives about scaling ozone impacts in
     trees. Environmental pollution 2001; 115, 373-393.

[9] UN/ECE, Mapping Manual Revision, 2004. UNECE convention on long-range trans-
     boundary air pollution. Manual on the Methodologies and Criteria for Modelling
     and Mapping Critical Loads and Levels and Air Pollution Effects, Risks and Trends.
     <www.icpmapping.org>.

[10] Loibl, W., Smidt, St.. Potential Ozone Risk for Selected Tree Species in Austria, Envi-
     ronmental Science and Pollution Research1996; 3(4) 213-217.

[11] Loibl, W., Modelling Tropospheric Ozone Distribution considering the spatio-tempo-
     ral dependencies within complex terrain; in J. Kraak, M. Molenaar (eds.) Advances in
     GIS ReseachII.Taylor& Francis Ltd. London. 1997; 667 - 678.

[12] Gerosa, G., Ballarin Denti, A..Regional scale risk assessment of ozone and forests. In:
     Karnosky D.F., Percy K.E:, Chappelka A.H., Simpson C., Pikkarainen J. (Eds). "Air
     Pollution, Global Change and Forests in the New Millennium", Elsevier Ltd., 2003
     119-139

[13] Gerosa,G.,Finco A.,Marzuoli R.,Ferretti M. and GottardiniE.. Errors in ozone risk as-
     sessment using standard conditions for converting ozone concentrations obtained by
     passive samplers in mountain region. Journal of Environmental Monitoring 2012;14,
     1703-1709.

[14] Gerosa, G., Ferretti, M., Bussotti, F. and Rocchini, D.. Estimates of ozone AOT40 from
     passive sampling in forest sites in South-Western Europe. Environmental Pollution
     2007: 145(3), 629-635.

[15] Cionni, I., Eyring, V., Lamarque, J. F., Randel, W. J., Stevenson, D. S., Wu, F., Bodek-
     er, G. E., Shepherd, T. G., Shindell, D. T., and Waugh, D. W.. Ozone database in sup-
     port of CMIP5 simulations: results and corresponding radiative forcing, Atmospheric
     Chemistry and Physics 2011 11, 11267-11292.

[16] Riahi, K., Rao, S., Krey, V., Cho, C., Chirkov, V., Fischer, G., Kindermann, G., Nakice-novic, N., Rafaj, P.,. RCP 8.5—A scenario of comparatively high greenhouse gas emissions. Climatic Change 2011;109, 33–57.

[17] Vuuren, D.P., Stehfest, E., Elzen, M.G.J., Kram, T., Vliet, J., Deetman, S., Isaac, M., Klein Goldewijk, K., Hof, A., Mendoza Beltran, A., Oostenrijk, R., Ruijven, B.,. RCP2.6: exploring the possibility to keep global mean temperature increase below 2°C. Climatic Change 2011; 109, 95–116.

Chapter 10

A New webGIS Platform Dedicated to Forest Extreme Events in the Alps: Aims and Functionalities

Stefano Oliveri, Marco Pregnolato and
Giacomo Gerosa

Additional information is available at the end of the chapter

1. Introduction

Many recent studies [1-6] show that, in the next future, climate change could represent a relevant threat for the state of health of Alpine forests. As a consequence, Alpine communities could be constrained to face an increase of both the frequency and the severity of damaging events affecting their forests (fires, pests attacks, windthrows, heavy snows). Although not all extreme damage occurrences are a direct consequence of extreme meteorological events, many of them have been linked to climatic phenomena. In some cases the relation is absolutely immediate: the storms Vivian (1990) and Lothar (1999), e.g., produced huge windthrows damages to the Alpine forests mainly in Germany, Switzerland and France with millions cubic meters of timber affected. The well-known heat wave registered during summer 2003 pro-duced, following the statistics, peaks in the frequency of forest fires both in the same season and in the next winter. Well the same, the extremely hot summer of 2003 has been correlated to several pests outbreaks [7,8] and several studies are conjecturing the possible influences of climate change on forest pathology. Hypothesis have been made about effects on the plant chemical defense mechanisms and on pests range expansions and diapauses, showing both a direct influence of temperatures and droughts periods on the biological cycle of the agents and an impact over the physiology of weakened host trees. Moving from these considerations, when MANFRED project was conceived partners recognized the need for a tool aimed at enabling Alpine communities to share their knowledge on past extreme events, in order to facilitate the diffusion and the sharing of practices which can be applied to prevent those extreme events, mitigate their effects and recover after extreme damaging occurrences. Whilst the greatest part of MANFRED activities have been intended to model and predict the future, the activities related to the development of the webGIS platform implied to look at the past, with the aim to derive the lessons learnt from a set of past extreme occurrences and use them as a tool to better manage the future.

2. Materials and methods

From an operative point of view, the work has been developed through the following steps:

- census of the events, with a special focus on extreme occurrences;

- identification of a set of relevant occurrences, which have been investigated in depth using a case histories approach;

- publication of results on a webGIS platform (available at the Internet address www.man-fredproject.eu/webgis) [9], giving open access to the whole set of data gathered and allowing to perform, on those data, both geographic and thematic queries.

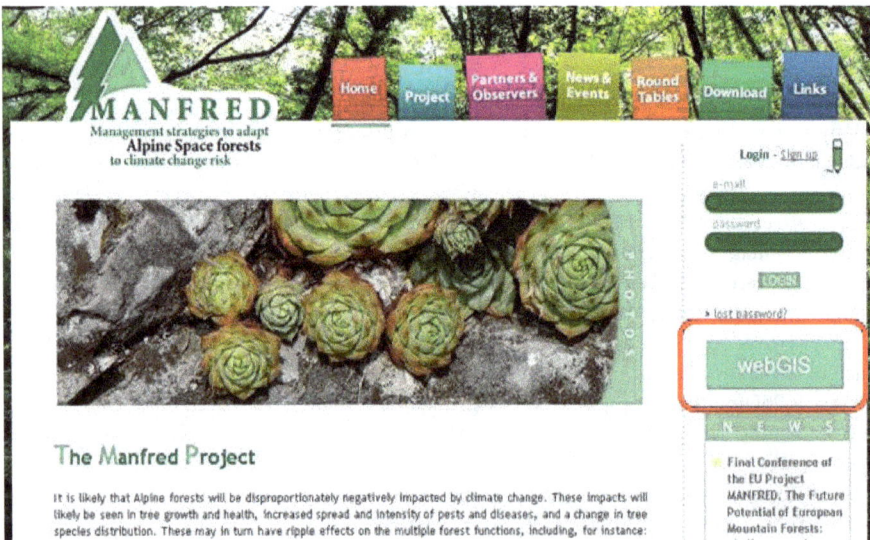

Figure 1. The project home page, with the button allowing the access to the webGIS platform

2.1. Definition of extreme events

The concept of extreme occurrence has to do with one's perception, familiarity and personal sensitivity. As a consequence, the definition of extreme event is associated with an intrinsic level of subjectivity. Yet, even sticking to the mere technical point of view, the interpretation of extreme stress event in a forest environment may dramatically depend on the different technical backgrounds and experience of the individuals giving the meaning. Foresters, landscape ecologist, conservationists, disaster and risk managers, statisticians and meteorologists would probably provide several keys, also referring to different forest-related benefits and roles. In fact, according to a multi-criteria and cross-sector approach, there could be more than one effective way to define an extreme event. Moreover, lack of numeracies and high heterogeneity of available data between the different Alpine regions, make it difficult the adoption of quantitative methods aimed at a standard identification of extreme occurrences. As a consequence, to warrant the selection of extraordinary events, a set of differentiated methods and approaches were applied based on the information and data available at the different areas of the Alpine arch. This kind of approach has been considered consistent with the aims of the activities carried out in the framework of MANFRED project, whose goal did not consist in a scientific definition of extreme event. It was considered much more suitable and useful to the aim of the project to describe an event as extreme mainly following the characteristic of being a non ordinary occurrence, the management of which requires going beyond present and local technical or theoretical know-how, to acquire new knowledge and to develop and implement ad hoc strategies.

2.2. Data on fires

A dedicated survey allowed the identification of a reference set of data on forest fires, available at Alpine scale. Based on those information, a pan-alpine database of fires has been generated, also in cooperation with another Alpine Space Programme project: ALPFFIRS (www.alpf-firs.eu). In the database, each fire is described through a broad set of attributes, listed in Table 1. The database contains information on about 82.000 forest fires, with a differentiated level of thematic contents based on regional availability of data. More than 26.000 occurrences are referred to the decade 2000 – 2009, the time span for which each Alpine region was able to provide information on fires. With specific regard to the identification of extreme events, it has to be highlighted that Total Burnt Area is the only quantitative attribute recorded, without significant lacks of numeracy, in all the fire dataset available at Alpine level. As a consequence, it was used as reference attribute for extreme events selection at Alpine scale. The 99th percentile of the values of the parameter was used for extreme events identification and 105 ha resulted as Alpine threshold for extreme occurrences selection. This kind of approach has a relevant drawback: extreme fires should be better identified in connection with an integrated evaluation of a broad set of variables (dealing, e.g., with territorial and environmental conditions, fire propagation and overall impacts). On the other hand, it allows to make comparisons between different Alpine regions, monitor the spatial and temporal evolution of extreme occurrences and make quantitative evaluations regarding ongoing trends. The analysis at Alpine scale were integrated with a set of regional deepening. It is well known, in fact, that forest fires have distinctive features depending on local conditions. A fire perceived as an exceptional occurrence in one region could fall into a normal regime of events in a different place. Therefore, the different co-operating Alpine regions were asked to make local selections of extreme fire occurrences, based on the whole set of data (both temporal and thematic) available and by adopting locally suitable methodologies.

Attribute	Notes	Domain
Year		
Date	Of the signaling	
Season	April-October: summer	
	November-March: winter	
X coordinate	Of the ignition point. UTM WGS84	
Y coordinate		
Cause		1. natural
		2. doubt
		3. anthropogenic
NUTS2	Of the ignition point	
NUTS3		
Altitude	Of the ignition point	1. 0 – 500m
		2. 500 – 1.000m
		3. 1.000 – 1.500m
		4. "/> 1.500m
Total Burnt Area		
Forest Burnt Area		
Non Forest Burnt Area		
Aspect	Of the ignition point. Derived from DTM	
Vegetation Unit	Of the ignition point. Derived from the "Map of Natural Vegetation of Europe"	The legend has been derived from the "Map of Natural Vegetation of Europe" report and structured in three different levels
Data Owner		
Mail of the Data Owner		
Data Provider		
Mail of the Data Provider		

Table 1. Attributes of the pan-alpine database of fires

2.3. Data on biotic and other abiotic factors

Whilst data on fires are basically comparable between the different areas of the Alps (at least for the core set of attributes gathered in the pan-alpine database), information on pathogens, windthrows and heavy snows are characterized by high heterogeneity in data availability between the different Alpine regions. In fact:

- some areas do not run dedicated monitoring systems at all;

- in other regions data are available only in paper format, which makes their acquisition and use difficult in the framework of a project with limited temporal duration;

- even in areas with information available, the categories of data acquired are highly heterogeneous both in terms of thematic contents (e.g. some regions monitor loggings and other regions monitor damages) and spatial/temporal resolution of the monitoring systems.

Given the high unhomogeneities in the data available at the different Alpine regions, data on biotic and abiotic factors have been introduced in distinct regional sub-sections. They are not conceived to build a comprehensive Alpine picture of the investigated phenomena and do not allow comparisons between the different regions. Moreover, whilst for fires the provided data allowed the definition of an overall Alpine threshold which was estimated with the aim to identify extreme events, information available on biotic and abiotic factors showed not to be suitable for a quantitative selection of extreme events. Besides the heterogeneity of the data, it must be highlighted that information on biotic and abiotic factors are often acquired on yearly basis (e.g. overall loggings or damages due to a certain disturbance), thus making difficult or impossible to link the values to one or more specific occurrences. With the aim to identify the most relevant occurrences a questionnaire has been produced, that partner spread, in their reference regions, to make a list of memory sifted events that, in the last decades, local population, authorities or practitioners perceived as "non-ordinary" or extreme. The questionnaire (Figure 2) asks for a basic set of information: area interested by the event, year or time-span in which the event occurred, stress category (biotic or abiotic), damage factor, main forest cover interested, overall damage (expressed as overall area or damaged volume), eventual further documentation and comments.

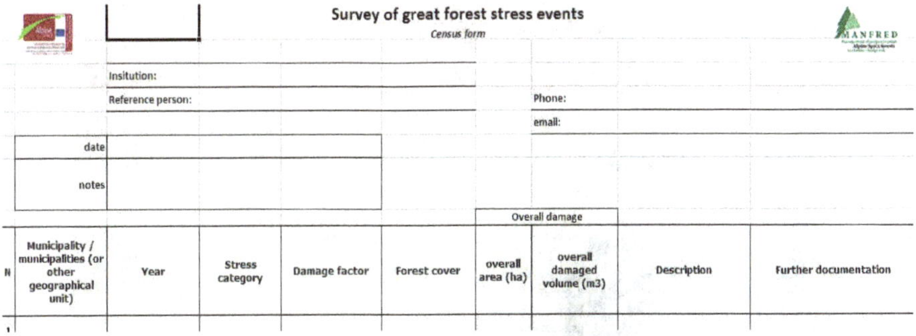

Survey of great forest stress events
Census form

Insitution:									
Reference person:					Phone:				
					email:				
date									
notes									

N	Municipality / municipalities (or other geographical unit)	Year	Stress category	Damage factor	Forest cover	Overall damage		Description	Further documentation
						overall area (ha)	overall damaged volume (m3)		

Figure 2. Structure of the questionnaire used to make a census of "non-ordinary" biotic and abiotic events occurred in the last decades in the Alps

In spite of the relevant drawback of such an approach, consisting in the high subjectivity in the process of signaling the events, it must be pointed out that it allowed to carry out the first geo-referenced census available at Alpine level related to 'non-ordinary' or extreme biotic and abiotic occurrences that affected forests in the Alps.

3. Structure and contents of the platform

From a technological point of view, the platform has been developed using J2EE (Java enterprise edition) technology. It runs on an Application Server Tomcat 6.0 and uses a db server PostgreSQL 8.4 and PostGIS 1.5

The webGIS platform consists of three distinct sections (Figure 3) dedicated to:

- fires;

- biotic factors;

- abiotic factors.

Figure 3. The main sections of the webGIS platform

Each section contains:

- a set of geographical data;
- Alpine and/or regional reports conceived to summarize and comment the data published;
- reports containing the results of the analysis on the case histories.

It must be pointed out that all the reports produced are based on reference schemes developed for the specific purposes of the project.

3.1. Fires section

It allows to work on three reference dataset:

- Alpine;
- Regional;
- Case histories.

When the Alpine option is activated from the Reference dataset box, the platform allows to query and map the data of the pan-alpine database of fires (decade 2000 – 2009) at three aggregation levels: NUTS2 (Region), NUTS3 (District) or it is possible to visualize the distribution of the ignition points of the single fires (Figure 4). Data can be queried based on many of the attributes of the database (Date, Cause, Class of altitude, Aspect, Class of extension) and it is possible to map the results of the queries both based on the overall number of fires or on the overall Burnt Area. In the resulting maps, grey areas evidence the territories for which a specific kind of information is not available. Besides working on the overall set of occurrences, the platform is also conceived to perform queries and produce maps specifically referred to extreme occurrences. The Identify button can be used to obtain, from the maps, analytical information on the results of the queries and the detail of both the Data Owner and the Data Provider. The Legend tool optimizes the visualization of data on the map, based on each query performed. When single fire events are visualized, the legend classes are pre-defined and intended to distinguish ordinary from "extreme" (>105 ha) Alpine occurrences. The Download area hosts a dedicated report summarizing the main results of the analysis carried out on the pan-alpine database. Specifically: state of the art on data availability, descriptive statistics of the overall dataset and of extreme occurrences selection, results of a fire selectivity analysis carried out with the aim to evidence ongoing trends in fire patterns at Alpine scale and results of a deepening performed to identify the Fire Danger Indexes better performing in predicting extreme fire occurrences in the Alps.

When the Regional option is selected in the Reference dataset box, the platform forces to choose (Zone box) a Region. Once the Region has been selected, the system allows to visualize and query (based on the same attribute categories of the previous section) the occurrences that the different regional partners have considered as extreme, using the whole set of data (both from a temporal and thematic perspective) locally available on fires (Figure 5). The Download area hosts the reports that regional partners have produced to comment the results of the analysis on local extreme occurrences and the set of data introduced in the webGIS platform.

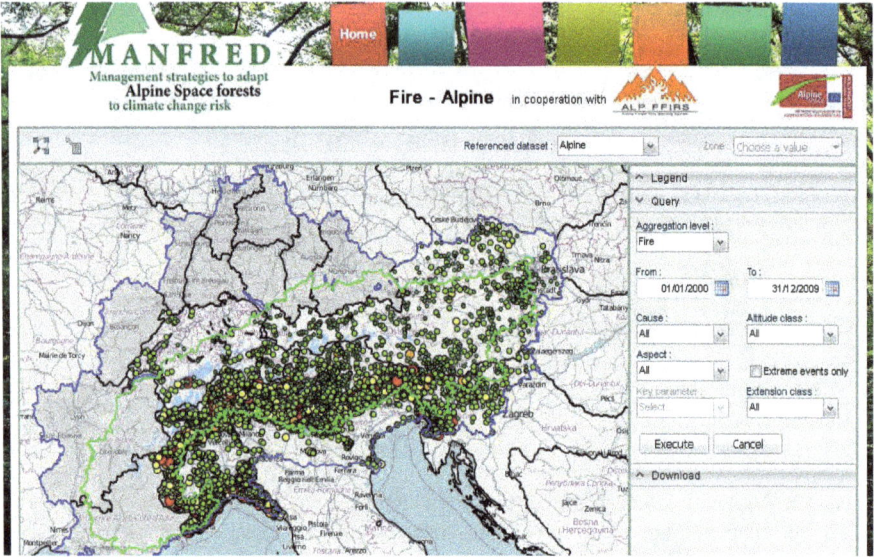

Figure 4. Fire section of the webGIS platform: querying and mapping environment of the Alpine sub-section

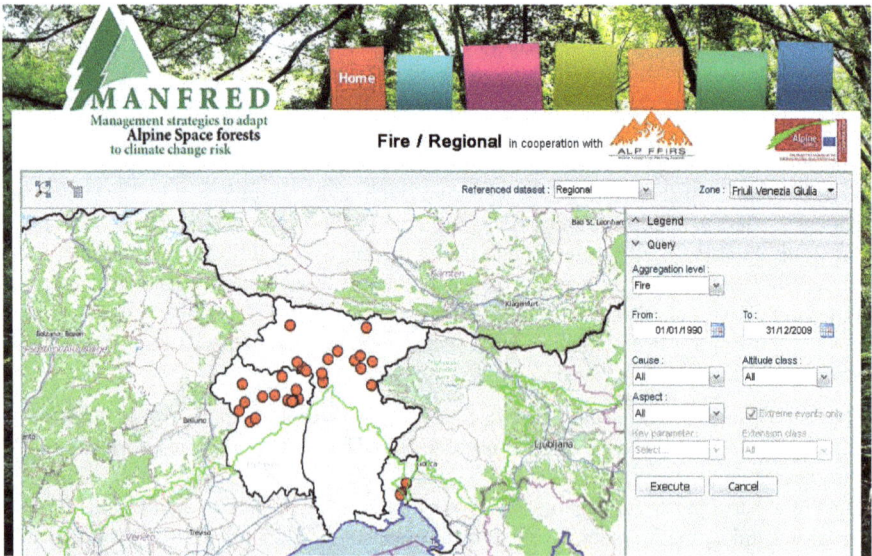

Figure 5. Fire section of the webGIS platform: querying and mapping environment of the Regional sub-sections

Last area of the fire section is the one dedicated to Case histories. When this option is activated, the platform opens an Alpine map, with the spatial distribution of the fire case histories investigated in the project. The Identify button can be used to obtain basic information on each case history and to download the reports on them carried out by the partners (Figure 6).

The following figures (Figure 7 and Figure 8) provide general schematizations of the Alpine and regional fire sections. For each of them, they give an indication of the main querying options and functionalities.

Figure 6. Fire section of the webGIS platform: querying and mapping environment of the Case histories sub-section

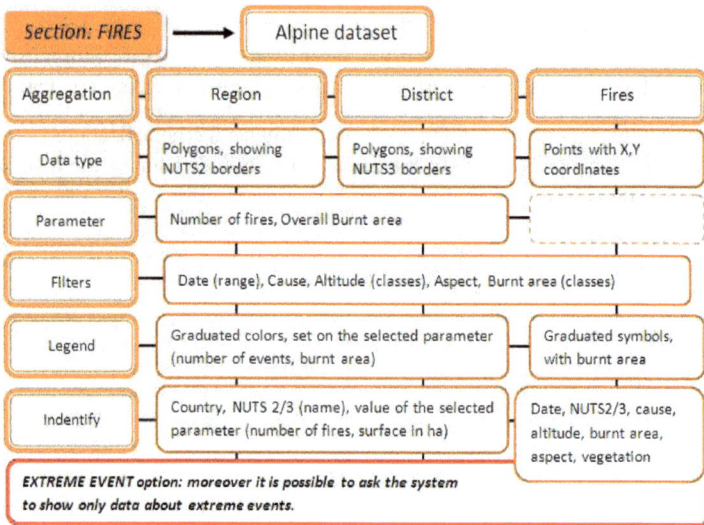

Figure 7. General schematization of the Alpine fire section, with its main querying options and functionalities

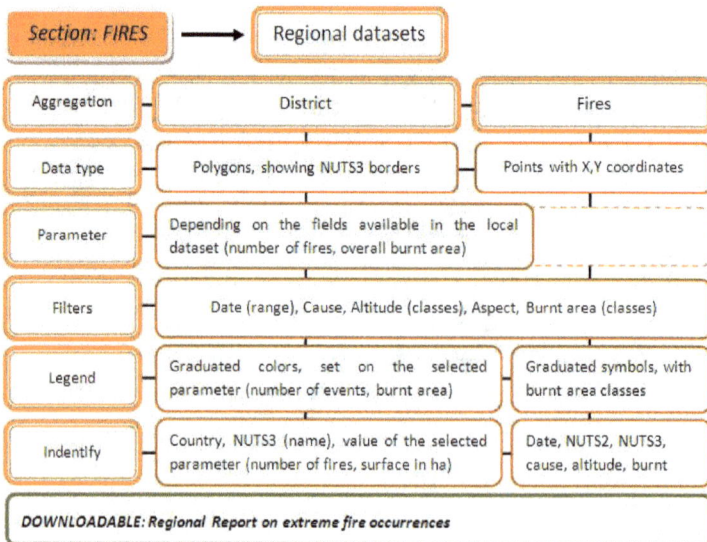

Figure 8. General schematization of the regional fire sections, with their main querying options and functionalities

3.2. Biotic and abiotic factors section

These sections have the same structure and can be considered comparable from an architectural point of view. They respectively contain data and information on pathogens, windthrows and heavy snows.

They work on four reference set of data:

- Loggings;

- Damages;

- Memory Sifted Events;

- Case Histories.

The Loggings sub-section is conceived to visualize, for the regions with available data on loggings at the level of forest parcel, the distribution of the parcels that, on a yearly basis, have been characterized by "non-ordinary" logging levels (Figure 9). For the selection process a dedicated procedure has been applied which was developed by partners in the framework of the project. The procedure integrates yearly data on loggings at parcel level and information on the area of the parcels. It has been intended to minimize the influence of parcels extension in the identification of outlying logging occurrences and to make the different regional distributions of logging data in fact comparable, by means of a normalization process. Besides allowing to show the distribution of outlying parcels, the platform gives the possibility to perform queries at District level, based on the date of loggings and, when available, on the specification of the damaging agent.

The Download Area hosts the reports produced by regional partners to summarize and comment the data introduced in the platform.

The following figure (Figure 10) provides a general schematization of the sub-sections dedicated to loggings both in the biotic and abiotic factors sections.

The Damages sub-section allows to map and query data on any Indicator (e.g. loggings, damaged area, number of signaling), provided by the partners regions, useful to describe the spatial and temporal (yearly basis) evolution of the "pressure" of biotic and abiotic agents on local forests. The platform is conceived to query and map the data, at NUTS2 (Region) or NUTS3 (District) level, based on a broad set of parameters: time-span, various levels of thematic information on the damaging agent (e.g., for pathogens, Category, Order, Family, Species) and querying Key parameter (Figure 11).

Like in the previous case, a dedicated area can be accessed to download regional reports commenting and summarizing the data hosted by the platform.

The Case Histories sub-section is intended to allow the user to visualize, on an Alpine map, the distribution of case histories investigated in the framework of the project. Through the Identify button, the user can obtain basic information on each case history and download the dedicated report produced on the specific occurrence.

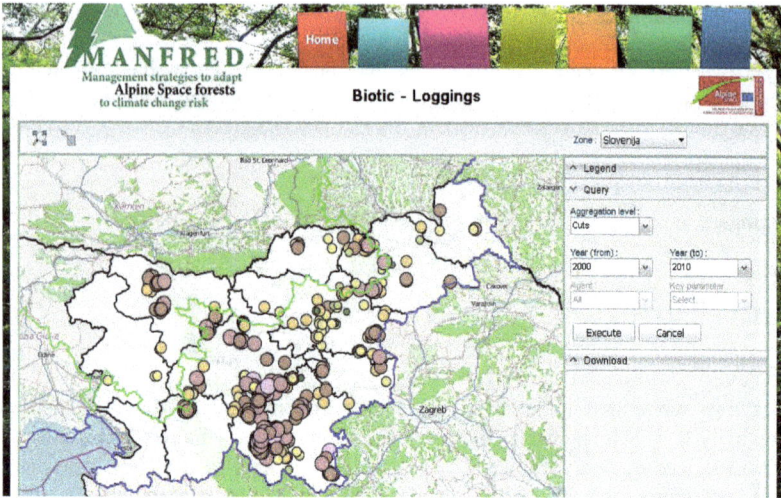

Figure 9. Sections dedicated to Biotic factors and Abiotic factors of the webGIS platform: querying and mapping environment of the Loggings sub-section

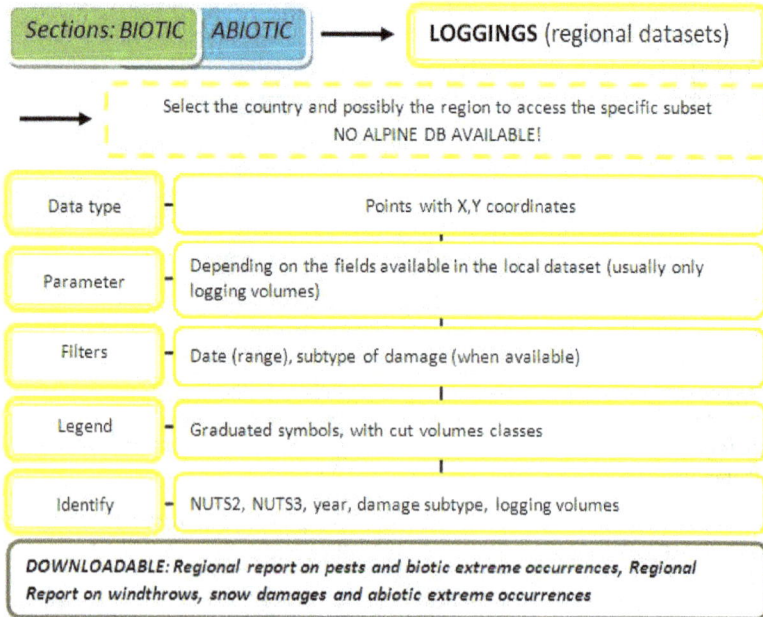

Figure 10. General schematization of the sub-sections on logging both in the section dedicated to biotic and abiotic factors. Main querying options and functionalities are evidenced

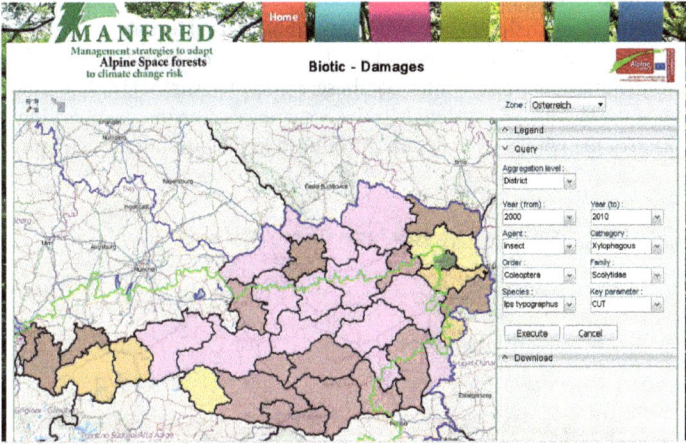

Figure 11. Sections dedicated to Biotic factors and Abiotic factors of the webGIS platform: querying and mapping environment of the Damages sub-section

The following figure (Figure 12) provides a general schematization of damages area, both in the biotic and abiotic factors sections.

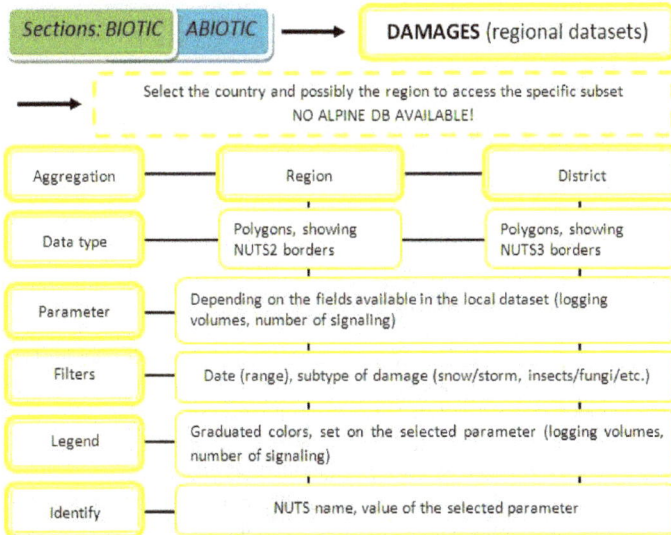

Figure 12. General schematization of the sub-sections on damages both in the section dedicated to biotic and abiotic factors. Main querying options and functionalities are evidenced

The Memory Sifted Events sub-section has been conceived to allow the user to query the database generated through the spread of the questionnaire used to make a census of the events perceived as "non-ordinary" or extreme. The sub-section gives access to a page conceived to allow the query (not the mapping) of the database, based on District, Damage factor and Year of occurrence of the events. The system lists the occurrences meeting the querying rules, with the specification of the whole set of details introduced in the questionnaire (Figure 13).

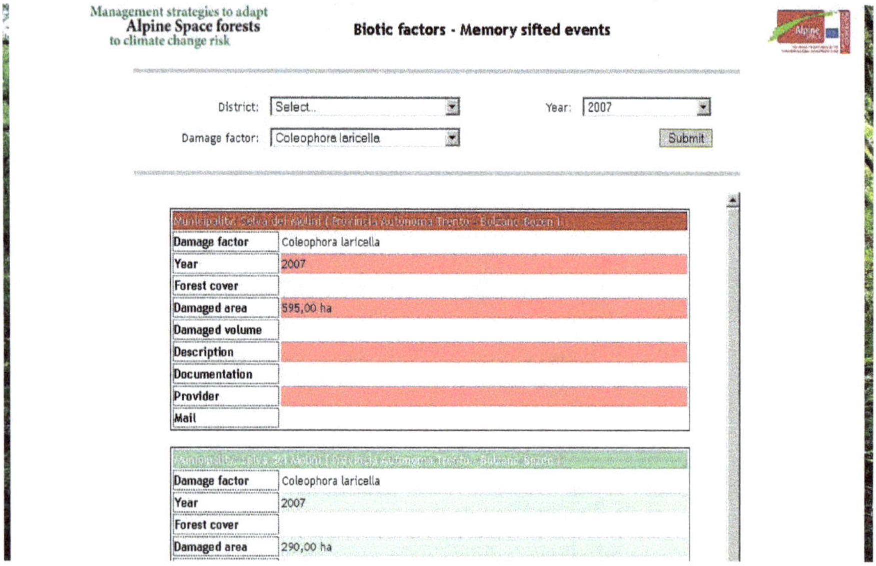

Figure 13. Sections dedicated to Biotic factors and Abiotic factors of the webGIS platform: querying environment of the Memory Sifted Events sub-section

4. Conclusions

Building a webGIS within the MANFRED project had different goals. Obviously, to contain and to make easily accessible the vast dataset gathered during the three years of the project among several partners, giving body to the effort of collecting a common knowledge base, provided by the several research centers and institutions of all the involved countries. The webGIS provides, in the form of an interactive reference book, the useful key to the consultation and interpretation of those data. It also gives an overall picture of the situation regarding forest damages monitoring procedures on the Alpine scale. Both the database and the webGIS are intended to be open-ended activities, to be further improvable, and to be run steadily as a web resource. The possibility to locate and retrieve information about extreme events occurred in the recent past can be particularly meaningful for local administrators and involved practitioners, who could find themselves in the necessity to deal with new similar events due to fire, snow, storm or pests within an Alpine forest environment. To this purpose, contact details of the main referents for the case histories have been made available in the webGIS. We remark here the importance to confront each others on our experiences from the recent past and on the strategies we developed to manage them, both assimilating the good practices and learning from errors. We believe, in fact, that the most effective way to face the uncertainties related to climate change is to have at disposal the widest and most various range of knowledge, enriched by the contribute of everyone between the two ends of the Alps and beyond. This research made the contacts between several experts and professionals from the six countries that have been involved. These actions have hopefully created a stable network of people and of shared knowledge: a wealth in resources that we wish to be durable and to be able to bear fruit in the preservation and the future management of our forests and our communities.

With specific regard to the data gathered and the investigations carried out, it is important to remark that:

- the pan-alpine database of fires is probably, today, the most complete set of information on forest fires available at Alpine level;

- the census of memory sifted events represents the first geo-referenced census available at Alpine level related to *'non-ordinary'* biotic and abiotic occurrences that affected forests in the Alps;

- all the case histories (#34) and the main contents of Alpine and regional reports (#20) have been integrated in a document (Handbook) titled *"Extreme stress events in the Alpine forests: management experiences based on recent occurrences"*, available in the MANFRED project website.

Based on the work done and taking the difficulties experienced in collecting and integrating the different set of data available at Alpine level as a starting point for mentioning some relevant future challenges, we can underline that it would be strategic, with the aim to build an integrated picture on the state of health of forests in the Alps, to improve the existing monitoring tools and steer the acquisition of data towards a higher degree of homogeneity between the different Alpine regions. Moreover, the opportunity to run permanent monitoring systems on the Alpine forests should be recognized and, hopefully, the platform generated in the framework of MANFRED project should be further integrated and updated in the future.

Acknowledgements

This work was conducted under the European Project MANFRED, funded by the European Regional Development fund of the Alpine Space Program, reference number 15-2-3-D (MANFRED).

A large set of partners co-operated in the activities: ERSAF. Ente Regionale per i Servizi all'Agricoltura e alle Foreste; FVA. Forstliche Versuchs - und Forschungsanstalt BW; Slovenian

Forest Service; Slovenian Forest Institute; Catholic University of Brescia; Italian Ministry of the Environment, Land and Sea; University of Camerino; CEMAGREF. Institut de recherche pour l'ingénierie de l'agriculture et de l'environnement; IPLA. Istituto per le Piante da Legno e l'Ambiente; Regione Autonoma Valle d'Aosta; LWF. Bavarian State Institute of Forestry; WSL. Swiss Federal Institute for Forest, Snow and Landscape Research; BOKU. University of Natural Resources and Life Sciences; BFW. Research and Training Centre for Forests, Natural Hazards and Landscape.

External contributions were provided by the University of Torino and by Regione Autonoma Friuli Venezia Giulia.

The research and its publication have been funded (partially or fully) by the Catholic University of the Sacred Heart, in the framework of its programs aimed at the promotion and the dissemination of scientific research.

Author details

Stefano Oliveri[1*], Marco Pregnolato[2] and Giacomo Gerosa[2]

*Address all correspondence to: stefano.oliveri@ecometrics.it

1 Ecometrics srl., Italy

2 Catholic University of Brescia, Mathematics and Physics Department, Italy

References

[1] Faccoli, M, & Battisti, A. in press) Climate change and forest pests: models from the Alps- Canadian Journal of Forest Research

[2] Marini, L, Ayres, M. P, Battisti, A, & Faccoli, M. Climate affects severity and altitudinal distribution of outbreaks in an eruptive bark beetle. Climatic Change (2012). DOIs10584-012-0463-z.

[3] Lindner, M, Maroschek, M, Netherer, S, et al. Climate change impacts, adaptive capacity, and vulnerability of European forest ecosystems. For Ecol Manag (2010). , 259, 698-709.

[4] Boisvenue, C, & Running, S. W. Impacts of climate change on natural forest productivity-evidence since the middle of the 20th century. Glob Change Biol. (2006). , 12, 1-21.

[5] Battisti, A, Stastny, M, Netherer, S, Robinet, C, Schopf, A, Roques, A, & Larsson, S. Expansion of Geographic Range in the Pine Processionary Moth Caused by Increased WinterTemperatures. Ecological Applications (2005). , 15(6), 2084-2096.

[6] Battisti, A. Forests and climate change- lessons from insects. Forest@ (2004). http:// www.sisef.it/accessed 15 June 2012), 1(1), 17-24.

[7] Faccoli, M. Effect of Weather on Ips typographus (Coleoptera Curculionidae) Phenology, Voltinism, and Associated Spruce Mortality in the Southeastern Alps. Environ. Entomol. (2009). , 38(2), 307-316.

[8] Battisti, A, Stastny, M, Buffo, E, & Larsson, S. A rapid altitudinal range expansion in the pine processionary moth produced by the 2003 climatic anomaly. Global Change Biology (2006). DOIj.1365 2486.2006.01124.x, 12, 662-671.

[9] MANFRED project Official websitewebGIS platform area. www.manfredproject.eu/ webgisaccessed 13 July (2012).

Chapter 11

Eco-Engineering and Protection Forests Against Rockfalls and Snow Avalanches

Frédéric Berger, Luuk Dorren, Karl Kleemayr,
Bernhard Maier, Spela Planinsek, Christophe Bigot,
Franck Bourrier, Oliver Jancke, David Toe and
Gillian Cerbu

Additional information is available at the end of the chapter

1. Introduction

The primary function considered to be of economic importance for forest stands has always been wood production. Today, society is aware of the necessity of forests' provision of additional functions: tourism, outdoor activities, fauna and flora protection, biodiversity, etc. In mountain areas, forests can also serve a protective function against natural hazards.

Many forests in the Alps cover steep to very steep slopes (gradients of 35 - 70 degrees) and so have an important protective function against natural hazards, such as: rockfall, snow avalanches, shallow landslides and erosion. The primary function of protection forests is to protect people or assets against the impacts of natural hazards. In order to provide this function,, the first 'products' of these forests are standing trees. Trees act as obstacles to mass movement hazards and/or to the propagation downslope of these hazards (within the MANFRED project we have focused on rockfall and snow avalanche risks). In the European Alps, mountain forests and the protection they provide have a long and distinguished history. This function has been recognized for centuries, as evident from logging bans declared from ~1350 onwards. However, in the last few decades, forest management has shifted its focus from timber management to multiple uses and forest ecosystem management. During this transition, there has been an increasing awareness of the need to manage the multiple functions of mountain forests aside from production and protection.

This protective function is also clearly identified in the first paragraph of the Mountain Forest Protocol of the Alpine Convention of 1996: "mountain forests ...provide the most effective, the least expensive and the most aesthetic protection against natural hazards." In Austria and Switzerland alone, approximately 50 million € are spent yearly to maintain or improve the protection provided by mountain forests [1,2].

In order to sustain the protective effect of these forests, they often have to be managed. Traditionally, mountain forests have been exploited for their timber and non-timber products, with the exception of forests on slopes above residential areas that present active natural hazards. A transition in forest management has occurred due to the increasing use of mountain areas and the diversification of Alpine economies, which have evolved from agricultural to tourism-based economies. During recent decades the damage potential, and therefore the importance of protection forests, has increased. Remote mountainous areas that were formerly avoided over the winter are now expected to be permanently and safely accessible to tourists. Moreover, settlements have been spreading into areas that were considered unsafe in the past, and infrastructure crossing the Alps (roads, power-lines, etc.) has greatly increased [3]. Thus, forests traditionally designated as protection forests have now gained wider recognition due to their increasing importance as well as their direct economic and social benefits [4]. The protective effect of forests against geo-hazards such as rockfalls and snow avalanches cannot be neglected in risk management. Furthermore, forest cover is constantly evolving, and targeted silviculture strategies are needed to maintain or increase forests' protective role. For this reason protection forest management has evolved with time from doing nothing ('banned forests'), to strictly silvicultural based management, e.g., selective cuttings to create openings in dense forests for the promotion of regeneration, to ecological engineering in mountain forests.

2. Eco-engineering: Definition and concepts

The term "ecological engineering" was originally described by Odum at the beginning of the 1960s, as "the management of nature" [5]. In the 90s, Mitsch and Jorgensen (see [5]) largely worked on this concept and finally defined it as "the design of sustainable systems, consistent with ecological principles, which integrate human society with its natural environment for the benefit of both" (see [5]). Bergen et al. In 2001 insist on the need for ecological engineering practices to be based on the science of ecology(see [5]). They also expand this to include all types of ecosystems and interactions between man and nature. More recently, Jones et al. highlighted the ethical, relational and intellectual dimensions of ecological engineering, therefore distinguishing it from environmental engineering, as explained before by Allen et al. (see [5]). Adapted eco-engineering practices are required to provide forests' ecosystems services over the longer-term including: timber production, wildlife habitat, water quantity and quality, hazard protection, recreation, inter alia that support quality of life and human well-being. Applying eco-engineering concepts can help to both protect forests and benefit humans.

Within the MANFRED project, eco-engineering refers to a mix of silvicultural measures and strategic creation of dead wood barriers such as higher left stumps and diagonally-placed felled tree stems on slopes. If required, real avalanche or rockfalls barriers can be also constructed with locally-felled wood. These actions mitigate natural hazards and promote forest renewal, enabling the forests to remain in a vital, stable state. Before eco-engineering practices are implemented, a series of analyses have to be carried out to determine the types, magnitudes and frequencies of the natural hazards that occur on site. Secondly, the current protective capacity of the forest has to be assessed. Thirdly, the developmental stage of the forest along with its future evolution will have to be evaluated. This evaluation needs to take into account the probable consequences of climate changes on both natural hazards and tree species' geographical distributions. Consequently, if the degree of protection does not suffice or tends to decrease in the future and if the forest has a potential for optimising its protective function, human intervention using eco-engineering can be a good option.

To support these interventions, protection forest researchers have been developing indicators and target values for specific mountain forests with different protective functions. Increasingly, the protective functions of mountain forests need to be evaluated alongside other functions and uses. This is difficult as foresters and forest researchers are still determining the characteristics of a forest that can provide optimal protection. Knowledge is limited due to the slow reaction of natural forests to the rapid environmental and socio-economic changes that have occurred in the Alps over the past 100 years. Fortunately, there has been considerable progress in research and practice on both the protective effect of forests against natural hazards as well as on the management of protection forests. The combination of this knowledge and experience provides a basis for balancing tradition and technology. This work resulted in practical guidelines in Switzerland, France and Italy as well as a variety of analysis tools for managing protection forests.

However, even if eco-engineering and silvicultural concepts, along with practical guidelines are based on forest ecosystems' natural cycles, adopting this type of management is impossible on all mountain forests due to financial restrictions. In this context, the knowledge of the spatial distribution of these protection forests and their differing abilities in preventing natural hazards becomes essential. Recognizing that forests offer protection against rockfall and snow avalanches is one matter, quantifying this effect is another one.

The three main objectives of the MANFRED project in the field of protection forest management in light of a changing climate have been to: 1) select and test a methodology for protection forest mapping, 2) propose an operational synthesis of protection forest management guidelines and, 3) upgrade these guidelines by integrating the findings from this project.

3. A methodology and tools for the mapping of protection forest against rockfall and snow avalanches

A forest has a protective function only if it protects society and facilities from hazards. Stemming from this, the 3 most important topics for protection forest mapping are: hazards, human infrastructure (at risk) and forests. This mapping can be done, depending on objectives of the exercise, at different scales. The regional scale is traditionally used for strategic forest planning and management, and the local scale (watershed, versant/slope, corridor) applied for silvicultural treatment. Mapping at the regional scale can be performed without accurate forest data; however, this is not the case for the local scale. In order to analyse the potential consequences of climate change on forests' protection function, here we have decided within to use a regional scale for analysis in order to use the same scale as the tree species distribution maps produced for the project. These maps are at the scale of the European Alpine Space with a resolution of 1km2.

The first regional protection forests for testing methodologies to combat snow avalanches and rockfall have been conceptualized in prior European projects: Rockfor, Provialp, Proalp and IFP. The general principle for this mapping is based on answering the following questions (Figure 1):

1. Where are the release areas?

2. What is the maximum propagation area envelope?

3. Is any human infrastructure located in the propagation area and if so, is it endangered?

4. Are any forest stands located in the release area and/or in the propagation area above the human infrastructure endangered?

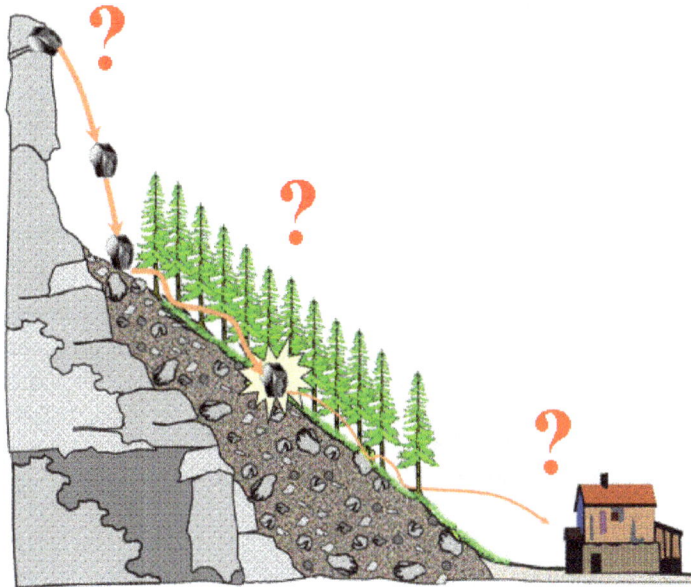

Figure 1. Schematic representation of the main questions to be addressed in mapping of protection forest for rockfall protection: Where are the release zones?, what is the maximum envelope of propagation, is there any infrastructure endangered and what is the role play by forest stands?

If the answer to the last question is yes, these forest stands serve a protective function, and an investigation at the scale of the stand is needed in order to qualify and/or quantify the efficiency of this protection.

3.1. Rockfall and snow avalanche potential maximal envelope of propagation mapping

The advantage of the regional scale is that models for identifying potential release and propagation zones can be used which do not requiring meteorological or geological data. The advantage of these models based on topographic criteria is that the only input datum is a digital terrain model. The calibration of these models has been made within the Interreg Alpine Space project Paramount. These models are: AvalForLIN for snow avalanches and RockForLIN for rockfalls. They have been developed by the French research centre, IRSTEA, and used to develop a method for protection forest mapping using Geographic Information Systems (GIS).

Firstly, all potential release points have to be mapped. 2D GIS models have been developed to localize these dependent on topographic conditions:

1. For rockfall, a simple slope threshold is applied to the slope surface raster (computed from the raster Digital Elevation Model [DEM]), according to the equation: $\alpha = 55° \times RES^{-0.075}$, where RES is the DEM resolution. All cells with values higher than the threshold α are qualified as potential release zones for rock fall.

2. For snow avalanches, curvature, slope, attitudinal and area criteria are chosen, depending on regional and geo-climatic conditions. Commonly in the European Alps, all cells in the raster with a slope of 28- 55 degrees, a convex form, a > 1000m altitude and > 500m² area are considered as potential release zones for avalanches.

Following, from each of the identified potential release points, 2D GIS models simulate the probable run out envelopes:

1. RockForLIN for rockfalls is based on the Energy Line principle [6], and can be used to compare rockfalls' run out envelopes to slope angles (Figure 2). The maximum spread of a block is determined by intersecting the ground and an imaginary line drawn from the release point with angle β. Different values for β were used: 32, 35 and 38°. Areas between 32 and 35° have a low but not 0 probability to be reached by rockfalls; between 35 and 38°, an intermediate probability; and higher than 38°, a high probability.

2. AvalForLIN for avalanches is also based on the Energy Line principle [7] (Figure 3). The maximal probable run out envelope is determined by the creation of an intersect between the ground and an imaginary line drawn from the release point with a calibrated angle. This angle is determined using the value of the energy line angle calculated for the point for which the slope angle is equal to 10°. AvalForLIN angles have been calibrated using snow avalanche cadastres (F, A, I, SLO).

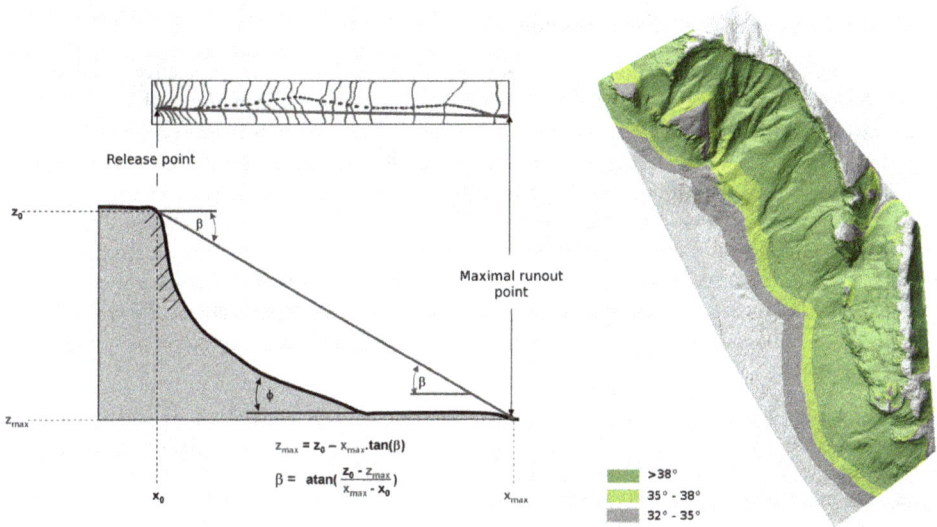

Figure 2. Rockfall Energy Line Principle and an example of a result obtained with it.

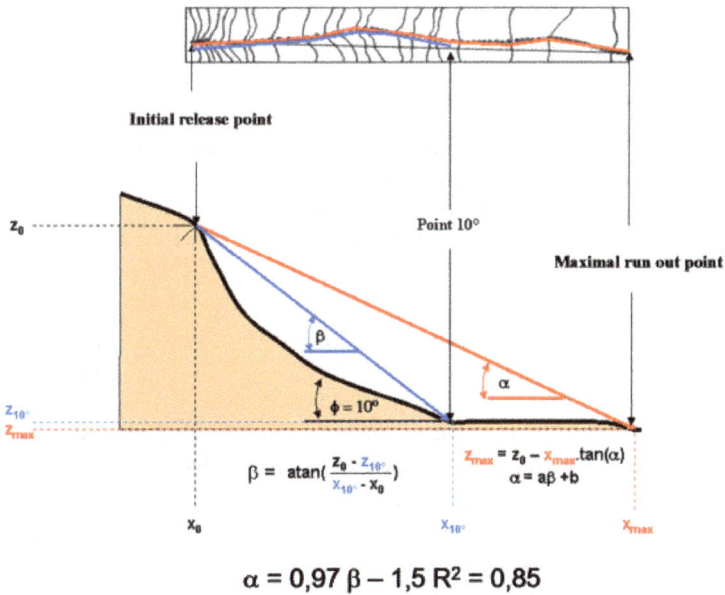

$$\alpha = 0,97\ \beta - 1,5\ R^2 = 0,85$$

Figure 3. Snow avalanches Energy Line Principle and the values of the model for the French Northern Alps.

3.2. Human infrastructure and protection forest mapping

Good information on the location of facilities is required to classify forests' protective functions and organize priority blocks for forest actions. At a regional scale, each of the Alpine Space countries can find this information in the geographic database of their respective national geographic institutes. Usually these databases list and correspondently map all human infrastructure: public facilities, dwellings, industries, as well as communication, electrical, gas and water infrastructure, etc. According to their importance or their extent, all items can be classified into 4 protection priority levels, from 0: low to 3: high level. This ranking is not obligatory, but facilitates the definition of priority levels for specific silvicultural actions depending firstly on the importance of the issues and secondly on forest stability. This ranking is required to be performed jointly with all actors involved in risk prevention policy of the study area.

By combining this map with the hazard run out envelope map, the potentially endangered infrastructure can be identified by selecting all the items located between release points and run out envelopes. The map obtained includes the endangered issues and the associated release and run out zones.

The last step of the process is then to cross this map of endangered issues with the map of the geographical extension of forest stands. This forest map can be the one provided by National Forest Inventories or the one available in the forest services. As the mapping is made at a regional scale, the dendrometrical description of the forest stands is not required. The information required is that of the surface covered by forest. Identification of forest stands potentially serving a protection function is then obtained by combining the endangered items map with the forest cover map, and by selecting all forested areas located above an endangered item and on/or between the associated release and run out zones.

This map of potential protection forest areas is required to be validated by a field survey. But before this, it can be used to define an area within which forest management dedicated to the improvement of the protection function needs to be applied. In other terms, this map defines the potential area of use of protection forest management guidelines. By using this methodology and these tools, policy- and decision- makers can then provided an efficient comparison between different regions.

The strength of this methodology lies in its ability to display the area within which forests are able to provide a protective function against rockfall and/or snow avalanches; often such areas are unknown having not been previously identified. A decrease in forest canopy in these protection forest areas could have dramatic consequences requiring adaptations to forest management to ensure the sustainability of this protective function.

Using this method, 43% of forests in Switzerland have a protective function, 42.7% in Val d'Aosta, and 29.5% and 24.7% in the French departments of Haute-Savoie (Figure 4) and Isère. In Austria and Germany, the area of forest providing a protective function is at 25% in Austria and of 34% in Germany. In Slovenia the only data available are for forests officially classified as protection forests: 9% of the forested area. For the Northern part of the Alpine Space, protection forests make up ~ 33% of the total forest cover.

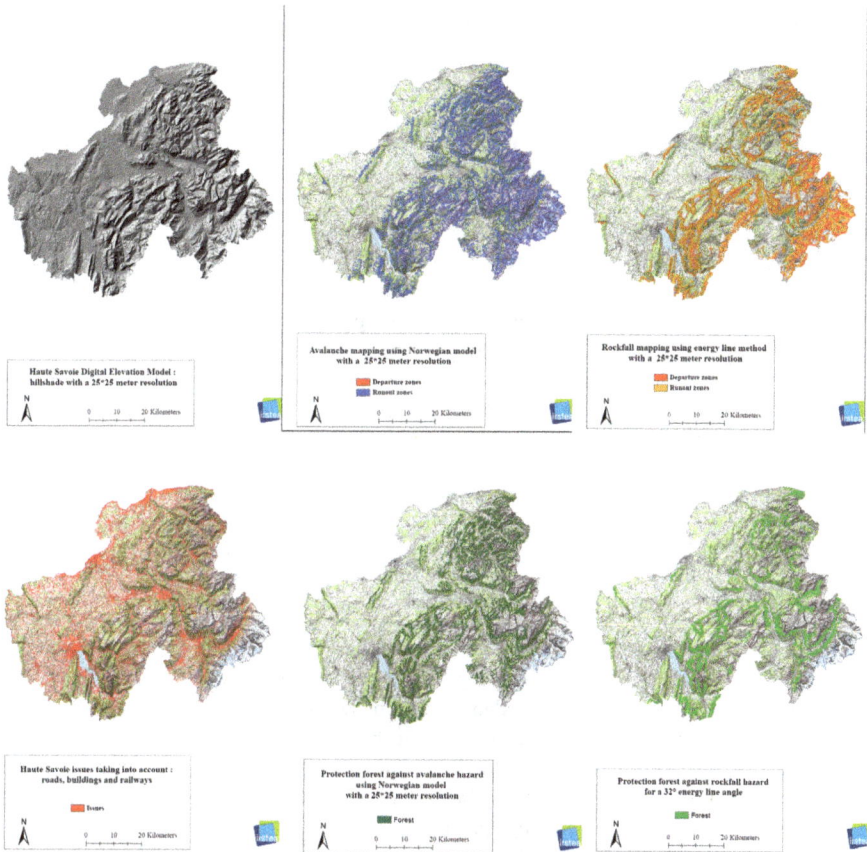

Figure 4. The principle steps for protection forest mapping. The example of the French Department of Haute-Savoie (one of the pilot area of the project MANFRED) : the Digital Terrain Model, the snow avalanche mapping results, the rockfall mapping results, the human infrastructure map, the forest's protection function in snow avalanche prevention, the forest's protection function in rockfall prevention

4. Silvicultural recommendations for the management of snow avalanche and rockfall protection forest

Protection forests act as natural obstacles which avoid or limit snow avalanche release and rockfall propagation. This protective function is created by the trunk and the crown of the trees present in these forests. In order to ensure the efficiency and the sustainability of this function, such forests have to be managed to ensure a stable and continuous forest cover. In other terms, silvicultural systems used and any natural disturbances that occur need to leave a sufficient amount of forest cover intact to maintain this protective function. Thus forest management plans have to be designed to take into consideration a long-term perspective for sustainable risk mitigation and the implementation of optimized management for protection forests. As concerns other forests, the primary objective of forest management should be ensuring the regeneration of forest stands. In the case of mountain forests, foresters have to create gaps to facilitate regeneration. Cutting trees may seem paradoxical in protection forests but is however necessary. In the context of risk prevention, these openings have to meet specific criteria to limit the negative impacts of cutting trees in protection forests. In order to help foresters define such adapted protection forest management, the consortium of the MANFRED project proposes a list of silvicultural recommendations and prescriptions. For the main silvicultural criteria, this list gives the thresholds values which facilitate the optimal protective effect of forest stands. These thresholds are mentioned in the synthesis of the guidelines currently used in the Alpine Space [8,9,10] and based on the expert knowledge contributed within the MANFRED project.

5. The use of stumps and felled trees in eco-engineering

In Austria, but also increasingly in Switzerland, France and Italy, the effects of the presence of couloirs or larger openings in protection forests are mitigated by cutting trees and 1) cutting tree stems during harvesting as high up as possible (leaving tree stumps at a height of > 1.3m) and 2) positioning the felled stems on the slope diagonal to the slope direction. Important criteria for selecting the trees to be cut are: the position and the growth tendency with respect to the corridor, the DBH (thicker stems, or if possible multiple trunks on top of each other, are clearly more effective barriers and should be left), tree instability, the effect on tree regeneration, the size of the gaps after cutting, as well as the shadow effect. The shadow effect is the phenomenon of in protection forests, of trees growing behind each other, or younger trees tending to grow downslope of older trees; i.e., the older tree protects younger trees downslope [11].

Another important aspect is the direction in which the felled trunk is positioned on the slope. In snow avalanche release zones trunks should be placed perpendicular to the steepest slope direction. One of the most important points is that these felled trees have to be kept in place on the slope in order to avoid any displacement or mobilization due to snow gliding or to the impact of a snow avalanche.

In the case of rockfall, a choice can be made to direct all rocks away from a channel, preferably into areas with a high stand density or a high surface roughness (e.g., depressions where many larger rocks have been deposited). Alternatively, if the corridor has become a real channel, in which forest regeneration is inhibited, all rocks can be directed into this channel by using accurately-positioned felled trunks that orient rocks towards this channel. A precondition for the latter case would be that there is sufficient protection at the end of the corridor, i.e. a rockfall net or rockfall dam.

	Recommendations	Thresholds
Release zone	Remove unstable trees	Coefficient of stability value (Height/Diameter at breast height = H/D) Coniferous: H / D ≤ 65 Broad-leaves: H / D ≤ 80
	The following effects: interception of snowfall, anchoring of the snowpack by punching, increasing soil roughness, are efficient when the height of the trees is twice higher than the height of the snow cover.	
	On the edges of gaps obtain trees with the greatest crown length as possible (more than 2/3 of the height of the tree)	
	Limit the proportion of deciduous trees and larch	Presence of deciduous trees should be maintained at < 30%
	Depending on the ecological conditions, a certain amount of broad-leaves are needed for the stability of stands	
	Deciduous trees are more suitable for the prevention of snow gliding during periods of smaller quantities of snow allowing more sunlight to reach the canopy floor and melt the snow. With large quantities of snow their effect is negligible. Fir and spruce needles on the ground enable good sliding which can cause avalanches.	
Transit and run out zones	Limit gap size (same thresholds as for transit and run out zones)	Length of gap (measured on the slope) along the line of the steepest slope (H = average height of trees) L ≤ 1.5 H The gap width <15m
	Maintain an effective winter canopy cover	Value of the canopy cover in winter according to the slope (in °) 30° >30% 35° > 50% 40° & more > 70%
	Harvest trees leaving stumps with a min. height of 1.30m or, if rockfall can occur, completely remove stump to ground-level (or screed in order to avoid a trampoline effect).	
	Fell/cut trees at an oblique angle to the slope leaving felled trees on the ground in a position that they cannot be easily moved.	
	Limit gap size	(same threshold as release zone)
	Promote stable deciduous trees to limit the effect of powder avalanches (decrease forest canopy permeability)	Corridor edge ≥ 70%, otherwise > 30%
	Remove unstable trees along corridors	Coefficient of stability value (Height/Diameter at Breast Height = H/D) Coniferous: H / D ≤ 65 Broad-leaves: H / D ≤ 80
	Harvest trees leaving stumps with a min. height of 1.30 m or, if rockfall can occur, completely remove stump to ground-level (or screed in order to avoid a trampoline effect).	
	Do not leave deadwood (i.e. logs) on the forest floor.	
	Remove trees that, when they fall, can reach issues	

Table 1. Silvicultural recommendations for the sustainable mitigation of snow avalanches by forest stands

	Recommendations	Thresholds
Release zone	Remove unstable trees (leverage effect due to the wind) at the top of cliffs or outcrops and in the release area	Coefficient of stability value (Height/Diameter at Breast Height = H/D) Coniferous: $H / D \leq 65$ Broad-leaves: $H / D \leq 80$
	Maintain a high basal area compatible with the sustainability of the stand at the foot of the release area	
	Whenever possible, limit the boulder's distance of entry in the stand.	
	Promote deciduous trees which are more resistant than conifers with equivalent diameter. Maintain more than 30% of deciduous trees among the largest trees. Depending on the ecological conditions, a certain amount of conifers are needed for the stability of stands.	
	Limit the size of gaps (same thresholds as for transit and run out zones)	
	Harvest trees leaving stumps with a min. height of 1.30 m or, if rockfall can occur, completely remove stump to ground-level (or screed in order to avoid a trampoline effect).	
	Fell/cut trees at an oblique angle to the slope leaving felled trees on the ground in a position that they cannot be easily moved.	
Transit and run out zones	If possible, increase the planimetric length of the wooded strip.	Recommended horizontal length of the wooded strip > 200m
	Limit the size of gaps	Length of gap (measured on the slope) along the length of steepest slope (H = average height of trees) High forest < 40m Coppice < 20m, In all cases, recommended value: $L \leq 1.3\,H$ with a wooded strip below the gap > 2 H (recommended 4H)
	Promote deciduous trees which are more resistant than conifers with equivalent diameter. Maintain more than 30% of deciduous trees among the largest trees. Depending on the ecological conditions, a certain amount of conifers are needed for the stability of stands.	
	Maintain an adapted basal area for the efficient trees	In the transit zone: the basal area of trees with a diameter of ≥15 cm is required to be ≥ 25 m²/ha
		In the run out zone, the basal area of trees with a diameter of ≥15 cm is required to be ≥ 20 m²/ha
	Maintain an adapted stem density for the efficient trees Maintain a high density in a band of 25m on either side of a corridor	In all cases the stem density for trees with a diameter of ≥20 cm is required to be ≥ 350 stems/ha
	Remove unstable trees along corridors	Value of the coefficient of stability (Height/Diameter at Breast Height = H/D) Coniferous: $H / D \leq 65$ Broad-leaves: $H / D \leq 80$
	Remove trees pose a threat to infrastructure	

Table 2. Silvicultural recommendations for rockfalls sustainable mitigation by forest stands

To test the efficacy and durability of such diagonally felled trees, within the MANFRED project specific research actions have been carried out in which 1) experiences from practitioners as well as resistance measurements on the durability of felled stems were collected and 2) full-scale rockfall experiments on felled stems conducted.

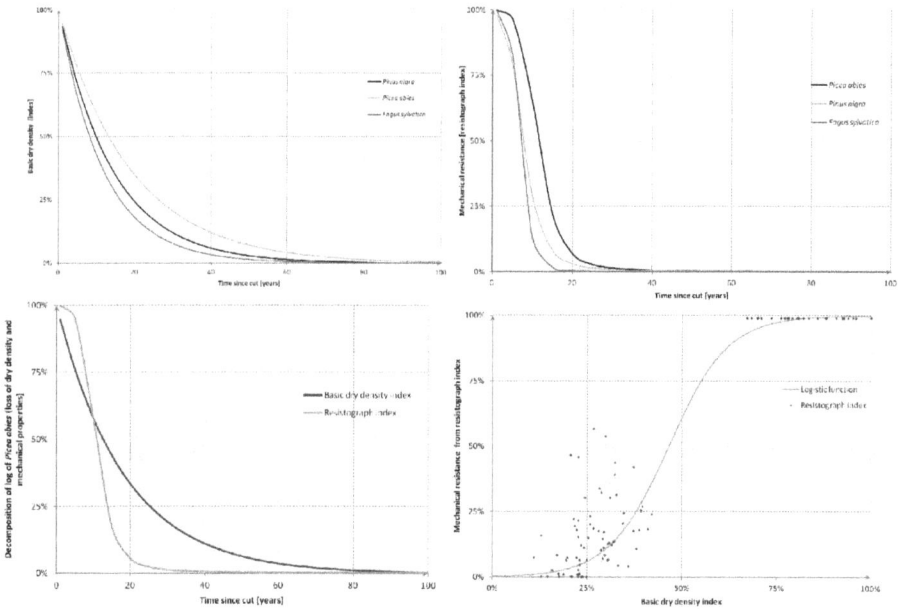

Figure 5. Graphical representations of the breaking dynamics of *Picea abies* (dark line), *Fagus sylvatica* (gray line) and *Pinus nigra* (soft grey line) relative to wood decay and its consequences on the loss of dry densities (stumps) and mechanical properties (stumps, log). The last graph represents the sigmoidal correlation between the loss of dry density and the loss of mechanical properties for *Picea abies* stumps. The loss of the mechanical properties has been measured using a resistograph.

The research has further shown that the felled trees have a significant effect on the energy loss (30% on average during an impact on a felled stem) and on the fall direction of the rock. If all the stems are oriented in the same direction, laboratory experiments showed that the rock will also change its direction. The results also indicate that the most optimal orientation of the stems would be between 45° -70° less than the steepest slope direction (slope aspect) (Figure 6).

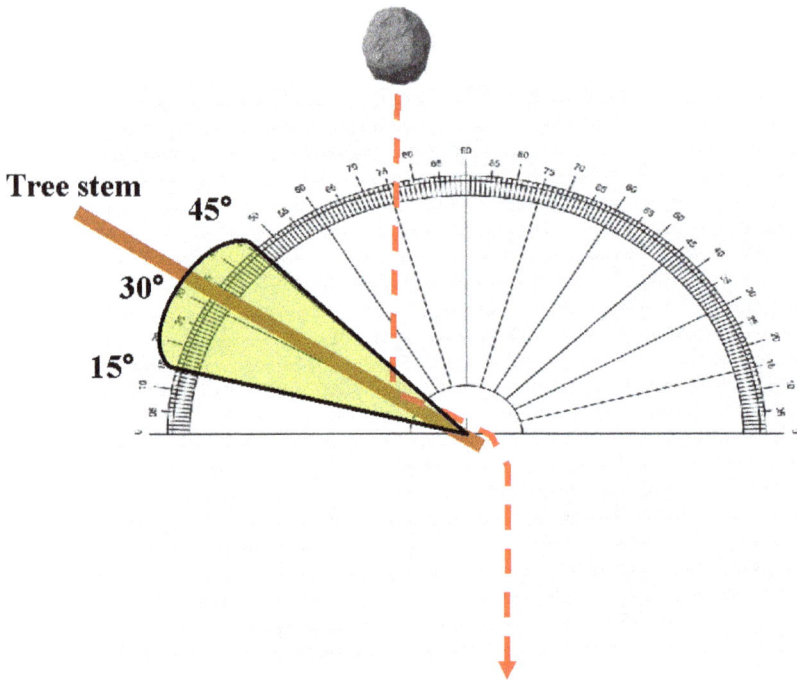

Figure 6. Representation of the range of optimal angles (green zone) for stems positioned obliquely to the steepest slope direction or the main fall direction of the rocks (dashed line).

6. Marteloscope: Definition, concepts and uses

Marteloscopes are training plots (approx. 1 ha) in protection forests where forest practitioners are asked to (virtually) manage the present forest in the best way to obtain an optimal long term protection function. They are asked when and how to apply silvicultural methods. After that, the notes of every participant are processed by a coupled growth dynamics (SAMSARA, PNN2) and protection function simulation model (for rockfalls: Rockyfor3D or RockforNet, for snow avalanches: SnowForIntercept). The next day, the best and other management options are discussed, especially in the light of the local context of the forest (ecological, economical and social functions).

The main objective of a marteloscope is therefore the training and education of forest practitioners.

There are multiple techniques involved in establishing a marteloscope. The first step is to select the plot. The plot's characteristics are the following: 1 ha (100 x 100m planimetric), uniform relief and slope, all trees ≥7.5 cm DBH must be recorded and their situation defined by absolute XYZ distances from the lower left corner of the plot. The selection of the site must be performed using the following check-list:

- Easy accessibility: Reasonable distance by car, Max. 30 min walking distance from the parking

- Infrastructure: Parking of adequate size, reasonably-priced accommodations nearby.

- Risk situation: The site should be situated in the transit zone of a real risk situation (greater scientific interest; far more demonstrative for the practitioners).

- Forest: Uniform characteristics on the entire marteloscope surface (±1 ha), the relief should also have uniform characteristics, it should be representative for a typical or specific forest type, the local context of the forest should be considered (ecological, economical and social functions).

- Data:It is advantageous if forest data is already available; i.e. information on previous forest management, regions where LiDAR or high resolution DEM data is available should be preferred, meteorological data, past events database.

- Safety: The safety of the practitioners should be considered (limited "exposition" time, limited number of persons, helmets etc.) as the safety of the issues located below the plot.

The choice of the procedure for mapping the trees is largely depending on the forest characteristics, on the available technical equipment and on the habits/competences of the field workers. It is possible with a compass, a clinometer and a decameter, but it will be faster and more precise using ultrasound or laser measuring devices or even a theodolite and high precision GPS.

9 marteloscopes have been implemented and used within the lifetime of the MANFRED project: 2 in France, 2 in Austria, 4 in Slovenia and 1 in Switzerland. Two sites deal with protection forest management against snow avalanches release and 7 with protection forest management in order to mitigate rockfall propagation. Four are covered by pure beech stands, 1 by a mixed stand of beech and silver fir, 1 with a pure Austrian black pine stand and 3 with pure spruce stands. All the data are available on MANFRED'S website. The figure below presents schematic representations of some of these training plots.

Within the framework of the activities of the MANFRED project, these 9 marteloscopes have served as a database for international know-how and knowledge exchange, as well as cases for the revision of regional protection forest recommendations and prescriptions.

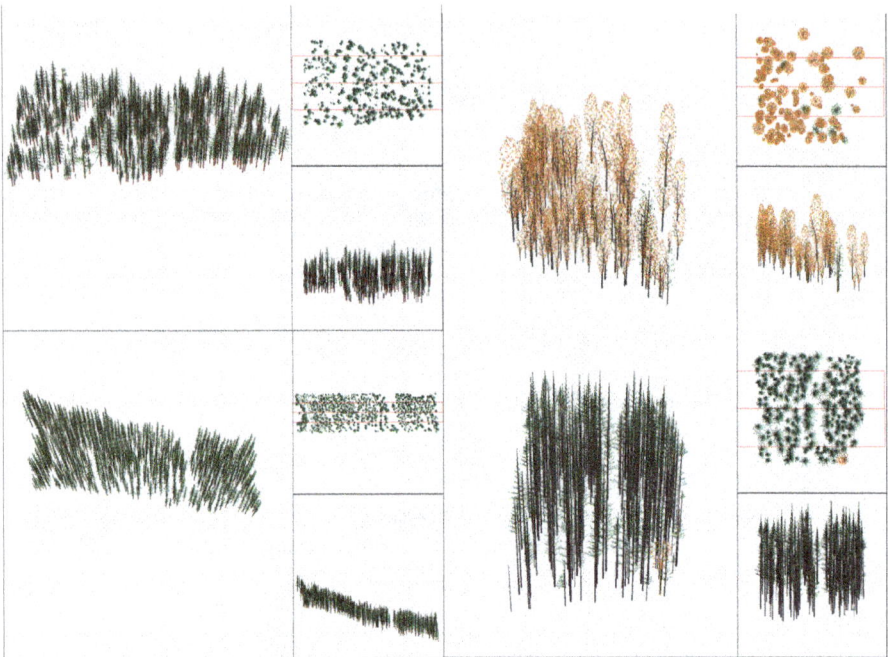

Figure 7. Schematic representations of 4 of the 9 marteloscopes implemented during the MANFRED project : Verbier (Spruce stand, CH), Solcava (Beech stands, SI), Valdrôme (Austrian black pine stand, F), Gashurn (Spruce stand, AT).

7. The probable consequences of climate change on protection forest

Snow avalanche occurrences are driven by 3 types of parameters: meteorological factors, topographic conditions and land use types in the release zones. To observe a snow avalanche release it's necessary to have a certain amount of snow height, a weak layer in the snow cover and to for the snow mass to overshoot a critical stability equilibrium value. The total snow mass is determined directly by the amount of snow precipitation, the presence of strong winds, changes in temperature and rainfall, along with land use types in the release area. The presence of forest vegetation limits the snow deposit on the slope and also modifies the snow quality. The efficiency of snow interception by the forest canopy is correlated with tree species: broad-leaf trees are less efficient than coniferous ones.

All of these parameters are directly influenced by the climate, and anthropogenic climate change thus will have an influence the frequency and severity of this hazard. The cumulative effects of possible climate variable changes (temperature, precipitation, wind...) on the release conditions of snow avalanches need to be evaluated. Until now, few studies are available on this subject. The two foremost studies have been performed within the Interreg Alpine Space project CLIMCHALP [12] and by the Cemagref [13].

The main results of these studies are 1) that there is no statistical correlation between the number of events and the evolution of the climate over the last century, 2) globally, experts agree that under warming climate conditions the snow cover stability will tend to increase, 3) observations show that over the past 30 years, snow avalanches have tended to stop at higher altitudes than usual, but for extreme events no differences in stopping altitude have been recorded.

Rockfall occurrences are also driven by 3 types of parameters: meteorological factors, topographic conditions and geology. One of the main parameters is the freeze-thaw-cycle. These cycles are directly related to weather conditions: high variation of temperature over a short time period, and duration and intensity of precipitation. In fact, the only certain change will be a temperature increase. The most probable consequences of climate change on rockfall will be the increase in occurrence. Up to now this trend has not been proven but some studies indicate an increase in rockfall occurrences for years with weather anomalies. The graph below shows recorded rockfall events from 1989 onwards in the French department of Haute-Savoie. The year 1999 is characterized by a strong winter and 2003 by a severe drought. These two years have the greatest number of recorded events (21 and 19 respectively) for the period 1989-2006. Between 1999 and 2004 the number of events is greater than the average number of occurrences. This is probably due to the consequences of the climatic conditions of the winter of 1999 on the stability of the geological facies.

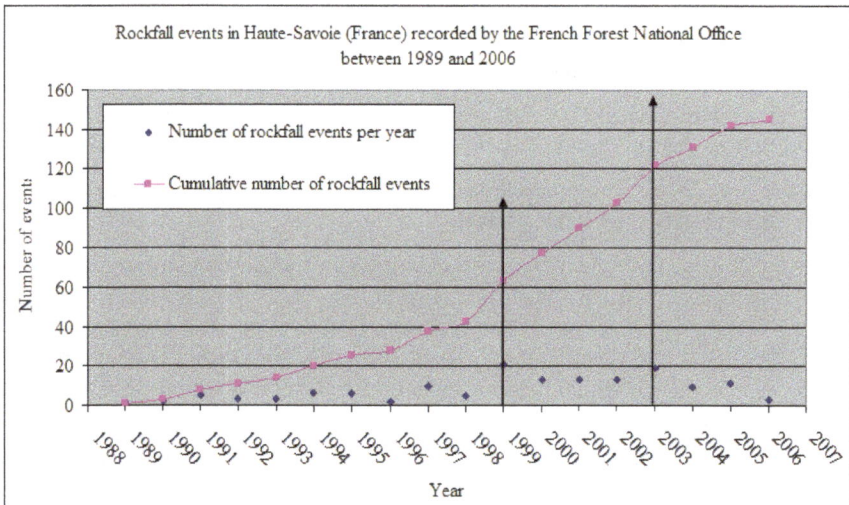

Figure 8. A chronology of the number of recorded rockfall events during the period 1989 – 2006 for the French Department of Haute-Savoie.

There is general consensus within the scientific community that climate changes will impact forest vegetation in 3 major ways:

- An upward attitudinal (and elevation shift) of the forest timberline and a shift in the distribution of species (already observed within Europe)
- An increase in forest growth (already observed within Europe)
- An increase in the development and impacts of pests and diseases

The study of the potential changes in tree species geographic distribution in the Alpine space has been provided by the MANFRED project. These distributions have been formalized via the probability of presence of tree species on 1x1 km raster maps. For the main tree species (*Fagus sylvatica, Picea abies, Abies alba, Pinus nigra* and *Pinus sylvestris*), and for the project's pilot areas for which the mapping of the protection forest has been completed, an analysis of the consequences of the evolution of the tree species distributions on the forests' protective function has been processed. The main conclusion of this study is that there will be an increase in the surface covered by forests potentially providing efficient protection against rockfall propagations and snow avalanche releases. This is due to 3 factors:

1. broad-leaf trees will progress upslope covering rockfall propagation areas previously covered by coniferous forests,
2. the coniferous trees will also progress upslope covering snow avalanche releases area currently not covered by forest,
3. there will probably be a transition step during which the snow avalanche release zones currently covered by coniferous forests will be covered by mixed forests, where these mixed forests will create conditions for efficient protection.

For the French departments of Haute-Savoie and Isère, the increase in protection forest will be 23% and 19% respectively, for rockfall, and 28% and 24% for snow avalanches (see Figure 9).

Due to these results, the most probable consequences of forest evolution in the field of snow avalanches and rockfall prevention are that:

- Over the next 100 years, the progression of broad-leaves vs. coniferous may decrease the current protective effect of mountain forests against snow avalanches to a minor extent. The snow interception of broad-leaf canopy is less than for a coniferous one; however, there will likely be a buffer zone characterized by the presence of mixed forest. This buffer zone will probably provide efficient protection if broad-leaves trees represent less than 30% of the forest canopy.
- The progression of broad-leaf vs. coniferous trees might increase the protective effect of mountain forests against rock falls. Broad-leaf trees have higher mechanical resistance to impact on their trunks than coniferous trees.
- The natural response of vegetation to an increase in temperature will be an attitudinal migration. Each degree of temperature increase will correspond to migration of 150m in altitude. Currently non forested snow avalanche release zones might be, in the future,

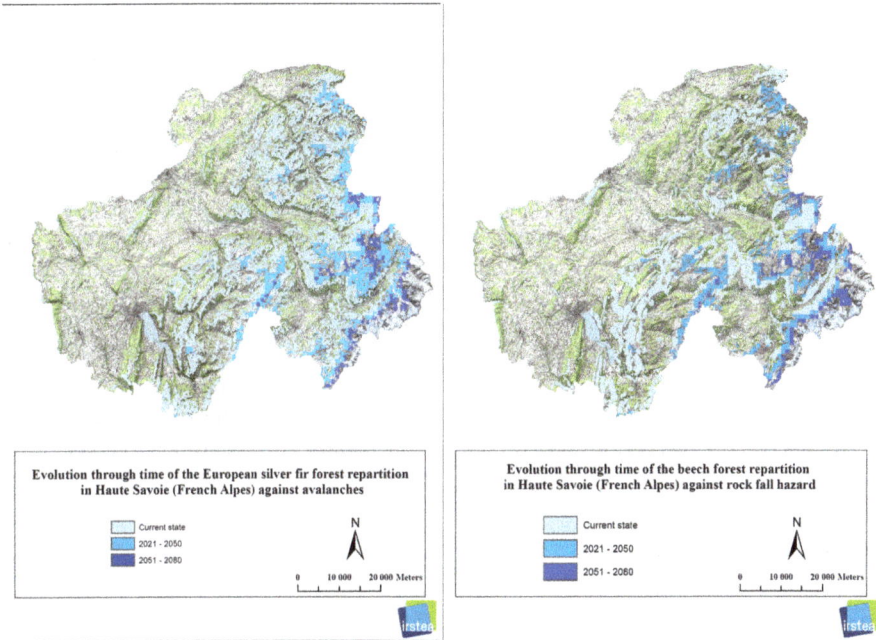

Figure 9. The evolution (2010 – 2080) for the French Department of Haute-Savoie of the geographic distribution of *Picea abies* in snow avalanche release zones and *Fagus sylvatica* in rockfall propagation zones. Light blue: zones covered in 2010, medium blue: zones covered as of 2050, dark blue: zones covered as of 2080.

covered by forest vegetation. If it's the case then these zones could be stabilized and fewer releases might be observed.

• By the end and due to these probable consequences, the major conclusion is that in the next 100 years the protection function of mountain forests will increase.

8. Conclusion and discussion

Because of the uncertainty of the climate change development foresters have to deal with new situations, which for protection forest management are not so easy to foresee. In fact the only certain changes will be the temperature increase and the modifications of the tree species geographic distribution. The results of the project Manfred show that the protection function of mountains forests will probably increase within the next century. For rockfalls mitigation, this protective effect will increase due to the progression on the upper slopes of broad-leaves trees ones. For snow avalanche mitigation, this protective effect will increase due to the colonisation of new releases zone by coniferous trees.

The complementary approaches developed within the Manfred project, the field observations, robust data acquisition, real size tests, synthesis of guidelines and modelling have proven to be an excellent basis for both research on the interaction between geo-hazards and protection forests and on the impact of climate changes on this function, as well as for the development of tools that are relevant for practitioners working in the field of natural hazard and risk management. With the help of the data and tools, we have been able to show that forests have the capacity to provide a degree of protection that is comparable to many other technical measures. They offer therefore possibilities for using mixed ecological and technical solutions for snow avalanches and rockfalls hazards management. It all depends on the characteristics of the phenomena (size of the release zone, size and energy of the boulders...), the length of the forested slope between release area and the elements at risk, and the structure of the forest. The protective effects of trees and stands can be very well quantified with the latest generation 3D simulation models (currently operational for rockfalls and under development for snow avalanches) that use Airborne Laser Scanning data both for mapping the forest and the terrain. Based on those analyses, the efficacy of the forest might be improved over the years by using the latest knowledge, as the tables of recommendations provided by the project Manfred, on protection forest management which includes the mentioned eco-engineering techniques.

Acknowledgements

This chapter is an outcome of the Interreg project Manfred, conducted within the Alpine Space programme.

Author details

Frédéric Berger[1*], Luuk Dorren[2], Karl Kleemayr[3], Bernhard Maier[4], Spela Planinsek[5], Christophe Bigot[1], Franck Bourrier[1], Oliver Jancke[1], David Toe[1] and Gillian Cerbu[6]

*Address all correspondence to: fredric.berger@irstea.fr

1 Irstea, National Research Institute of Science and Technology for Environment and Agriculture, Grenoble, France

2 FOEN, Federal Office for the Environment, Hazard prevention division, Bern, Switzerland

3 BFW, Federal Research and Training Centre for Forests, Natural Hazards and Landscape, Department for Natural Hazards and Alpine Timberline, Innsbruck, Austria

4 Stand Montafon, Forest division,Schruns, Austria

5 Slovenian Forest Institute, Ljubljana, Slovenia

6 FVA, Forest Research Institute of Baden-Wuerttemberg, Freiburg, German

References

[1] European Observatory of Mountain Forests 2000

[2] Swiss Federal Statistical Office 2002

[3] BUWAL Bundesamt für UmweltWald und Landschaft. Lawinenwinter 1998/1999. Bern, Switzerland; (2001).

[4] Brang, P, Schönenberger, W, Frehner, M, Schwitter, R, Thormann, J. J, & Wasser, B. Management of protection forests in the European Alps:an overview. For. Snow Landsc. Res. (2006). , 80, 23-44.

[5] Rey, F. Ecological engineering and engineering ecology: the case of rehabilitation of severely eroded mountainous catchments. Ecological engineering (accepted to be published in (2013).

[6] Heim, A. Bergsturz und Menschenleben. (1932). Beiblatt zur Vierteljahrsschrift der Naturforschenden Gesellschaft in Zürich, Fretz & Wasmuth, p., 77

[7] Lied, K, & Bakkehoi, S. Empirical calculations of snow-avalanche run-out distances based on topographic parameters.(1980). Journal of Glaciology, , 26(94), 165-176.

[8] Frehner, M, Wasser, B, & Schwitter, R. Nachhaltigkeit und Erfolgskontrolle im Schutzwald- Wegleitung für Pflegemassnahmen in Wäldern mit Schutzfunktion. Bern: Bundesamt Umwelt Wald Landschaft. (2005). p.

[9] Compagnia delle Foreste Sr.l. : Selvicoltura nelle foreste di protezione. Esperienze e indirizzi gestionali in Piemonte e Valle d'Aosta. (2006). 139-7-88890-122-3

[10] Gauquelin, X, et al. Guides des sylvicultures de montagne- Alpes du nord françaises. (2006). Cemagref, CRPF, ONF: 289 p.

[11] Dorren, L. K. A, & Berger, F. Le Hir C., Mermin E., Tardif P.: Mechanisms, effects and management implications of rockfall in forests.(2005). For. Ecol. Manag., 215, 1-3:183-195.

[12] www.climchalp.org

[13] Eckert, N, Deschatres, M, & Bélanger, L. Analyse des fluctuations spatio-temporelles des nombres d'avalanches dans les Alpes du Nord à partir de l'EPA. (2010). Sciences Eaux & Territoires, , 02, 16-25.

Chapter 12

Drought in Alpine Areas Under Changing Climate Conditions

Ernst Gebetsroither, Johann Züger and
Wolfgang Loibl

Additional information is available at the end of the chapter

1. Introduction

Forest ecosystem sensitivity to climate change is a result of direct impacts from climate (e.g. changes of temperature and precipitation) and indirect impacts from several biotic (e.g. pests) and abiotic factors (CO_2 and Ozone concentration) influenced by climate change [12]. Within this chapter we will concentrate on possible changes for drought hazards due to climate change scenarios. Drought generally has a negative impact on ecosystem productivity and increases mortality. Species adapted to cold and wet conditions with low reproduction rates and limited mobility seem to be most affected. It was found that beech, and surprisingly the broadleaved Mediterranean forests are highly sensitive to drought [1]. The drought of 2003 was in some areas, especially in Germany and France, the strongest drought during the last 50 years. The analysis showed that some time lag effect can occur, thus e.g. for beech the growth reduction was stronger in the following year 2004 after the drought event of 2003. Besides the strong impact of extreme years as 2003 experts assume that in the long run, a change in the frequency of hot and dry years could affect tree species composition and diversity more than one single event [6]. The hydrological cycle at the local scale might superimpose the influences from climate change on a broader scale but extreme events such as droughts cause growth reductions across many site conditions [4]. For the drought analysis within the MANFRED project and because of the information gap[1] on local site conditions for the whole study area we concentrated our analysis on the predicted input changes from precipitation and temperature.

---

1 like information about the soil, groundwater, surface runoff etc.

2. Method

2.1. Regional climate modelling

Global Circulation Models (GCMs) simulate large scale features of global atmospheric and ocean circulation, based on physical principles. A relevant disadvantage is the coarse horizontal resolution of app. 100-120km in mid-latitudes. This resolution is inadequate to reflect the orographic influences and effects of land cover and soil on atmospheric and hydrological processes at regional scales (Figure 1). Therefore a Regional Climate model (RCM) is embedded into a global model to calculate the atmospheric and hydrological processes on finer scales down to 10x10 km (Figure 2). Along borders the RCM is supplied with data from the GCM, inside the model domain the RCM develops its own dynamics. While the GCM simulates the response of the climate system to global driving forces (greenhouse gas, solar activity, volcanoes …) the RCM simulates additionally the impact of regional factors like terrain, soil and land cover.

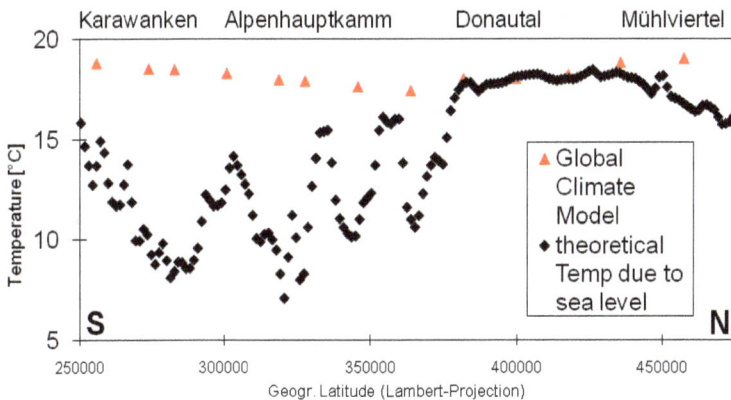

Figure 1. Temperature values in a N-S cross section through the Alps as calculated by GCMs (red triangles) and as expected due to elevation (black squares) (AIT, figure based on ECHAM4 results and a temperature-elevation gradient based on observation data)

Another important point is to distinguish between weather prediction models and climate models. A weather prediction model starts from actual state of the atmosphere and calculates a few days into the future, but after 10-14 days the model is no longer correlated to the real state of the atmosphere. On the other hand the simulation time in climate models may be up to centuries. Therefore climate models do not describe actual weather, but a potential realisation of weather conditions. Only in long term statistics like mean values or variability a climate model should be similar to real weather behaviour. Weather prediction like: „It`s raining on April, 24th 2045" is not possible, but statistical information similar to: „The probability of precipitation in June in the 2040s is higher than today" can be derived.

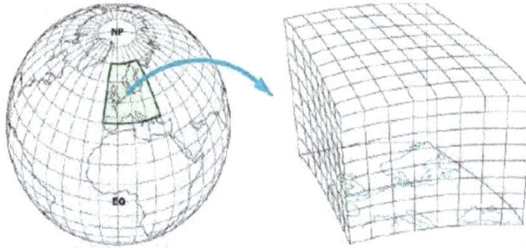

Figure 2. RCM domain (Δx= 10 – 50 km) embedded into a GCM (Δx= 100 – 300 km), Wegener Center Univ.Graz)

2.2. Three steps to regional climate scenarios

Regional climate is influenced by global atmospheric conditions and the global climate processes. Hence, for calculations of regional climate scenarios it is necessary to use a stepwise approach, consisting of three main steps:

Step 1: Selection of a particular emission scenario for consideration of atmospheric changes.

Step 2: Selection of a global climate model to calculate the scenarios for global climate processes.

Step 3: Selection of an appropriate regional climate model and calculation of the regional climate scenarios.

2.2.1. Step 1: Selection of a particular emission scenario for consideration of atmospheric changes

The Intergovernmental Panel on Climate Change (IPCC) provides a wide range of Green House Gas scenarios based on assumptions of the future development of technologies and society (Figure 3).

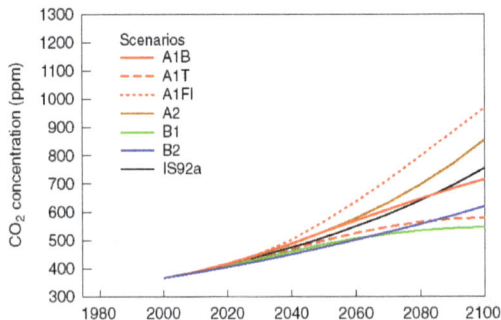

Figure 3. CO_2 concentrations in the atmosphere as predicted by different IPCC scenarios (IPCC 3rd Assessment Report – Climate Change, 2001[2])

The selection of an appropriate scenario is the first important step towards the estimation of climate change signals. The A1 storyline for example describes a future world of very rapid economic growth, global population that peaks in mid-century and declines thereafter, and assumes rapid introduction of new and more efficient technologies. Major underlying themes are convergence among regions, capacity building, and increased cultural and social interactions, with a substantial reduction in regional differences in per capita income. The A1 scenario family can be divided into three branches describing alternative directions of technological change in the energy system. They are distinguished by their technological emphasis: fossil intensive (A1FI), non-fossil energy sources (A1T), or a balance across all sources (A1B)[3]. Out of this the scenario A1B was selected because it represents a moderate increase of Green House Gases (GHG) and is located in the centre of all covered assumptions (Figure 4). The predicted temperature boost for different emission scenarios additionally depends on the used GCM and covers a range from app. 1°C to more than 5°C. The estimated global temperature increase for the scenario A1B until the end of the century is in the range of +1.5°C to +2.5°C, which underlines its conservative position as it does not tend to any extreme at the upper or lower bounds.

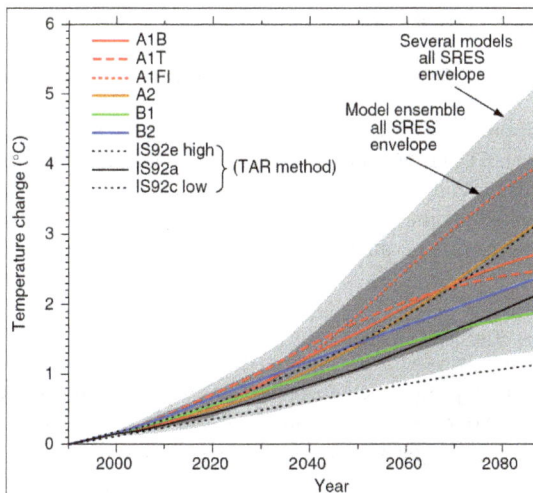

Figure 4. Increase of the global temperature according to different IPCC scenarios[4], (IPCC, 2001, 3rd Assessment report)

2 http://www.grida.no/publications/other/ipcc%5Ftar/?src=/climate/ipcc_tar/wg1/figspm-5.htm , accessed 15.10.2012

3 http://www.ipcc.ch/pdf/special-reports/spm/sres-en.pdf , IPCC SPECIAL REPORT, EMISSIONS SCENARIOS, 2000

4 http://www.grida.no/publications/other/ipcc%5Ftar/?src=/climate/ipcc_tar/wg1/fig9-14.htm, accessed 15.10.2012

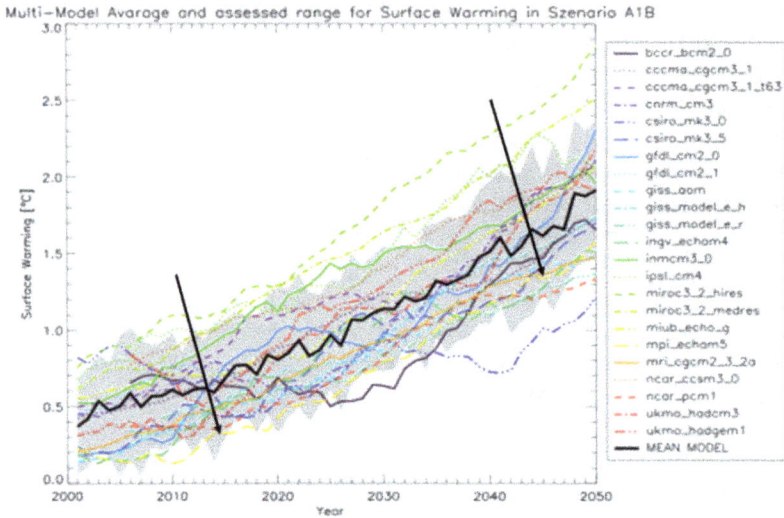

Figure 5. Increase of the global temperature according to the IPCC scenario A1B predicted by different GCMs. The arrows mark the temperature timeline given by the ECHAM5 model (dashed yellow line), [15]

2.2.2. Step 2: Selection of a global climate model to calculate the scenarios for global climate processes

The 2nd step in the approach is the choice of the GCM providing the driving data. In our case results from the German ECHAM5/MPI-OM ([17], [18]) model have been used. The temperature increase towards the end of the century predicted by the ECHAM5 model tends to be at the lower end compared to other models (black arrows in Figure 5). Further information on the ECHAM5 model can be obtained from http://www.mpimet.mpg.de/en/wissenschaft/modelle.html[5].

2.2.3. Step 3: Selection of an appropriate regional climate model

The 3rd step includes the appropriate choice of the RCM providing high resolution data needed for regional analyses. In our case the German "Consortial Runs" [8] were chosen. They were carried out with the RCM Cosmo-CLM[6]. The Grid on which these model runs are based consists of 257x271 cells with a resolution of 0.165° which corresponds to app. 18x18km in mid-latitudes (Figure 6). The time range of these simulations starts in 1961 and covers 140 years up to 2100. Many atmospheric parameters (like temperature, precipitation, snow, wind components, radiation, cloud cover, etc.) are available on a daily or even hourly basis from the German Climate Data Center (DKRZ) via the Climate and Environmental Retrieval and Archive (CERA[7]).

---

5 http://www.mpimet.mpg.de/en/science/models/echam/echam5.html

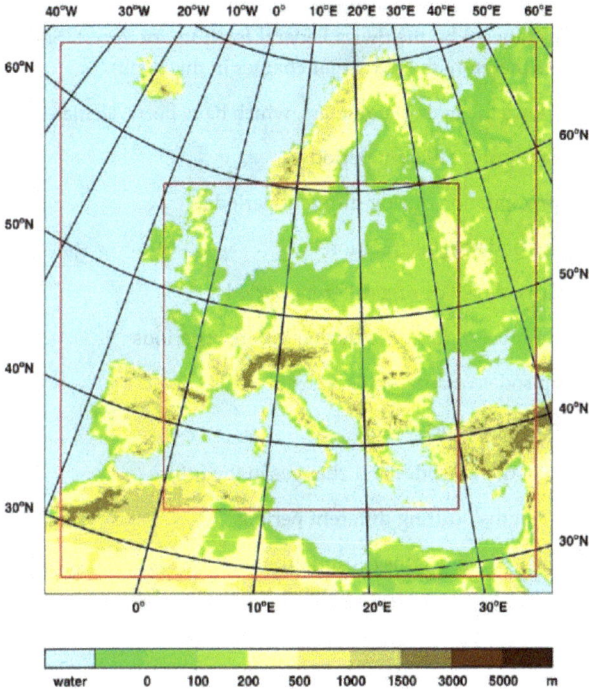

Figure 6. Model domain for Cosmo-CLM Consortial Runs (257x271 cells, resolution: 0,165°), [8]

2.3. Drought hazard calculation with regional climate input data

As there is no single drought parameter which can summarize the impact of climate change on drought for forests, it has been decided to calculate several parameters to provide as much information for forest practitioners as possible. For the current study results of the IPCC climate scenario A1B[8] downscaled to a spatial resolution of 1x1km and daily time resolution have been applied. Temperature as well as precipitation data for every day have been analysed within the time period from 1961 to 2100. As the scenario A1B is rather moderate concerning emissions and temperature increase (cf. Figure 4&5) the results according to climate change impacts on drought presented within the MANFRED project can be seen, despite of the uncertainty of the climate change scenarios, as a quite probable development.

---

6 http://www.clm-community.eu/, accessed 3.9.2012

7 http://cera-www.dkrz.de,accessed3.9.2012

8 http://www.ipcc.ch/ipccreports/tar/wg1/029.htm (accessed 3.9.2012) and Figure 3&4

All calculations have been performed for the whole study area of the MANFRED project with a horizontal resolution of 1x1km. Climate data where statistically downscaled from the original 18x18km simulation results and investigated for 10 case study regions. The downscaling exercise has been conducted by the Swiss Federal Institute for Forest, Snow and Landscape Research (WSL)[9] (see therefore the particular chapter in this book).

The following list shows the main parameters, which have been calculated as 30-year means:

- Total precipitation during different periods[10]
- Change of total precipitation during different periods
- Mean yearly cycles of daily precipitation
- Mean number of dry days[11] during different periods
- Mean change of number of dry days during different periods
- Number of dry episodes of different lengths
- Change of number of dry episodes
- Maximum length of dry episodes and changes in the future
- Mean heat wave lengths[12] during different periods
- Mean change of heat wave lengths during different periods
- Mean daily temperature

3. Results for the entire Greater Alpine Region (GAR)

The Figures 7 and 8 show that precipitation in general is reduced in the whole area at any time. For regions south of the Alps a stronger reduction is estimated and the decrease after 2050 is more significant than till 2050.

To calculate the number of dry days we have used a value used a threshold of 1mm precipitation per day as threshold for a dry day. This was done because 1mm of rainfall usually will not reach the soil in forested areas because it is intercepted in the canopy. Furthermore, climate models tend to overestimate the precipitation and values of zero precipitation occur almost never. The calculation of the number of dry days can be performed with different thresholds, but this will not significantly change the results as we focus on the change signals[13] and not the absolute values of dry days (Figures 9 and 10).

---

9 http://www.wsl.ch/

10 vegetation period (April to September), annual and seasonal results

11 a dry day is a day with less than 1mm precipitation

12 heat waves are: 3 consecutive days with Temp. > 30°C & until mean max. Temp. < 30 °C & not max. Temp. of one day < 25°C

Change of Total Precipitation during Vegetation Period
2021/2050 - 1971/2000

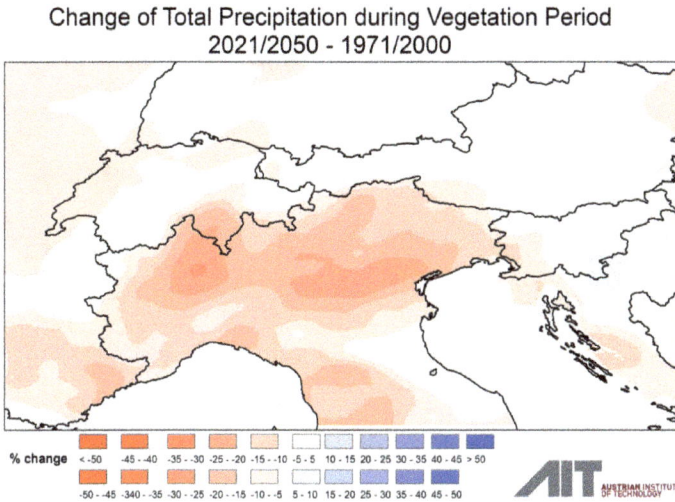

Figure 7. Relative change of total precipitation between 1971/2000 and 2021/2050 (30 yr mean) (Source: ECHAM5/CLM A1B Consortial run, 18x18km resolution, 1x1km downscaling: WSL. Data compilation and spatial analysis: AIT)

Change of Total Precipitation during Vegetation Period
2071/2100 - 1971/2000

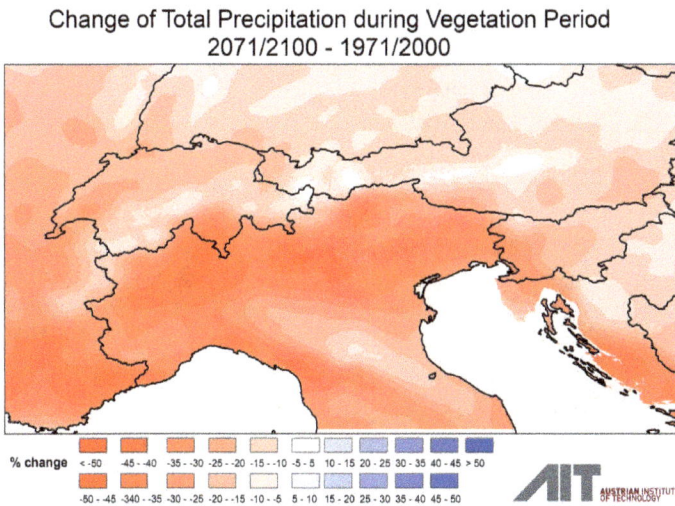

Figure 8. Relative change of total precipitation between 1971/2000 and 2071/2100 (Source: ECHAM5/CLM A1B Consortial run, 18x18km resolution, 1x1km downscaling: WSL. Data compilation and spatial analysis: AIT)

---

13 the relative differences not the absolute values

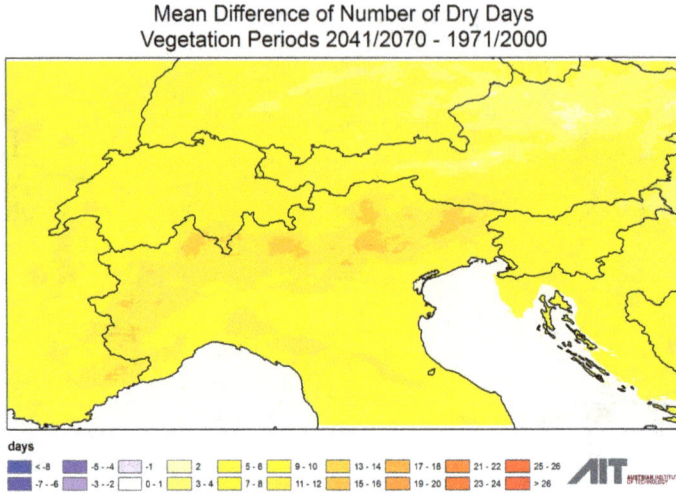

Figure 9. Change of number of dry days during vegetation period in 2041/2070 compared to 1971/2000 (Source: ECHAM5/CLM A1B Consortial run, 18x18km resolution, 1x1km downscaling: WSL. Data compilation and spatial analysis: AIT)

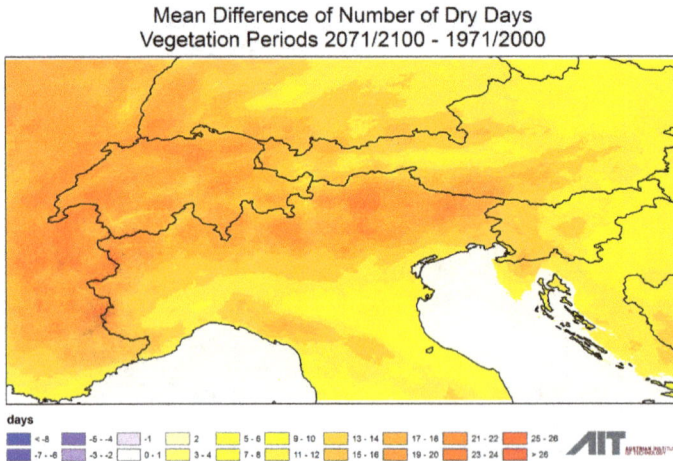

Figure 10. Change of number of dry days during vegetation period in 2071/2100 compared to 1971/2000 (Source: ECHAM5/CLM A1B Consortial run, 18x18km resolution, 1x1km downscaling: WSL. Data compilation and spatial analysis: AIT)

The amount of total precipitation will be reduced in the future, and the number of dry days and also the maximum length of dry episodes will increase. But also the distribution of rainfall during the year will change which may be assumed to have a stronger impact on vegetation (Figure 11).

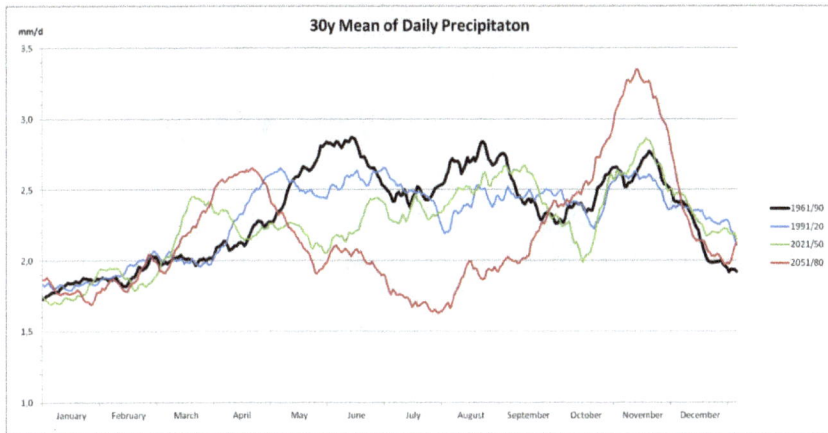

Figure 11. Mean annual cycle of daily precipitation for different periods (30 yr means). (Source: ECHAM5/CLM A1B Consortial run, 18x18km resolution, 1x1km downscaling: WSL. Data compilation and analysis AIT)

The 30y mean of daily precipitation behaviour for the entire GAR shows a shift from main precipitation during the period from May to August to a maximum in April and November. Major reductions may occur from May to end of September so that during summer the drought hazard will increase significantly and also be strengthened due to the temperature increase in this period of the year, as predicted for the future.

3.1. Development of heat waves

As additional parameter we calculated a Heat Wave Index (HWI) according to [7] and [9], which is defined as at least 3 consecutive days with max. temperature above 30 °C and each further day with max. temperature above 25 °C as long as the mean of the max. temperatures over the whole period is not below 30°C. Usually, temperature decreases with higher altitudes, which means that a threshold of 30°C will never be reached in these regions. Nevertheless, some "heat waves" will also occur at higher elevations. To be able to calculate a HWI in a mountainous region, all temperature values had to be normalized to sea level. To achieve this, an average temperature gradient of 0.7°C for each 100m of altitude was assumed. Thus, from a Digital Elevation Model (DEM) of the Greater Alpine Region with a spatial resolution of 1km a theoretical temperature at sea level was calculated. These normalized temperature values were used to determine the HWI for the vegetation period from April to September (Figure

12 and 13). As shown in Figure 14 the maximum length of heat waves increases over the entire GAR, with significant stronger rise south of the Alps.

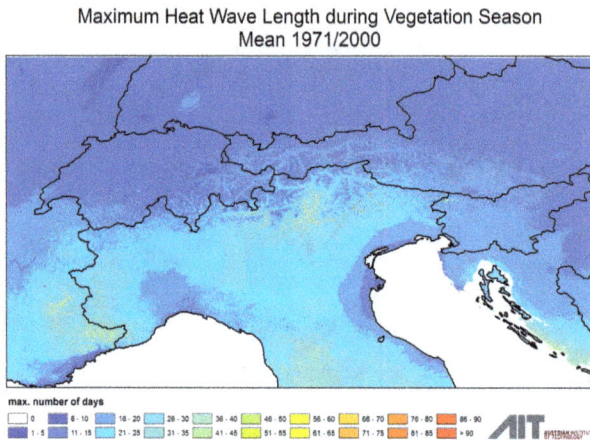

Figure 12. 30-year mean of maximum heat wave length during vergetation period vegetation season 1971/2000 – temperature normalized to sea level. (Source: ECHAM5/CLM A1B Consortial run, 18x18km resolution, 1x1km down-scaling: WSL. Data compilation and analysis: AIT)

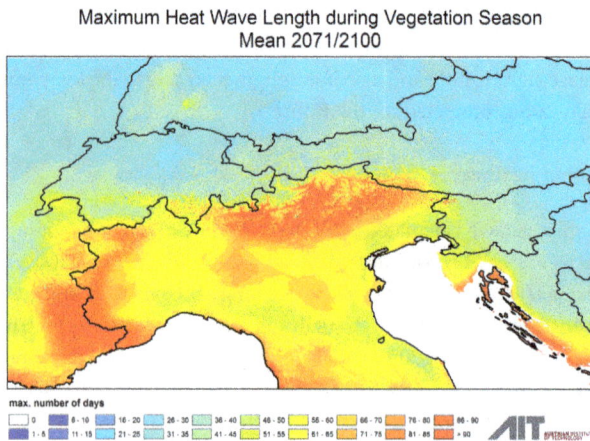

Figure 13. 30-year mean of maximum heat wave length during vergetation period vegetation season 2071/2100 – temperature normalized to sea level. (Source: ECHAM5/CLM A1B Consortial run, 18x18km resolution, 1x1km down-scaling: WSL. Data compilation and analysis: AIT)

Mean Difference of max. Heat Wave Lengths
Vegetation Periods 2071/2100 - 1971/2000

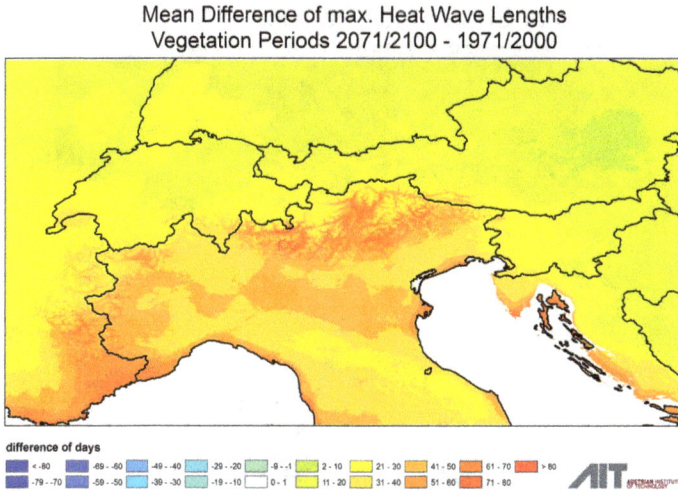

Figure 14. Change of maximum heat wave length during vergetation period for 2071/2100 compared to 1971/2000 – temperature normalized to seal level (Source: ECHAM5/CLM A1B Consortial run, 18x18km resolution, 1x1km downscaling: WSL. Data compilation and analysis: AIT)

4. Presentation of major results for the case study regions

This section discusses exemplarily major results for five of the ten case study regions. The different location of the regions can be seen in Figure 15. Location and extent of the regions have been provided by the project partners of MANFRED. The criteria for taking the five samples out of the ten was to select one from each country providing sample regions as well as covering the climate characteristics from north to south and from west to east.

All results represent statistical means of the entire regions of interest. Larger regions – overall the Italian region I1 (Lombardy) – show a wider range of climate characteristics as the area reaches from low elevated areas with Mediterranean influence to high Alpine areas. The differences in the area mean and area max values give some hint regarding the range of the values.

The following table of diagrams (Figure 16) presents the annual variation of precipitation for the selected case study regions. They all show declining precipitation during the summer months. But from north to south the changes somehow deviate. In the northern regions – influenced by Atlantic climate ("Nordstaulagen") – the case study regions D1, A1 may expect a growing peak of spring precipitation and from May to September a high decrease during the future decades. The regions, which are influenced by the Mediterranean Sea (I1 and S1) will face a decline of summer precipitation and the evolution of two distinct peaks: a smaller spring

Locations of Case Study Regions in Alpine Area

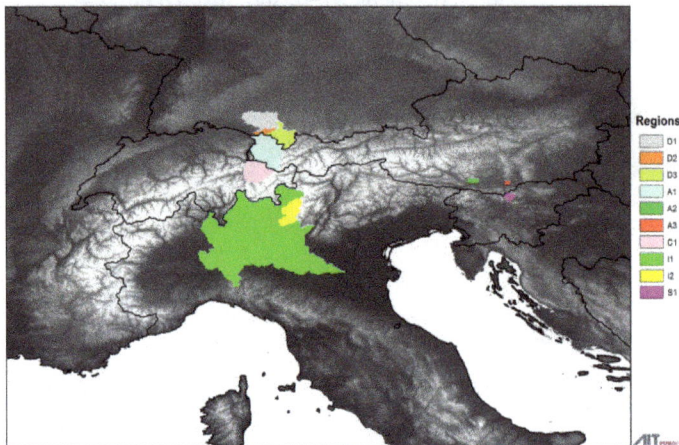

Figure 15. Location of the ten case study regions explored in MANFRED

peak and a high peak in November. The Swiss region (C1) seems to be affected by both influences – the precipitation decline in summer is still high, two precipitation peaks occur – a higher spring peak which is characteristic for the northern regions and a smaller autumn peak which is typical for Mediterranean influence. In all areas there is a significant secondary sink in October.

The diagrams in Figure 17 show the evolution of the total precipitation during the growing season for the selected case study regions: All regions show a moderate decline of the average and the maximum rainfall sum. The German (D1) as well as the Slovenian (S1) case study region show moderate precipitation ranges between 600 and 800 mm which declines in Germany to 600 for max values and 550 mm for average values, in Slovenia the max values remain almost the same, the average values decline from almost 700 mm to 350 mm. All other case study regions show an absolute decline of about 100 - 200 mm of the max values in the high alpine areas and again an absolute decline of about 50 - 100 mm of the average numbers referring more to the low elevated areas. For the lower elevated areas like the Italian region I1 the decline may turn out to be severe as the precipitation sum during the vegetation period is between 500 and 700mm and the temperature is going to be higher which results in higher evapotranspiration.

While in Figure 17 the total precipitation during the growing season is shown, Figure 18 depicts the seasonal precipitation sums during different 30 year periods. In all regions a significant decrease of the summer precipitation is obvious: the total rainfall during summer declines from 400 to 300 mm in the northern regions, from app. 400 to 250 mm in the Slovenian region (S1) and from 250 to 150 mm in the Italian areas with Mediterranean influences (I1).

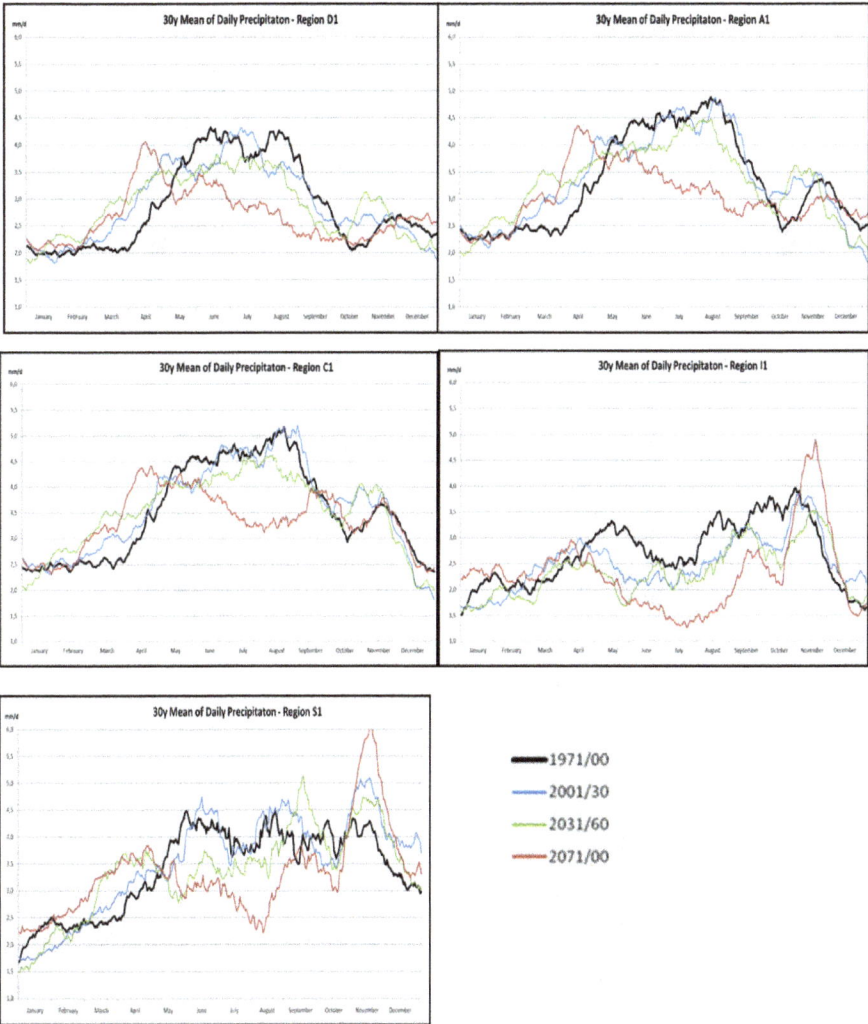

Figure 16. Mean yearly distribution of precipitation from 1971/2000 to 2071/2100 for case study regions D1, A1, C1, I1 and S1 (Source: ECHAM5/CLM A1B Consortial run, 18x18km resolution, 1x1km downscaling: WSL. Data compilation and analysis: AIT)

The following table of diagrams (fig. 19) indicates the evolution of the number of dry days during the vegetation season for the selected case study regions. All regions show a distinct increase of the average and the maximum number of dry days: the less alpine German region (D1) starting from 69 (average) / 73 (max) dry days may expect an extra of about 20 dry days

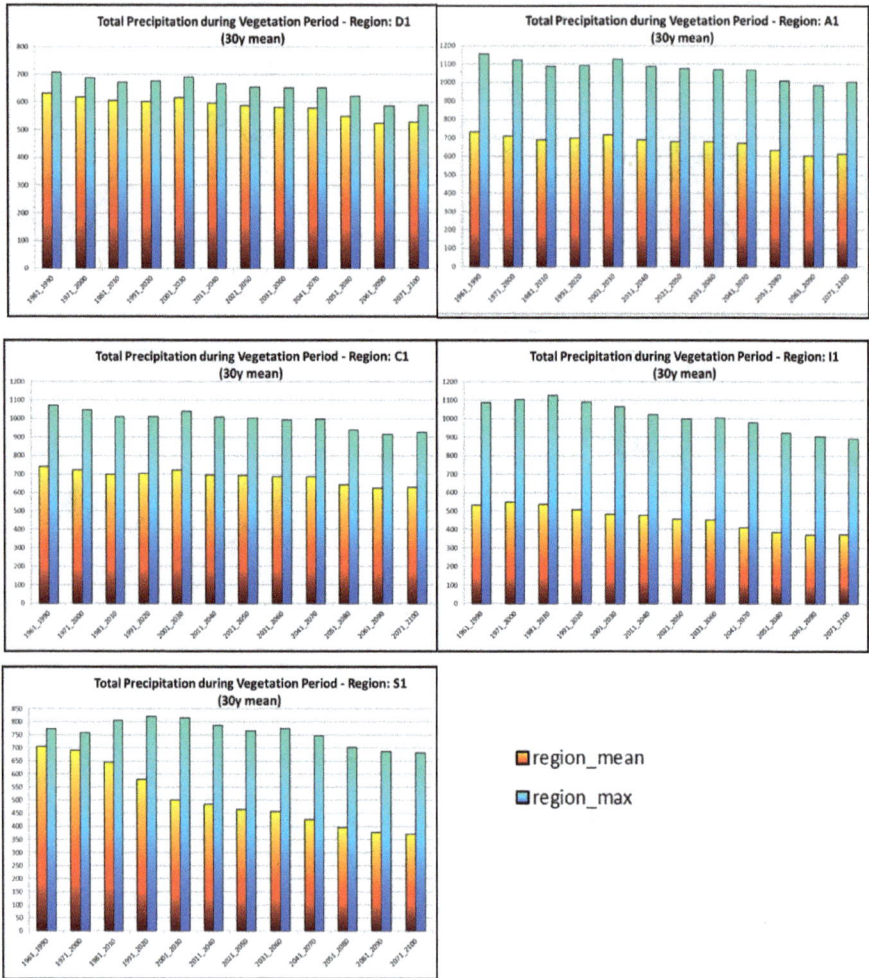

Figure 17. Mean total precipitation during the vegetation period from 1971-2000 to 2071/2100 for case study regions D1, A1, C1, I1 and S1 (Source: ECHAM5/CLM A1B Consortial run, 18x18km resolution, 1x1km downscaling: WSL. Data compilation and analysis: AIT)

till 2100, the Swiss region (C1) starts from 48 (average) / 55 (max) dry days and will face an increase of app. 15 dry days till 2100. The Vorarlberg Region (A1) with alpine climate and some moderate influence of Lake Constance starts with 80 (average) / 90 (max) dry days and may expect another 20 dry days till 2100. In the small Slovenian region (S1) the number of dry days will increase from app. 120 by 10 to 20 days till 2100. The large Italian region (I1) shows the widest range because of the Alpine and the Mediterranean influences with hottest tempera-

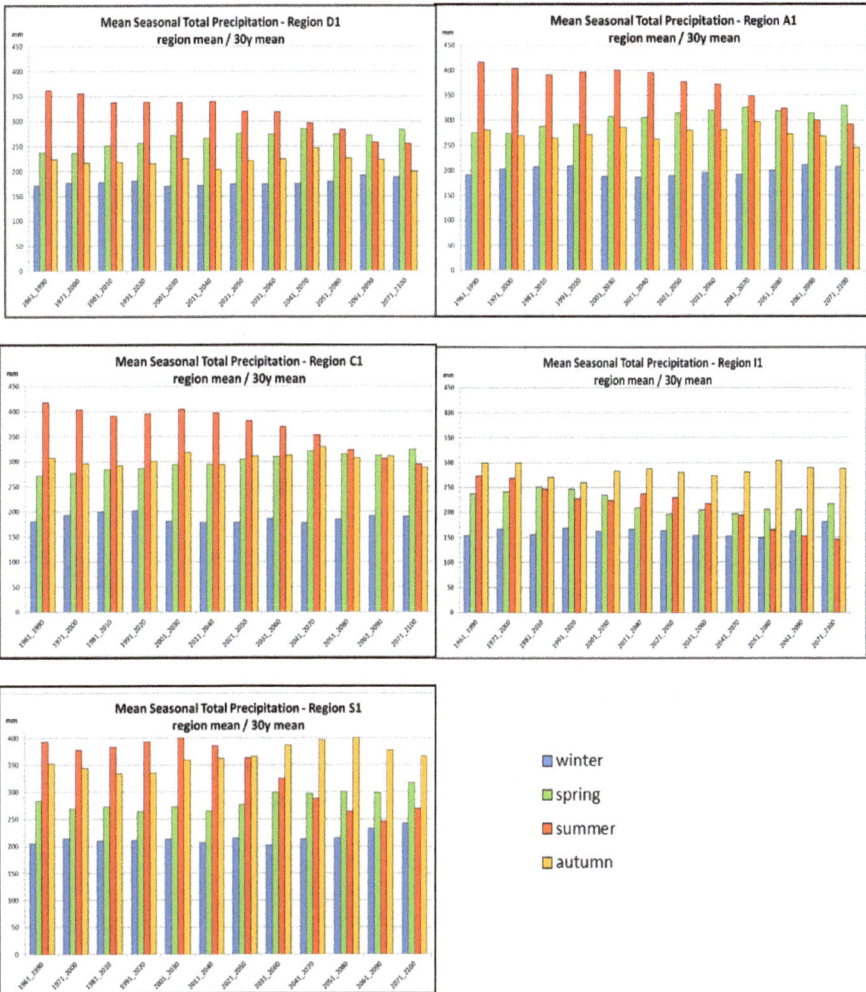

Figure 18. Mean seasonal total precipitation sums from 1971-2000 to 2071/2100 for case study regions D1, A1, C1, I1 and S1 (Source: ECHAM5/CLM A1B Consortial run, 18x18km resolution, 1x1km downscaling: WSL. Data compilation and analysis: AIT)

tures in the Po Valley: 110 (average) and 135 (max.) dry days with an expected increase of 15 dry days.

The following table of diagrams (fig. 20) shows the evolution of the seasonal 30 year mean number of drought periods greater than 10 days for the selected case study regions. In general the number of more than 10-day dry periods for current climate accounts to 1-2 events per

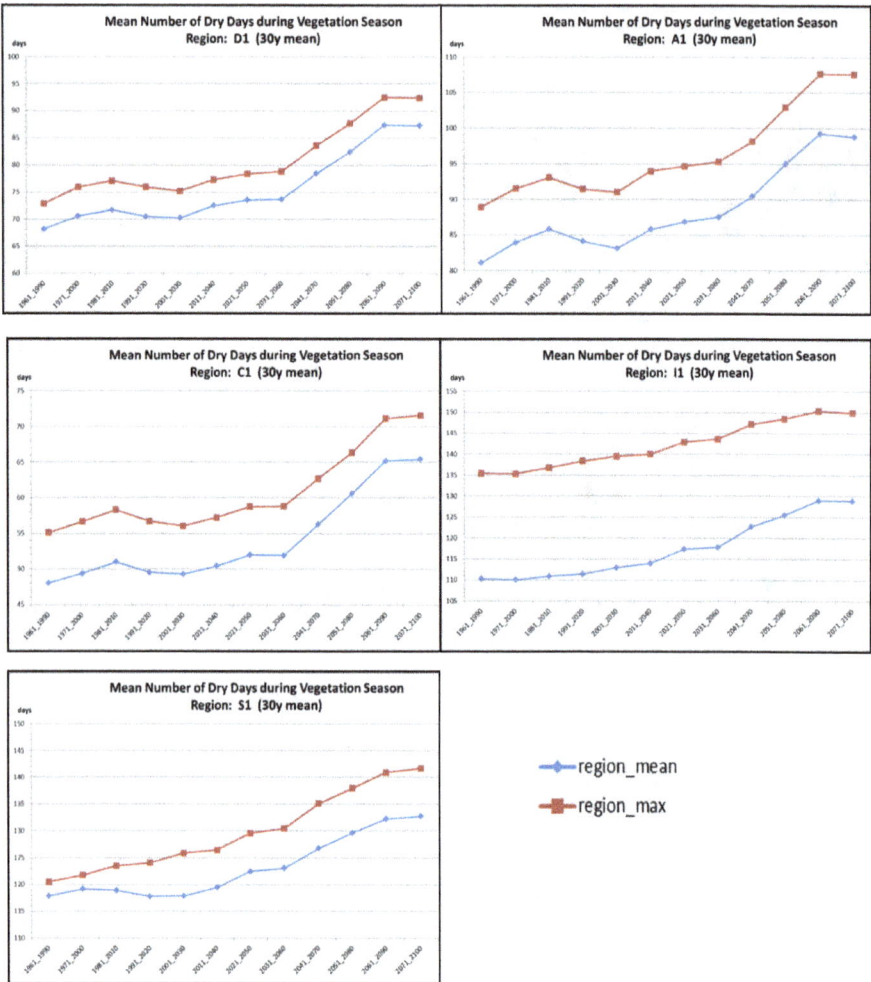

Figure 19. Mean number of dry days during vegetation season from 1971/2000 to 2071/2100 for case study regions D1, A1, C1, I1 and S1 (Source: ECHAM5/CLM A1B Consortial run, 18x18km resolution, 1x1km downscaling: WSL. Data compilation and analysis: AIT)

season and is expected to grow by app. 1 event during summer months till the end of the century. All regions show a significant increase of such drought periods during summer. The northern regions (A1 and D1) show a rise up to nearly 200 %, the Slovenian region up to 90 % and the Italian regions – due to higher occurrences at present – app. 40 %. At most locations there may be some reduction in spring, but overall changes during the other seasons except summer are not very significant.

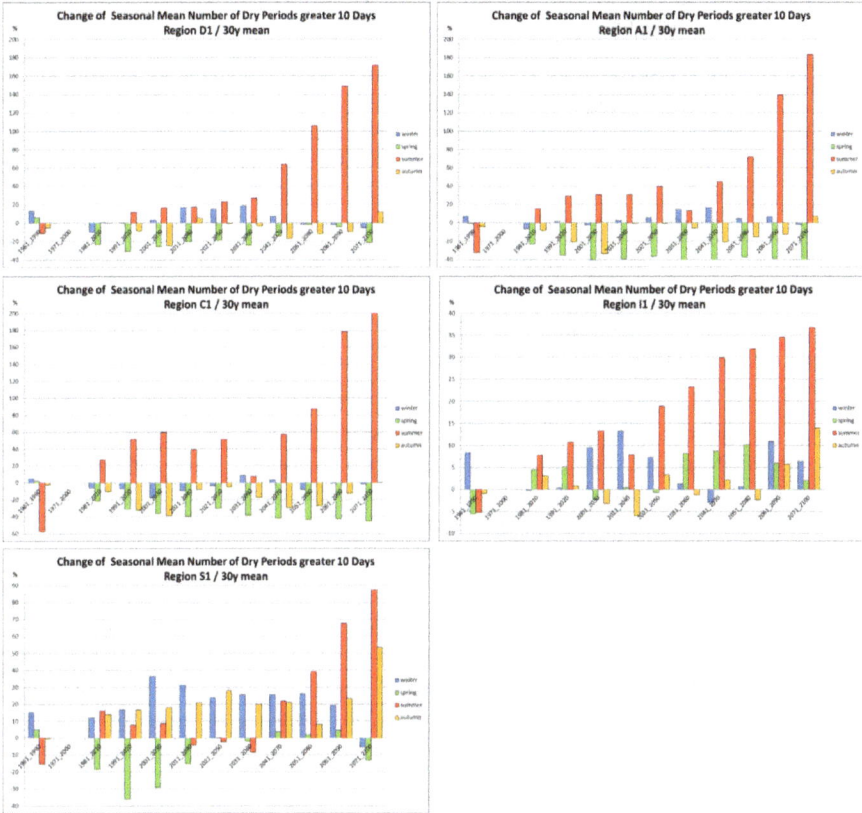

Figure 20. Relative change of mean number of dry periods greater than 10 days compared to period 1971/2000 for case study regions D1, A1, C1, I1 and S1 (Source: ECHAM5/CLM A1B Consortial run, 18x18km resolution, 1x1km downscaling: WSL. Data compilation and analysis: AIT)

5. General result discussion

The Figures above let expect on the one hand a decrease of total precipitation within the vegetation period during the coming decades till 2100 and on the other hand a substantial increase of dry day periods – both aspects may lead to increasing drought stress in the particular forest areas in the greater alpine area.

- The tendency to significant lower rainfall in summer can be found in all areas. The decline of precipitation south of the Alpine main ridge (Northern Italy, Slovenia) is much stronger due to the Mediterranean influence. The applied climate scenario assumes a decline of summer precipitation up to 50%. This surely will affect forests in southern regions and may lead to extended drought stress and fire hazard.

- A similar picture gives the comparison of the number of dry days in the future with the 30 year mean for 1971/2000 (Figure 20). The number of dry days north of the alpine main ridge is 1/3 less than south of it. The increase of dry days is estimated to achieve 15 to 20 additional days, which is quite severe in the southern regions with numbers above 120 dry days for current climate. The number of dry day episodes as well as the length of the periods is expected to increase in all regions by 2 or 3 events per length class for the shorter episodes. The long periods (20 to 30 days) occur only in the southern regions and are also expected to increase.

Thus the combination of precipitation decrease, increase of the length of dry day periods as well as the increase of temperature may result in severe drought stress effects on the forest areas.

Investigations of forest growth and dendroecology compiled from tree ring chronologies have shown that differences between tree species to cope drought events in wide range occur. Over a longer period after an intense drought event[14] Norway spruce tend to be more effected as common beech and Scots pine. Pedunculate and sessile oak species showed very little growth losses in the investigations by Beck [1].

In general the estimation of the local risks for forest and forest management needs local forest experts and stakeholders which are able to balance the different demands on forests to sustain in the future. The presented results of the single case study regions of the project MANFRED can be used as input for these analyses by local experts. The MANFRED project provides much more information than discussed within this book chapter. Detailed data have been provided to the stakeholder of the different case study regions and all MANFRED project partners.

6. Uncertainty and spatial variability of regional climate projections

Uncertainty in regional climate projections can be roughly divided into four components: (1) Natural variability[15], (2) uncertainty in external forcing (mainly anthropogenic forcing like greenhouse gas emissions and land use change)[16], (3) uncertainty due to imperfect simulation of the climate system (model uncertainty)[17], and (4) downscaling uncertainty. Prein et al. ([15, 16]) analysed the relative contributions of natural variability, emission scenarios and models to total uncertainty over Europe. The uncertainty components for *air temperature* and *precipi-*

---

14 e.g. the drought of 2003 in middle Europe

15 The term natural variability refers to deterministic and random fluctuations in the climate system, that occur on various spatial and temporal scales,[3].

16 GHG emissions depend on many factors including growth of population and economy, energy prices and energy source selection, cultural and social interactions, technological development and land use, the latter serving either as GHG sink or as GHG emission accelerator. Since it is impossible to strictly predict the future development of these factors, the pathway of future emissions remains uncertain. The IPCC Special Report on Emission Scenarios (SRES) [14] provides different storylines of how the world might develop and the associated trajectory of future GHG emissions.

tation over Europe due to internal variability, model formulation, and emission scenario assumptions until the mid and the end of the 21st century are regarded in Prein's study, based on 23 different GCMs driving 84 simulations of the CMIP3[18] ensemble.

They found that uncertainty due to the formulation of the climate models describing atmospheric processes is largest for temperature as well as precipitation regime changes. The following Figure 21 presents a comparison of simulation results for seasonal change of temperature and precipitation in the south western part of the greater alpine Region (GAR) – the region, where most of our case study regions are located.

Changes of precipitation amount are calculated relative with respect to 1961-1990. The black lines indicate the median changes, [5].

The cross in the diagrams marks the median of precipitation change (horizontal) and temperature change (vertical): The temperature change medians of the four seasons oscillate between 1.2 and 2 °C while the seasonal precipitation change medians shows rates of 0 to 7% of the particular precipitation total. The changes of all seasonal temperature means show the widest range between +0.5°C and 3°C during summer. The ECHAM5-driven simulation marked with green dots show always the smallest temperature gain, while the HADCM3-based[19] results marked with blue and purple dots show constantly the highest increase. For precipitation the single models show high deviations during the seasons, ranging from -20 to +20 %. ECHAM-based models show declines in spring and summer, and (little) precipitation increase in autumn and winter, while the HADCM3-scenarios show no clear trend indicating some increase as well as some decrease during the seasons. These results give an idea of the uncertainty range which can be expected also for the model output applied in MANFRED.

But uncertainty shall not be mixed up with spatial variability unless the variability itself shows uncertainty: A rough exploration of the spatial variability – and the uncertainty - has been conducted in the project reclip:century carried out by AIT ([13], http://reclip.ait.ac.at/reclip_century).

Besides uncertainty, there are also distinct location related differences between the regions, which are effects of terrain and position within the global atmospheric pressure and humidity motion patterns. Thus the simulations show a clear influence of the Alpine main ridge on the spatial distribution of climate change signals, with e.g. precipitation increase north of the Alps in spring, summer and autumn, and some decline in the southern and western parts. The temperature change pattern show rather an east-west- than a north-south-gradient. For Austria regional change trends for temperature and precipitation have been extracted from three regional climate simulation runs (based on 2 GHG scenarios and 3 RCM/GCM-combinations[20])

---

17 Uncertainties due to climate models arise from incomplete understanding and simplified formulation of climate processes in the models [19]. In order to quantify these uncertainties, ensembles of different or modified climate models are used.

18 Coupled Model Intercomparison Project Phase III, http://cmip-pcmdi.llnl.gov/, accessed 08.04.2013

19 "HadCM3 stands for the Hadley Centre Coupled Model version 3. It was developed in 1999 and was the first unified model climate configuration not to require flux adjustments (artificial adjustments applied to climate model simulations to prevent them drifting into unrealistic climate states)".http://www.metoffice.gov.uk/research/modelling-systems/unified-model/climate-models/hadcm3, accessed 5 September 2012

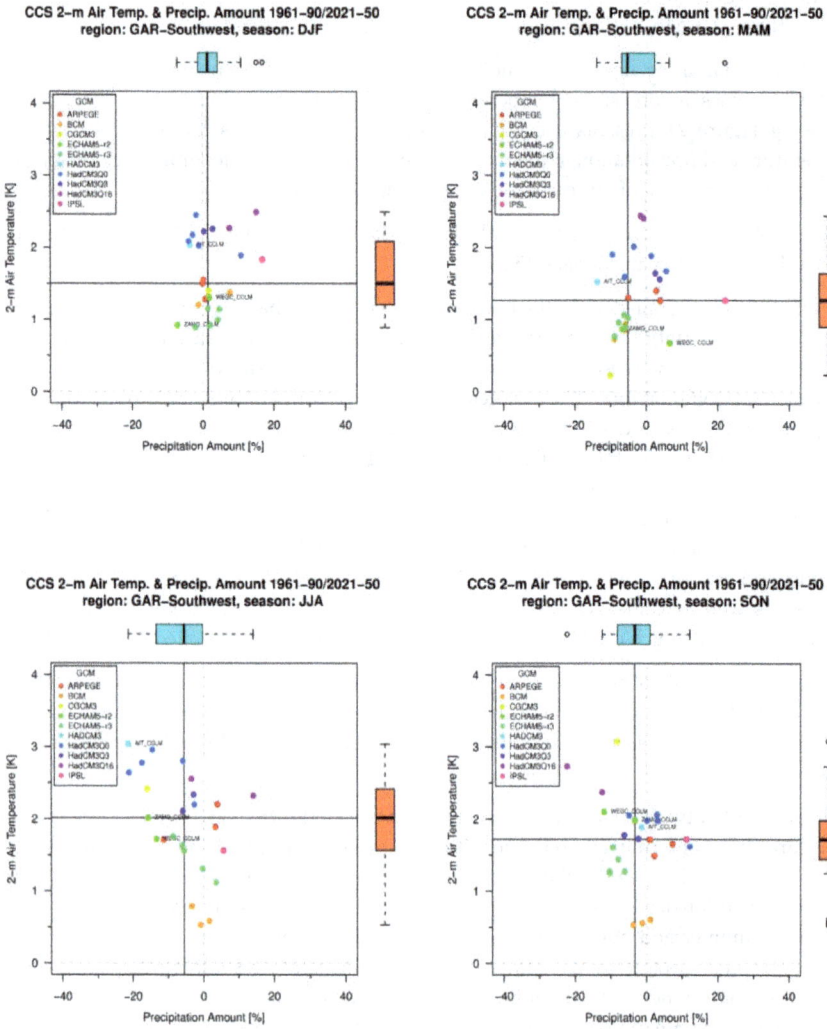

Figure 21. Scatter plots for the changes in 2-m air temperature and precipitation amount between 2021/2050 and 1961/1990 for GAR-South-west for 22 ENSEMBLES simulations (http://www.ensembles-eu.org) [10, 11].

---

20 The observed seasonal changes till 2050 as explored in these three reclip:century simulations are integrated in the diagrams in Fig. 21 and marked by names: AIT-CCLM is based on a A1B/HADCM3/CCLM combination, WEGC-CCLM is based on an A1B/ECHAM5/CCLM combination and ZAMG-CCLM is based on a B1/ECHAM5/CCLM model combination.

for the four seasons and related to Austrian climate regions. The range of climate change signals within a region refers to uncertainty, the range of climate change signals between the regions refers to local effects of spatial variability[21].

Winter:		
	temperature: +1.6 to +2.2 °C,	prec.: +8 to +13% increase
	more in E	less increase in the S and W
Spring:		
	temperature: +1.0 to +1.2 °C.	prec.: constant to light decreases,
	more distinct in the E	
Summer:		
	temperature: +1.0 to +2.5 °C,	prec.: little loss,
	scenarios disagree!	more distinct in the S
Autumn:		
	temperature: +1.7 to +2.3 °C,	prec.: little loss,
	W, S: higher increase, N: divergent	more distinct in S, SE and eastern Alps

Such general location based variability/uncertainty in climate signal change can be expected for the entire Greater Alpine Region.

7. Further research and conclusion

The drought analysis within the MANFRED project where concentrated on the analysis of predicted input changes from precipitation and temperature. The local effects of changes in precipitation, drought length and frequencies or heat waves have to be estimated with local experts including local site conditions. The results show different manifestations of the assumed climate change scenario (A1B) within different areas, especially north or south of the alpine main ridge. It has to be noticed that the analysed climate change scenario is not an extreme scenario, thus the real changes in temperature and precipitation might lead to more severe hazards for forests in the future (see Figure 4 and 5). Above we briefly mentioned the uncertainties of climate change scenarios, which especially referring to precipitation are not neglectable, thus reducing this uncertainty should be one important research task for the future. Beyond this more ample spatial referenced information about site conditions as soil texture, surface runoff, evapotranspiration etc. would be very important to develop a closer understanding of the local hydrological conditions for the whole analysed Greater Alpine Region. A general increase in winter precipitation (mainly in northern and central Europe) and

---

21 This Tabble below referes to results for Austria and it is for the year 2050, not further

decreases in summer precipitation (mainly in central and southern Europe) can be found in different climate scenario simulation. Analysis of extreme events (as drought periods, length and frequencies) have more drastic consequences on tree growth than gradual changes in mean climate conditions [12]. In the future more advanced extreme event analyses have to be developed corresponding with reduced uncertainty in climate change scenarios.

Author details

Ernst Gebetsroither*, Johann Züger and Wolfgang Loibl

*Address all correspondence to: Ernst.Gebetsroither@ait.ac.at

AIT Austrian Institute of Technology GmbH, Foresight & Policy Development Department, Vienna, Austria

References

[1] Beck, W. Auswirkungen von Trockenheit und Hitze auf den Waldzustand in Deutschland- waldwachstumskundliche Ergebnisse der Studie im Auftrag des BMELV: DVFFA- Sektion Ertragskunde, Jahrestagung 2010 http://www.nw-fva.de/ ~nagel/SektionErtragskunde/band2010/Tag2010_07.pdfaccessed on 6 September (2012).

[2] Déqué, M, et al. An intercomparison of regional climate simulations for Europe: assessing uncertainties in model projections, Clim. Change., 81, suppl. 1, DOIs10584-006-9228-x, (2007).

[3] Ghil, M. Natural Climate Variability in Encyclopedia of Global Environmental Change, The Earth System: Physical and Chemical Dimensions of Global Environmental Change. John Wiley & Sons, Ltd: Chichester, (2002). , 1

[4] Granier, A, Reichstein, M, Bréda, N, Janssens, A, Falge, E, Ciais, P, Grünwald, T, Aubinet, M, Berbigier, P, Bernhofer, C, Buchmann, N, Facini, O, Grassi, G, Heinesch, B, Ilvesniemi, H, Keronen, P, Knohl, A, Köstner, B, Lagergren, F, Lindroth, A, Longdoz, B, Loustau, D, Mateus, J, Montagnani, L, Nys, C, Moors, E, Papale, D, Peiffer, M, Pilegaard, K, Pita, G, Pumpanen, J, Rambal, S, Rebmann, C, Rodrigues, A, Seufert, G, Tenhunen, J, & Vesala, T. and Wang, Q: Evidence for soil water control on carbon and water dynamics in European forests during the extremely dry year: 2003. Agricultural and Forest Meteorology, (2007). , 143, 123-145.

[5] Heinrich, G, & Gobiet, A. reclip:century Uncertainty Assessment. Report Part D: Wegener Center, 2011, Graz (http://reclip.ait.ac.at/reclip_century,accessed 28 September (2012). , 1.

[6] Hemery, G. E. Forest management and silvicultural responses to predicted climate change impacts on valuable broadleaved species. Short-Term Scientific Mission report for Working Group 1, COST Action E42, (2007). refs. www.ForestryHorizons.eu., 196.

[7] Hollweg, H, Böhm, D, Fast, U, Hennemuth, I, Keuler, B, Keup-thiel, K, Kysely, E, Kalvova, J, & Kveton, J. V. Heat Waves in the South Moravian Region during the Period 1961-1995. Studia geoph. Et geod., (2000)., 44, 57-72.

[8] Keuler, K, Lautenschlager, M, Wunram, C, Keup-thiel, E, Schubert, M, Will, A, Rockel, B, & Boehm, U. Climate Simulation with CLM, Scenario A1B run Data Stream 2: European region MPI-M/MaD. World Data Center for Climate, (2009). (1)

[9] Kysely, J. Mortality and displaced mortality during heat waves in the Czech Republic. Int. J. Biometeorol, (2004)., 49, 91-97.

[10] Lautenschlager, M, Legutke, S, Radtke, K, Rockel, B, Schubert, M, Will, A, Woldt, M, & Wunram, C. Ensemble Simulations over Europe with the Regional Climate Model CLM forced with IPCC AR4 Global Scenarios. Model and Data group, Max Planck Institute for Meteorology, Hamburg, (2008). Technical Report 1619-2257(3), 150.

[11] Lautenschlager, M, Keuler, K, Wunram, C, Keup-thiel, E, Schubert, M, Will, A, Rockel, B, & Boehm, U. Climate Simulation with CLM, Climate of the 20th Century run Data Stream 2: European region MPI-M/MaD. World Data Center for Climate., (2009). doi:DOI:10.1594/WDCC/CLM_C20_1_D2.http://dx.doi.org/DOI:10.1594/WDCC/CLM_C20_1_D2(1)

[12] Lindner, M, Garcia-gonzalo, J, & Kolström, M. (2008). Impacts Of Climate Change On European Forests And Options For Adaptation." European Forest, 2008; http://www.metla.eu/tapahtumat/2009/JFNW2009/Lindner.pdf.doi:DOI:10.1594/WDCC/CLM_A1B_1_D2.http://dx.doi.org/DOI:10.1594/WDCC/CLM_A1B_1_D2, accessed 28 September 2012), 40ff.

[13] Loibl, W, Züger, J, & Köstl, M. eclip:century 1. Report Part C: Climate Scenarios: Comparative Analysis, AIT, Vienna, 2011; (http://reclip.ait.ac.at/reclip_century,accessed 28 September (2012).

[14] Nakicenovic, N, Alcamo, J, Davis, G, De Vries, B, Fenhann, J, Gaffin, S, Gregory, K, Grübler, A, Jung, T. Y, & Kram, T. La Rovere EL, Michaelis L, Mori S, Morita T, Pepper W, Pitcher H, Price L, Raihi K, Roehrl A, Rogner H-H, Sankovski A, Schlesinger M, Shukla P, Smith S, Swart R, van Rooijen S, Victor N, Dadi Z. IPCC Special Report on Emissions Scenarios. Cambridge University Press: Cambridge, United Kingdom and New York., 2000; http://www.ipcc.ch/pdf/special-reports/spm/sres-en.pdfaccessed 28 September (2012).

[15] Prein, A. Uncertainties in the driving data of regional climate models in the Alpine region. Wegener Center Scientific Report Wegener Center Verlag: Graz, Austria, (2009). (30-2009), 30-2009.

[16] Prein, A. F, Gobiet, A, & Truhetz, H. Analysis of uncertainty in large scale climate change projections over Europe Meteorol. Z., doi:(2011)., 20, 383-395.

[17] Roeckner, E, Bäuml, G, Bonaventura, L, Brokopf, R, Esch, M, Giorgetta, M, Hagemann, S, Kirchner, I, Kornblueh, L, Manzini, E, Rhodin, A, Schlese, U, Schulzweida, U, & Tompkins, A. The atmospheric general circulation model ECHAM 5. PART I: Model description, (2003).

[18] Roeckner, E, Lautenschlager, M, & Schneider, H. . IPCC-AR4 MPI-ECHAM5_T63L31 MPI-OM_GR1.5L40 SRESA1B run no.1: atmosphere 6 HOUR values MPImet/MaD Germany. World Data Center for Climate, 2006; DOI:10.1594/WDCC/EH5-T63L31_OM-GR1.5L40_A1B_1_6H.   http://dx.doi.org/10.1594/WDCC/EH5-T63L31_OM-GR1.5L40_A1B_1_6H (accessed 28 September 2012)

[19] Stainforth, D. A, Allen, M. R, Tredger, E. R, & Smith, L. A. Confidence, uncertainty and decision-support relevance in climate predictions. Phil. Trans. R. Soc, (2007). A 365: 2145-2161. DOI:rsta.2007.2074.

Chapter 13

Coupling a Forest Growth Model with a Soil Carbon Simulator

Klaus Dolschak, Robert Jandl and
Thomas Ledermann

Additional information is available at the end of the chapter

1. Introduction

The impact of climate change on forests can be assessed with simulation models. The obvious advantage is the possibility of running a large number of case studies, without the need for field trials for each specific case. Especially in the context of climate change effects simulation models have become a standard tool. The alternative approach of evaluating the forest growth at sites with site conditions that may resemble the future conditions of the target site has numerous weaknesses. Therefore, models are deemed superior for the interpretation of the consequences of climate change. Simulation models are representative for the state-of-knowledge. Upon their development they are validated against data of existing forests, *i.e.* either data from national forest inventories or long-term forest productivity experiments [1–4].

The available models vary widely with respect to the model structure and the embedded modules. *Productivity models* are often conceived as a practical tool where forest practitioners find the opportunity to directly define specific silvicultural treatments such as thinning interventions, and the model provides information on the consequence of these actions on the temporal trend of stem growth. An important incentive for the development of this model type was that formerly used yield tables proved to be less reliable for several reasons. Yield tables were often derived from a rather limited number of field experiments and neither their species composition nor their management is reconcilable with present concepts and requirements of forestry [5, 6].

Succession models place their emphasis on the regeneration success of different tree species under future climate conditions. The focus is on the identification of site conditions, with the potential of changing the tree species combination. The strong part of these models is the description of the competition between the different tree species within a forest and the success of trees for natural regeneration [7–10].

Biogeochemical models such as BIOME-BGC are large scale models that model primary productivity [11]. The simulation results mostly apply for large regions and are skillfully broken down for smaller regions. Although the models are generally not conceived for the description of silvicultural treatments, there are efforts underway to introduce options of forest management [12].

We chose a productivity model, because we wanted to properly depict management options that are expressed in terms of generally understandable silvicultural interventions. In productivity models the emphasis is typically on the effect of silvicultural interventions, but not necessarily on the provision of ecosystem services such as carbon sequestration.

The sequestration of carbon in forest soils is often not integrated in the productivity models. Instead, the carbon dynamics are calculated separately. A notable exception is a Swiss study were the impact of storm damage on the soil carbon pool had been evaluated [13]. The effect of tree species on the forest soil carbon pool has been shown [14–17]. Therefore, we are convinced that a full assessment of forest management options needs to comprise both the carbon pools in the total biomass and the soil.

Our objective was the linkage of a soil productivity model with a soil carbon model in order to emphasize the implication of forest management on the soil carbon pool.

2. Sites and methods

2.1. Site

The chosen site for our simulation exercise in the Ossiacher Tauern. It is a montane forest area on silicatic bedrock. Although the site would be naturally dominated by deciduous trees, it is currently ideal for growing Norway spruce (*Picea abies* (L.) Karst.) [18].

2.2. The forest productivity model CALDIS

CALDIS is derived from the celtic term "Caldis Vâtis" translating to "forest prognosticator". The core of CALDIS is a module simulating the basal area increment of individual trees of the dominating tree species in Austria, *i.e.* Norway spruce, fir, larch, three pines, European beech, oak, and other deciduous species. In recognition of the scarcity of georeferenced single trees the model was intentionally developed as distance-independent model. The data for model development were taken from the Austrian National Forest Inventory. The model handles competition between trees with the basal area as the parameter. The basal area increment of individual trees; BAI, was derived from ?tree size?, ?competition? and ?site properties?. The by far most important parameter for the basal area increment is 'tree size", represented by stem diameter and size of the canopy. The competitive situation of a single tree is captured as the sum of the basal area of all trees with a larger diameter than the sample tree. Site factors such as topography, elevation, aspect, depth of the organic soil layer, and soil type contribute a small, yet significant, improvement of the model. Besides the diameter and height growth module, the model contains a mortality module and a regeneration module. The model has been used with the acronym "PrognAus" [6]. The productivity model has been amended with several additional modules. In order to capture the long-term development of a forest, an ingrowth module was added describing the recruitment of naturally regenerating trees. "Recruitment" is the estimation of the number of trees exceeding a predefined threshold value of height and diameter. The recruitment of trees was estimated from the appearance of new small trees in the fixed plots of the Austrian National Forest Inventory that were observed in consecutive inventories. A model component estimates the probability of the successful regeneration for 13 tree species. An important site factor for most tree species was soil moisture. Several additional site parameters such as exposition and elevation are have a significant influence. The module describes the regional situation of Austria particularly well, but may be less valid in other countries, because the impact of site factors on the natural regeneration is derived from the data set and therefore the amplitude of Austrian forests is well described [19]. A climate-sensitive part of the model was developed in order to make the productivity programme available for long-term scenarios. Additional climatic parameters used were the annual temperature sum (sum of daily mean temperatures above 3 °C), the annual precipitation, the temperature sum and the precipitation during the growing season. In order to capture the impact of the past, the precipitation during the growing season of the previous year and the average of temperature sum and precipitation in the past 30 years are used as parameters. Dryness is derived from a transformation of the monthly precipitation and temperature. The response of different tree species to climate is embedded in the model. The basal area increment of European beech responds strongly to increases in the temperature sum, and is greatly reduced at lower precipitation rates. Oak responsds similarly to the temperature sum, and is less affected by declining precipitation rates. At lower temperature sums, the basal area increment of Norway spruce is higher than the increment of beech and oak. Nevertheless, the increment rate remains high, even under warmer conditions. The dependence on the precipitation shows that spruce can tolerate dry conditions rather poorly [20].

An important element of forest development is the disturbance regime. Prominent factors in the Alpine Space are insect attacks, storm damages, and possibly fire [21–23]. It has been shown that the pressure is not solely caused by climate change effects, but also by forest management effects, that potentially can be modified by means of strategies for adaptive forest management [24, 25].

The influence of storm events on forests was derived from field evidence for tree mortality due to disturbance, and wind speed data from the Zentralanstalt für Meteorologie und Geodynamik. A challenge was linking forest damage to a particular storm event, because the data of the Austrian Forest Inventory allow to assign a damage event to a period of a few years whereas the wind-speed is recorded with a resolution of 2 seconds. The pragmatic assumption was that the strongest storm in the observation period of the Forest Inventory was responsible for the observed forest damage. It was shown that Norway spruce is particularly vulnerable to storms during the dormant season, whereas the fully developed canopy of beech forests makes them more vulnerable during the growing season. The storm module represents the strong influence of tree size parameters. Larger trees and trees with a high height/diameter ratio are more vulnerable to storm damages [26, 27].

The future pressure from insects is difficult to predict for a particular site in a certain time span, although clear indicators for the potential danger are available [22, 28, 29].

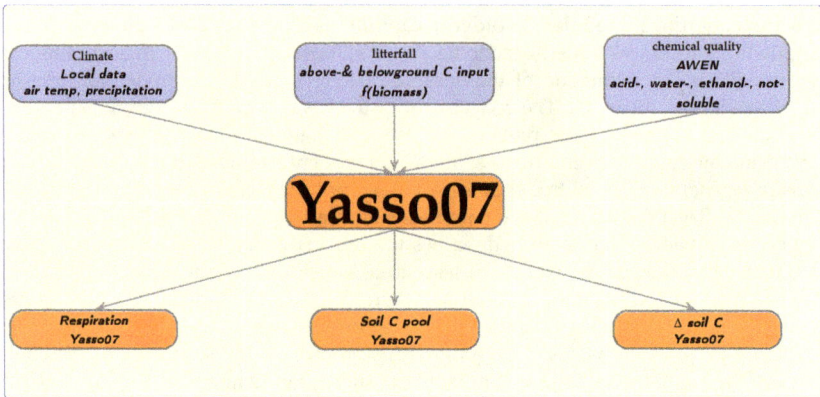

Figure 1. Yasso07 is a simulation model for the decomposition of soil organic matter. The input parameters are the annual climate, represented by the mean annual temperature, the difference of temperatures between the coldest and the warmest month, and the annual precipitation, the total carbon input to the soil, and its chemical quality.

The annual output of CALDIS comprises stem volume, tree height, and stem diameter of the individual trees of stand. The disturbance regime is focused on storm damages, because the storm-inflicted damages are well described in measurable parameters. Disturbances by an inadequately high density of the deer population and by pests and pathogens are not incorporated in the simulation model and need to be taken into account in the interpretation of the simulation results.

The CALDIS model was fed by climate data from A1B and B1 scenarios. The A1B scenario foresees a stronger warming trend, the B1 scenario is more optimistic with respect to controlling the warming trend [30].

2.3. The soil carbon simulator Yasso07

Yasso is an intentionally simple soil carbon simulation model. Its general concept is shown in Figure 1: The user supplies air temperature and precipitation as climate parameters, the quantity of the above- and belowground influx of carbon to the soil, and information on the chemical quality of the incoming carbon. The model can deal with different temporal resolutions of the input information. The core of Yasso is a decomposition model of soil organic material [31, 32]. In our simulation Yasso was run 10 times for each data set in order to account for the uncertainty about the parameter values. These repetitions do not account for the variability in the user-provided input data.

As output Yasso gives a time series of the total soil C pool, which is divided into carbon in woody matter, non-woody matter and the acid-, water-, ethanol and insoluble fractions. It can be understood that carbon in woody material describes the coarse woody debris and the forest floor material and the sum of the remaining fractions are carbon in the mineral soil. Yasso does not distinguish between soil horizons and does not explicitly state a soil depth. Operationally, a soil depth of 1 m can be assumed, although the reference to the "total soil C pool" is more appropriate, because all incoming carbon is processed by Yasso07.

3. Building a bridge between CALDIS and Yasso07

3.1. Data and folders

The CALDIS forest management model produces output in annual time steps. Folders, named corresponding to the year of the simulation, are located in one major CALDIS output folder. Each of these folders contains yet another set of subfolders which, at last, comprise the data: Content of folder [sim] describes the standing biomass, folder [mort] contains data about trees discarding from the simulation due to natural mortality events, and folders [VN] and [EN] comprise tree data about harvesting removals. Folder [verb] contains information about remaining trees, after a disturbance event or harvest has taken place.

Every line in an output file represents one tree of a sample plot. At the time this work was conducted, the forest management model CALDIS was still under development. Forty-two comma-separated output parameters described each tree. Depending on the number of model runs and the timeframe of the simulation the output can become quite extensive and may therefore present considerable data-handling challenges. A strategy for accessing the data generated by CALDIS and for skimming the data in order to assemble the parameters for the Yasso07 input file is shown in figure 1.

In this approach advanced options of the SAS® 9.2 datastep were used to access and process the data generated by CALDIS [33, 34]. If more than one model run was conducted, the program which is presented here is able to access all model runs at once.

3.1.1. The import of CALDIS output data into a SAS®9.2 dataset

First, a list is created, storing all data-paths and filenames of CALDIS output files in a SAS®9.2 dataset. The SAS Command [pipe] enables the SAS application to access MS-DOS/UNIX commands [35]. The output of the command is accessed by SAS again, and written to a dataset. In this case, it is the common MS-DOS® command [dir] which displays the content of a defined folder. Three sub-options are used, in order to prevent the input of non-relevant information.

```
filename DIR_FILE pipe "dir /b/s d:\XYZ\YOUR_CALDIS_OUTPUT\*.cds";
```

/s causes the inclusion of the content of all subfolders.

/b Only filenames and the full data-path, but no additional information are displayed.

*.cds Only files which have [cds] (relevant CALDIS output files) as file type extension are displayed.

Thereby a dataset which lists all CALDIS output files inside a stated folder is created. Its only parameter is the full data-path plus the filename stored as a character string. To avoid truncation of data path descriptions, the length of the variable is set to a maximum of 256 characters.

```
data FILE_LIST;
length data_path $256;
infile DIR_FILE length=reclaeng;
input data_path $ varying256. reclaeng ;
run;
```

In the next step, this list is used to import the data files into one major data record, consisting of the output of all model scenarios and runs. This is achieved by running a macro loop from the first observation of the generated list to the last. Each cycle of the loop the character string parameter at the current observation is stored as a global macro variable. Then this string is used in a data step behind an [infile] statement. A temporary data record is created which contains data of the current CALDIS output file. The global variable, in which location and filename are defined, is also stored as parameter making it possible to identify the origin of the data. If this is the first iteration of the loop, a new dataset is created by storing the temporary one. For all following iterations the temporary dataset is appended on this new dataset. After this event the next run of the loop starts. After the last iteration of the loop the data import is complete. This way all observations of all CALDIS output files are gathered in one dataset.

```
%macro data_import;
    %do o=1 %to &Num_Of_Obs;
        data _NULL_; set FILE_LIST;
            if _n_ = &o; call symput('path_to_file',data_path);
        run;
        data temp;
            infile "&path_to_file"
            DLM=',';
            INPUT all input parameters;
            data_path_key ="&path_to_file";
        run;
        %if &o ne 1 %then %do;
            proc append base=full_dataset data=temp; run;
        %end;
        %else %do;
            data full_dataset; set temp; run;
        %end;
    %end;
%mend data_import;
```

3.1.2. The break-up of the data location into key parameters

The labeling of the storage location provides information about the comprised data itself. In this next step, the path to the output data is used to create key parameters, which can be used as filter criteria at a given point of time. This is achieved by 'breaking down' the variable [data_path] into single 'words'. Every element of the location is stored as an own parameter. In our case numerous CALDIS runs of different base conditions were conducted. It seems reasonable to store the output with consistent labeling. The name of the subfolder, which comprises the output of one single run, was formatted containing information about CALDIS model setting details. An example is *P_12_S_3_C_2_R_30* describing:

- Plot 12
- Scenario 3
- Climate 2 [e.g. IPCC - B1]
- Run 30

This string is also split into its 'words' which are used to create a numeric key variable. Then the dataset which contains information that is gained from labeling of subfolders [FILE_LIST_key] is merged with the dataset which contains all the simulation results [full_dataset].

3.1.3. Compressing the dataset

Every CALDIS output file represents a sample plot, which was examined by the angle-count method. The parameter [nrepjeha] tells how many trees of the same characteristics exist on one hectare and serves as a multiplier for scaling up the sample plot to one hectare. The content of folders [sim], [verb], [mort], [VN] and [EN] differs exclusively in this parameter. We merged corresponding datasets of folders [sim], [verb], [mort], [VN] and [EN] via previously generated key variables, introducing [nrepjeha_sim], [nrepjeha_verb], [nrepjeha_mort], [nrepjeha_VN] and [nrepjeha_EN] as new parameters.

3.1.4. Editing climate data according needs of Yasso07

In the framework of this project, a regional downscaling of 2 different climate scenarios was conducted for the region of Ossiach (Carinthia). 100 Years were modeled, starting 2001. Furthermore daily time-step weather parameters were generated. These results were edited for the forest management model CALDIS as well as for the soil carbon model Yasso07.

In the Yasso07 model, the only factors which are driving litter decomposition are the composition of the material and the climatic conditions that can be described simply by using temperature and precipitation [36]. Climate data is assessed in quite a simple manner. Different ways of formatting are possible. The software accepts monthly or yearly time-steps. In our simulation exercise we used yearly time-steps. The input data consists of the time-interval, yearly annual mean temperature, the amplitude of mean temperatures of the coldest and the hottest month and the annual precipitation sum.

The data input of different climate parameter files was conducted analogue to the input of CALDIS data (see Chapter 3.1.1). Via the command [pipe] a list is generated. Afterwards the data locations (which are stored as macro variables) are passed to a data step, where they are used behind an [infile] statement to read in the files. One precipitation and one temperature file per climate scenario is generated. Annual mean temperature, the amplitude of the mean values of coldest and hottest month and the annual precipitation sum are calculated. Then they are stored in a permanent library for later access.

3.2. Estimating biomass compartments

In this step, different biomass estimations, according to tree species and compartment, are applied. These estimations are used to assess annual biomass fluxes, which are used later on as Yasso07 input parameters. Some biomass parameters which are going to be output parameters in the final release of CALDIS, were not yet defined and were computed externally. The parameters "stem volume" [m^3] and "mass of dry branches" [kg] were already defined output parameters of the forest management model.

For calculating other compartments the "diameter at breast height" (dbh) [cm] and the "height of the modeled tree" (h) [m] was used. A wet-to-dry stem volume shrinkage factor 0.881 was applied for all tree species [37]. A stem wood dry density of 430 and 680 kg/m^3 was assumed for coniferous and deciduous trees, respectively. For calculation of foliage biomass a discrimination was made between deciduous and coniferous trees:

Needle mass was calculated as

$$\text{needle weight} = \exp^{-1.6386} +2.0005 \times \log \text{bhd} - 0.5944 \times \log \text{h} \qquad (1)$$
$$\text{leaf weight} = 0.0057 \times \text{bhd}^{1.9836} \qquad (2)$$

where the needle and leaf weight are given in kg [37–39].

The estimation of the root biomass is supported by only a few field measurements. In a Finnish 35-year-old and in a mature 100-year-old Scots pine stand (*Pinus sylvestris L.*) the belowground biomass (located in the organic surface layer and the upper 30 cm of the mineral soil) accounted for 21% and 13.2% of total biomass, with fine roots, defined as roots with a diamterof the mineral soil) accounted for 21% and 13.2% of total biomass, where fine roots (defined as roots with a diameter < 2 mm contributed 31.8% and 15% to the belowground biomass, respectively [40]. In a modeling exercise of coupling the EFIMOD model with the module of soil organic matter ROMUL, the biomass of the fine roots was set to 10% of the coarse-root biomass [41–45].

In our model root biomass was set to 21% of the aboveground compartments, accounting for 17.4% of the total biomass. The ratio between the coarse root fraction (> 2 mm) and the fine root fraction was set to 20:1 [46].

Among uncertainties are wood density and the standard deviation of the utilized biomass expansion functions. The wood density might vary between sites; furthermore a single wood density might not adequately represent an entire tree species. The wood density is often not measured for individual sampling sites. Hence, a reference value is chosen, based on the judgment of the modeler. – The inherent uncertainty of biomass-expansion functions, as reflected in the standard deviation of its parameters, is often not propagated in the modeling exercise.

3.3. Estimating annual carbon fluxes

The input to Yasso07 calls for estimates of carbon fluxes, due to the aboveground and belowground litterfall of the standing tree biomass, mortality and harvesting (Figure 1). Especially for underground biomass compartments there are only vague figures to assess these fluxes. The ones which are applied here might be adapted intuitively for model calibration. For Scots pine and Norway spruce stands in southern Finland an annual needle turnover of 21% (turnover time 4.76 years) and 10% (turnover time 10 years), respectively, was measured, accounting for 49 to 75% of the total aboveground litter production of Scots pine stands [47–49]. For northern Germany spruce stands a lifespan of 4.2 to 5.7 years, and a positive correlation between lifespan and altitude was reported [50, 51] . For the coniferous tree species except for larch (*Larix decidua*), we set the annual needle turnover rate to 20%; resulting in a lifespan of 5 years [52]. The foliage of deciduous trees and of larch needles are recycled annually, and the turnover time was set to one year.The possibility, that site conditions might influence the lifespan of tree needles [50], was neglected in this model exercise. The foliage of deciduous trees as well as the needles of larch are recycled annually; their turnover time was set to one year.

The belowground plant structures contribute an important fraction to annual carbon fluxes [53] Yet the magnitude of these fluxes is difficult to quantify [54]. It was claimed that the belowground net primary production of trees might exceed its aboveground counterpart and that the fine root turnover is playing a major role [55]. The fine root biomass in a boreal forest accounted for 32% of the annual net primary production. Fine roots were turned over at an average of 1.07 times per year leading to a turnover time of 0.93 years [56]. For approximately 40-year-old Norway spruce stands in Germany the turnover time of fine roots in the upper 20 *cm* of the soil ranged from 1.18 to 2.29 times a year, resulting in turnover times of 0.85 and 0.42 years, respectively [57]. – Without discrimination between either stand-age or tree species we set the fine root turnover to 0.9 years.

The modeled annual branch litterfall for Norway spruce and Scots pine in Finland was 1.25% and 2.7% of the total branch biomass, respectively. Averaging branch turnover rates of European forest sites from the IBP Woodlands Data Set [58], the annual turnover of branches was set to 2.5% [52]. – In this work the annual branch litterfall was intuitively set to 3% of the total branch mass, leading to a lifespan of 33.3 years. Only little is known about the senescence rate of coarse roots [59]. Following a common practice we set it to the same level as branch litterfall [52].

Possible biomass inputs, originating from an understory shrub- and an herbaceous plant layer, were not taken into account, because no data was available to assess these fluxes.

Without discrimination between, either tree species, or biomass compartment, the carbon content of dry matter was assumed to be 50% [52].

3.3.1. Treatment of mortality events

The National Forest Inventory and regional repeated forest resource assessments are capable of informing about tree mortality. Trees are recorded as "dead" when they hold no living needles or leaves during the growing season, or when they have been removed and only a stump is left behind. In the simulation model CALDIS mortality is driven by a submodule. For consistently dealing with mortality events, the following assumptions about biomass fluxes were made:

- We made no distinction between natural mortality and harvesting events. Stem biomass is withdrawn from the stand, leaving behind a stump.

- The stump was assumed to have the shape of a cylinder with a diameter of bhd and a height of 20 *cm*. The stump volume vol_{stump} was calculated as

$$vol_{\text{stump}} = r_{d/w} \times \left(\frac{\text{bhd}}{2}\right)^2 \times \pi \times 0.2 \qquad (3)$$

where vol_{stump} is reported in $[m^3]$ and the bhd is entered in $[m]$.

- 60% of the branch biomass and 100% of the needles/foliage remain on the plot.
- 100% of the belowground biomass remains on site.

3.4. Assigning chemical characteristics to the carbon fluxes

The Yasso07 model is based on the assumption, that the components of organic litter can be classified in 4 types, according to their decomposition characteristics [36] These components can be either totally decomposed; releasing CO_2, or can be transferred into humus. The classes are

- components soluble in a non-polar solvent, ethanol or dichloromethane (waxes),
- water soluble components (simple sugars),
- acid hydrolysable components (cellulose),
- insoluble and non-hydrolysable components (lignin).

The chemical qualities are model input parameters and differ between tree species and litter type. In case the variability of the chemical qualities within litter types of single tree species is known, the mean value and its standard deviation can be used in the model in order to reflect the uncertainty of the model parameters. A database of chemical qualities of litter types is available from the Yasso07 manual [31, appendix]. In cases were no matching data for a certain tree species were available in the database, the respective parameter was represented with values that seemed most suitable. For the chemical composition of coarse roots and branch litter the same parameters were used as for decaying stems. – In our modeling exercise, no information on the standard deviation of the parameter values was available. The standard deviation was therefore set to zero.

Considering the size of the litter particles as a physical attribute, which is affecting their decomposition [43], the diameter of stump carbon is set to the mean breast height diameter of tree species per plot. For coarse roots and branch biomass we assume a diameter of 1 *cm*. The size of all other biomass compartments is set to zero. The size limit of woody litter was set to 3 *cm*. Hence, only stump necromass with a diameter exceeding 3 *cm* was treated as woody litter by the model.

3.5. Exporting Yasso07 input files

The final product of the last 4 steps is a dataset [Influx] which comprises storage locations of processed data, the height of annual carbon fluxes [*ton/ha*] of every tree and biomass compartment, plus the associated decomposition qualities. Via the parameter data_path this dataset is merged with FILE_LIST_key in order to retrieve required key parameters, resulting in a dataset named combine.

Yasso07 input files start with time-step '0' or '1'. If a zero time-step is stated, Yasso07 uses input from time '0' to derive initial steady-state soil conditions. If the input file contains no time step '0', but starts with year '1' the input from the first year of the simulation is used to estimate the initial soil carbon pool, irrespective whether year '1' represents steady-state conditions or not [31]. – In our modeling exercise we chose the option of estimating steady-state conditions with a time step '0'.

Yasso07 simulations require only one single input file which is providing carbon flux and climate information. These two data blocks are headed by captions [Yearly soil carbon input] and [Yearly climate]. The macro export_single_runs is the centrepiece of the output procedure. It exports data from SAS files to text files that are directly accessible by Yasso07.

```
%macro export_single_runs;
%do s=&firstscenario %to &lastscenario;
  %do c=&firstclimate %to &lastclimate;
    %do p = &firstplot %to &lastplot;
      %do r =&firstrun %to &lastrun;

        data temp;
      set combine
      (where=(Plot_ID = &p and
              Scenario_ID = &s and
      Climate_ID = &c and
      Run_ID =&r));
          call symput('lenght_Influx',_N_);
        run;

        %LET DSID=%SYSFUNC(OPEN(temp,IN));
        %LET Observations=%SYSFUNC(ATTRN(&DSID,NOBS));
        %IF &DSID > 0 %THEN %LET RC=%SYSFUNC(CLOSE(&DSID));

          %if &Observations ne 0 %then %do;

            data _NULL_; set temp climate.IPCC_&c;
              file
        "D:\your_YASSO_input\SINGLERUN_Scn&s._Clm&c._Plt&p._Run&r..in";

              if _N_<=&lenght_Influx then do;
                if _N_=1 then put "[Yearly soil carbon input]";

                put NEW_year flux flux_std acid a_std watr w_std
      ethn e_std nsol n_std hums h_std size "# "
      comp species "Scen " Scenario_ID
      "Plot " Plot_ID "Clim " Climate_ID "Run " Run_ID;

                if _N_=&lenght_Influx then do;
                  put " ";
```

```
                    put "[Yearly climate]";
                  end;
               end;
               else do; /* if _N_>&lenght_Influx */
                  put simyear annual_mean precsum
      month_diff "# " IPCC;
                  end;
               run;

            %end;
         %end;
       %end;
     %end;
  %end;
%mend export_single_runs;
```

The macro consists of one loop comprising 3 nested loops, iteratively running through every possible combination of the incrementers scenario, climate, run, and plot. A temp dataset is created out of the content of combine where the values of the incrementers match scenario, climate, run, and plot. The observation length of this dataset is stored as a macro variable [length_Influx]. – In our case, not all possible combinations of the incrementers represent valid data in combine. In this case the group of functions starting with [%LET DSID] returns zero as a value for &Observations [34], i. e., the combination of scenario, climate, run, and plot does not represent an existing data record. Hence, the next step is skipped, the next iteration of the loops starts. If the created dataset contains valid data (more than zero observations) the next [%IF &END] clause becomes active. A _NULL_ dataset is created which consists of the previously created temp dataset, and appending, the final product of the climate calculations. So flux data and the corresponding climate data exist in one record! Then a file is created. Its name contains the current values of the incrementers. The file extension [.in] marks Yasso07 input files. If a put statement is set after the file statement, the output is written not to the SAS®9.2 log-file, but to the defined external output file.

The headlines, used by Yasso07 for the denomination of the subsequent data records, need to be supplied by the code. This is implemented by adding the text line [Yearly soil carbon input] above the first observation record. The parameters, which are subsequently output to file, are the annual carbon inputs at each time step. The information contains the mean values and their standard deviation, followed by the decomposition characteristics and their standard deviation, and finally the diameter of the litter compartments. Eventually added comments to the input lines enable an easier interpretation of the input files by the model user. Such comments can be added after the # character.

After writing out the last observation of the carbon input subset (at position length_Influx), a blank line is inserted, marking the end of carbon flux data. In the following line the next headline, i.e. [Yearly climate], is added. It indicates the beginning of the climate input data. The variables which are then written to external output file are the annual mean temperature, the annual precipitation sum, and the temperature difference between the hottest and the coldest month. The data are written out for each time step. – After the last iteration of the central loop of the script, the export routine is completed and the macro is terminated. Thereby, individual Yasso07 input files, for each case that has been modeled in CALDIS, are created. The Yasso07 modeling runs need to be run individually, by loading the respective external files into the Yasso07 interface.

In the current modeling exercise the soil model Yasso07 was run with the following settings:

Initial soil carbon stock: The chosen option was 'steady state'. In this mode the model estimates the initial carbon stock from the incoming amount of carbon by above- and belowground litterfall. The model does not use eventually measured soil carbon stocks. The field data can therefore be used for the validation of the modeling results.

Sample size: The sample size defines how often one particular case (site) is simulated. If a sample size of '1' is chosen the model uses the most likely parameters of the decomposition model. Increasing the sample size invokes additional simulation runs where less likely parameter values are chosen. The sample size therefore allows the representation of the model uncertainty. – In our modeling exercise the sample size was set to 10.

Time: The model was run 100 time-steps with a step length of 1 year.

Size: The size of the incoming aboveground and belowground litter is affecting the decomposability of the organic matter. Smaller particles are decomposing quicker; we set the threshold to woody litter at 3 *cm*.

4. The outcome of the modeling exercise

The used regionalized climate scenarios show a warming trend in the next 100 years between 3 and 5 °C. The amplitude between the coldest and the warmest month remains constant (Figure 2).

Running the growth model CALDIS several times yields widely different results, because the probability of being affected by storms is variable. In addition, the ingrowth module contains a probability element. The by far largest impact on the standing volume has the storm module. The variability due to these stochastic processes by far exceeds the impact of the chosen climate scenarios on forest growth (Figure 3); in order to obtain a representative stem volume, the individual model runs were averaged with the statistical package SAS®.

Introducing the output of CALDIS into the soil model Yasso07 allows the estimation of the anticipated development of the soil carbon stock. Figure 4 shows an example for the temporal trend of soil carbon over 100 years and compares the effect of two climate scenarios and two dominant tree species. - In the presented example the soils would gain organic carbon. In a more realistic model run a major harvesting operation would reset the stand conditions and the soil carbon stock would be rather stable. Nevertheless, the comparison suggests that Norway spruce is able to accumulate substantial amounts of organic carbon. The climate effect is very strong. The warmer A1B scenario is reflected by smaller soil carbon stocks, because the increasing temperature stimulates the heterotrophic respiration in the soil and leads to elevated rates of CO_2-emissions. – More detailed interpretations of the modeling outcomes are given in the description of the case study area [18].

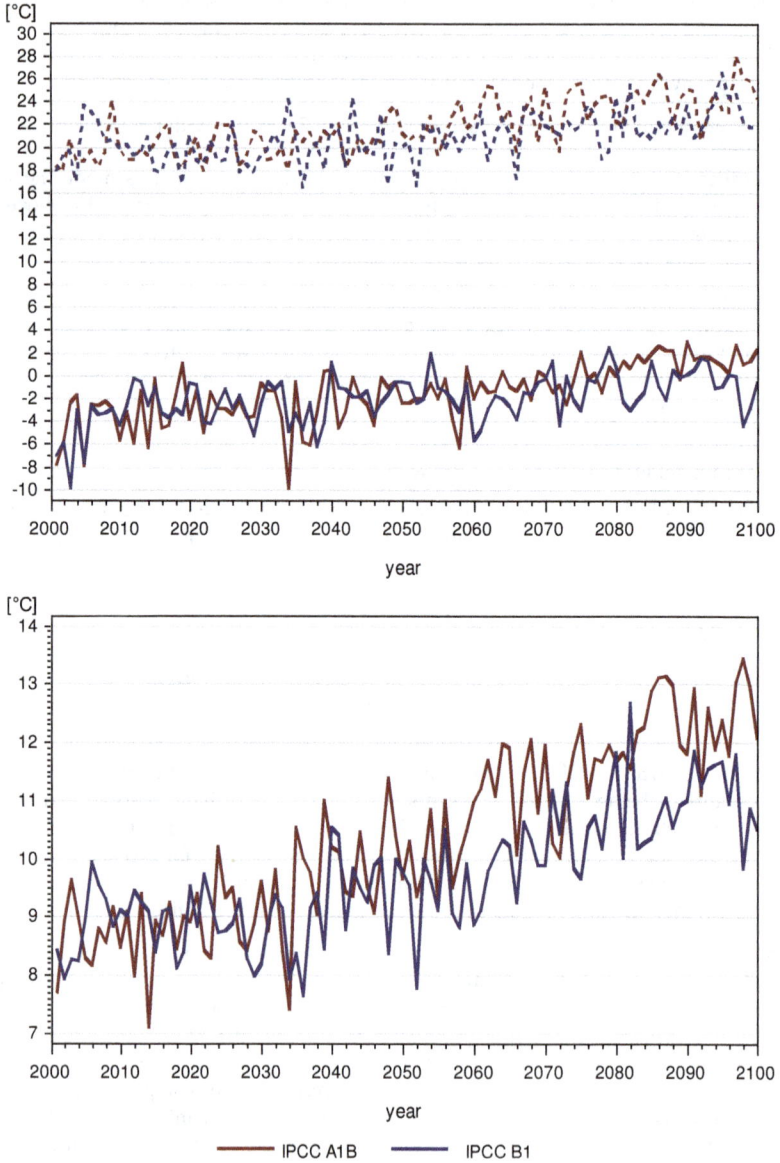

Figure 2. Difference between the coldest and the warmest month (upper panel) and mean annual air temperature (lower panel) for the test region Ossiach between the years 2000 and 2100. Blue: The IPCC climate scenario B1; red: the scenario A1B.

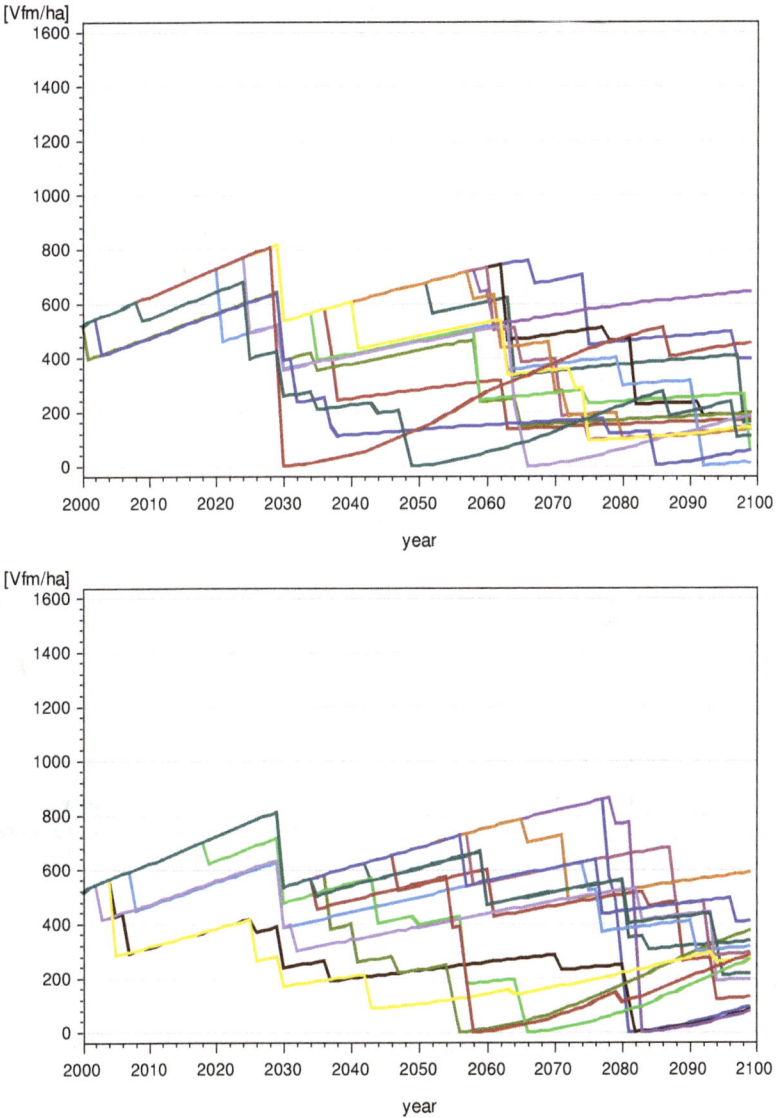

Figure 3. Simulated stem volume [m^3/ha] of an example stand in the Ossiacher Tauern. Different runs of simulation model CALDIS yield widely different stocks of the stem volume at one particular site for the simulation run from year 2000 to 2100. Upper panel: forest stand experiences climate according to the IPCC scenario A1B; lower panel: forest stand experiences climate according to the IPCC scenario B1.

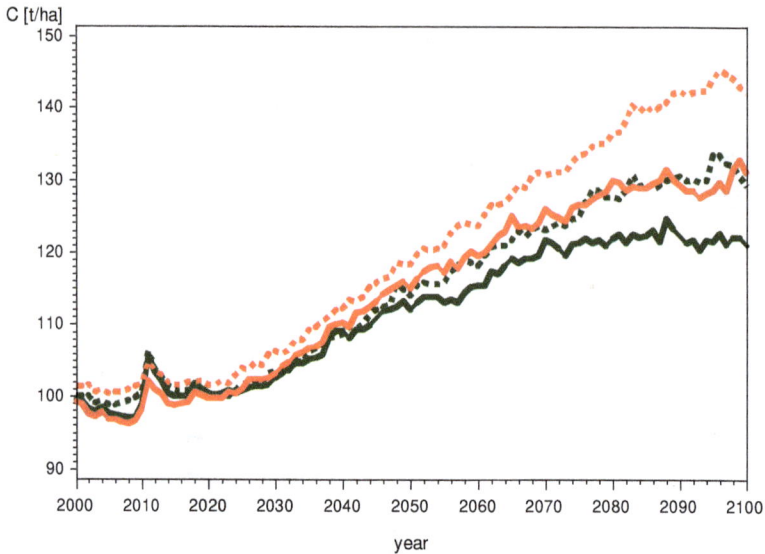

Figure 4. Simulated temporal trend of the soil carbon stock (C $[ton/ha]$) for an example stand in the Ossiacher Tauern between the year 2000 and 2100. Red: beech dominated forest, green: spruce dominated forest; full lines: climate scenario A1B; dashed lines: climate scenario B1.

Acknowledgments

This chapter is an outcome of Interreg Manfred. The modeling exercise was partly supported by a grant of the Section Environment of the Austrian Ministry of Agriculture, Forestry, Environment and Water Management.

Author details

Klaus Dolschak[1],
Robert Jandl[2] and Thomas Ledermann[2]

1 Dept of Forest Ecology, University of Applied Life Sciences (BOKU), Vienna, Austria
2 Forest Research Center (BFW), Vienna, Austria

References

[1] MJ Lexer, K Hönninger, and M Englisch. Schätzung von chemischen Parametern für Waldstandorte am Beispiel der Österreichischen Waldinventur. *Forstwissenschaftliches Centralblatt*, 118:212–217, 1999.

[2] Hubert Hasenauer, editor. *Sustainable Forest Management: Growth Models for Europe*. Springer, Berlin, 2010.

[3] Hans Pretzsch. GesetzmäSSigkeiten der Kronenformentwicklung und Wuchsraumbesetzung. Querschnittsanalyse auf der Basis langfristiger Versuchsflächen. In *DVFFA – Sektion Ertragskunde, Jahrestagung 2010*, pages 82–95, 2010.

[4] Aaron R. Weiskittel, David W. Hann, John A. Kershaw Jr., and Jerome K. Vanclay. *Forest Growth and Yield Modeling*. Wiley, Chichester, United Kingdom, 2011.

[5] Julius Marschall. *Hilfstafeln für die Forsteinrichtung*. Österreichischer Agrarverlag, Wien, 1975.

[6] Robert A. Monserud and Hubert Sterba. A basal area increment model for individual trees growing in even- and uneven-aged forest stands in austria. *Forest Ecology and Management*, 80(1-3):57 – 80, 1996.

[7] Herman H Shugart jr. and Darrell C West. Long-term dynamics of forest ecosystems. *American Scientist*, 69(6):647–652, 1981.

[8] Harald Bugmann. A review of forest gap models. *Climatic Change*, 51:259–305, 2001.

[9] Jacqueline Gehrig-Fasel, Antoine Guisan, and Niklaus E. Zimmermann. Evaluating thermal treeline indicators based on air and soil temperature using an air-to-soil temperature transfer model. *Ecological Modelling*, 213(3-4):345 – 355, 2008.

[10] Niklaus E. Zimmermann, Janine Bolliger, Jacqueline Gehrig-Fasel, Antoine Guisan, Felix Kienast, Heike Lischke, Sophie Rickebusch, and Thomas Wohlgemuth. Wo wachsen bäume in 100 jahren. *Forum für Wissen*, pages 63–71, 2006.

[11] J. S. Kimball, AR Keyser, SW Running, and SS Saatchi. Regional assessment of boreal forest productivity using an ecological process model and remote sensing parameter maps. *Tree Physiology*, 20:761–775, 2000.

[12] Hubert Hasenauer, Richard Petritsch, Maosheng Zhao, Celine Boisvenue, and Steven W. Running. Reconciling satellite with ground data to estimate forest productivity at national scales. *Forest Ecology and Management*, 276(0):196 – 208, 2012.

[13] E Thürig, T Palosuo, J Bucher, and E Kaufmann. The impact of windthrow on carbon sequestration in Switzerland: a model-based assessment. *Forest Ecology and Management*, 210:337–350, 2005.

[14] Daniel D Richter jr. and Daniel Markewitz. *Understanding soil change - soil sustainability over millenia, centuries, and decades*. Cambridge University Press, Cambridge, 2001.

[15] Robert Jandl, Marcus Lindner, Lars Vesterdal, Bram Bauwens, Rainer Baritz, Frank Hagedorn, Dale W Johnson, Kari Minkkinen, and Kenneth A Byrne. How strongly can forest management influence soil carbon? *Geoderma*, 137(3-4):253–268, 2006.

[16] H Fischer, O Bens, and RF Hüttl. Veränderung von Humusform, -vorrat und -verteilung im Zuge von Waldumbau-Massnahmen im nordostdeutschen Tiefland. *Forstwissenschaftliches Centralblatt*, 121:322–334, 2002.

[17] E. Detlef Schulze and Annette Freibauer. Carbon unlocked from soils. *Nature*, 437:205–206, 2005.

[18] Robert Jandl, Andrej Breznikar, Marko Lekše, Christian Tomiczek, Silvio Schüler, Klaus Dolschak, and Hans Zöscher. *Case Study Carinthia / Slovenia – Productive Forests Affected by Climate Change*, chapter 17, page this volume. InTech, 2013.

[19] Thomas Ledermann. Ein Einwuchsmodell aus den Daten der Österreichischen Waldinventur 1981-1996. *Centralblatt für das gesamte Forstwesen*, 119(1):40–76, 2002.

[20] Georg Kindermann. Eine klimasensitive Weiterentwicklung des Kreisflächenzuwachsmodells aus PrognAus. *Centralblatt für das gesamte Forstwesen*, 127(3/4):147–178, 2010.

[21] Bruna Comini, Elena Gagliazzi, and Giampaolo Cocca. *Abiotic Stressors: Fire Hazard and Risk*, chapter 5, page this volume. InTech, 2013.

[22] Holger Griess, Holger Veit, and Ralf Petercord. *Risk assessment for biotic pests under prospective climate conditions*, chapter 5, page this volume. InTech, 2013.

[23] Stefano Oliveri, Marco Pregnolato, and Giacomo Gerosa. *A new webGIS platform dedicated to forest extreme events in the Alps: aims and functionalities*, chapter 10, page this volume. InTech, 2013.

[24] Rupert Seidl, Mart-Jan Schelhaas, and Manfred J Lexer. Unraveling the drivers of intensifying forest disturbance regimes in Europe. *Global Change Biology*, 17(9):2842–2852, 2011.

[25] Peter Brang, Marc Hanewinkel, Robert Jandl, Andrej Breznikar, and Bernhard Maier. *Managing Alpine Forests in a Changing Climate*, chapter 20, page this volume. InTech, 2013.

[26] WAMOD. Auswirkungen des Klimawandels auf Österreichs Wälder - Entwicklung und vergleichende Evaluierung unterschiedlicher Prognosemodelle (WAMOD). Endbericht, Institut für Waldbau, Institut für Waldwachstumsforschung (BOKU); Institut für Waldwachstum und Waldbau, Institut für Waldinventur (BFW), Wien, 2010.

[27] Bin You and Mitja Skudnik. *Abiotic stressor: storms*, chapter 7, page this volume. InTech, 2013.

[28] Beat Wermelinger. Ecology and management of the spruce bark beetle *Ips typographus* - a review of recent research. *Forest Ecology and Management*, 202:67–82, 2004.

[29] Lorenzo Marini, Matthew P. Ayres, Andrea Battisti, and Massimo Faccoli. Climate affects severity and altitudinal distribution of outbreaks in an eruptive bark beetle. *Climatic Change*, 2012.

[30] Niklaus E. Zimmermann, Ernst Gebetsroither, Johannes Züger, Dirk Schmatz, and Achilleas Psomas. *Future Climate of the European Alps*, chapter 3, page this volume. InTech, 2013.

[31] Jari Liski, Mikko Tuomi, and Jussi Rasinmäki. Yasso07 user-interface manual. Technical report, Finnish Environment Institute, March 2009.

[32] Mikko Tuomi, T Thum, H Järvinen, S Fronzek, B Berg, M Harmon, JA Trofymov, S Sevanto, and Jari Liski. Global patterns of leaf litter decomposition. Technical report, Finnish Environment Institute, 2008.

[33] SAS. *SAS - STAT Users's Guide*. SAS - Institute, Inc., Cary, NC, 4 edition, 1992.

[34] SAS. *SAS 9.2 Macro Language: Reference*. SAS - Institute, Inc., Cary, NC, 4 edition, 2009.

[35] B Varney. Check out these pipes: Using Microsoft Windows commands from SAS®. SAS Global Forum 2008, Applications Development. Paper 092-2008, 2008.

[36] M. Tuomi, T. Thum, H. Järvinen, S. Fronzek, B. Berg, M. Harmon, J.A. Trofymow, S. Sevanto, and J. Liski. Leaf litter decomposition–estimates of global variability based on yasso07 model. *Ecological Modelling*, 220(23):3362 – 3371, 2009.

[37] Thomas Ledermann and Markus Neumann. Biomass equations from data of old long-term experimental plots. *Centralblatt für das gesamte Forstwesen*, 123:47–64, 2006.

[38] Otto Eckmüllner. Allometric relations to estimate needle and branch mass of Norway spruce and Scots pine in Austria. *Austrian Journal of Forest Science*, 123:7–16, 2006.

[39] C Wirth, ED Schulze, G Schwalbe, S Tomcyk, G Weber, and E Weller. *Dynamik der Kohlenstoffvorräte in den Wäldern Thüringens*, volume 23 of *Mitteilungen Thüringer Landesanstalt für Wald, Jagd und Fischerei*. Thüringer Ministerium für Landwirtschaft, Naturschutz und Umwelt, Gotha, 2004.

[40] H.-S. Helmisaari, K. Makkonen, S. Kellomäki, E. Esko Valtonen, and E Mälkönen. Below- and above-ground biomass, production and nitrogen use in Scots pine stands in eastern Finland. *Forest Ecology and Management*, 165:317–326, 2002.

[41] O.G. Chertov, A.S. Komarov, M. Nadporozhskaya, S.S. Bykhovets, and S.L. Zudin. Romul - a model of forest soil organic matter dynamics as asubstantial tool for forest ecosystem modeling. *Ecological Modelling*, 138:289–308, 2001.

[42] A Komarov, O Chertov, S Zudin, M Nadporozhskaya, A Mikhailov, S Bykhovets, E Zudina, and E. E. Zoubkova. Efimod 2 - a model of growth and cycling of elements in boreal forest ecosystems. *Ecological Modelling*, 170:373–392, 2003.

[43] T Palosuo, M Peltoniemi, A Mikhailov, A Komarov, P Faubert, E Thürig, and M Lindner. Projecting effects of intensified biomass extraction with alternative modelling approaches. *Forest Ecology and Management*, 255:1423–1433, 2008.

[44] NI Kazimirov, AD Volkov, SS Ziabchenko, AA Ivanchikov, and RM Morozova. *Mass and energy exchange in Scots pine forests of the European North. Nauka, Leningrad, Russia (in Russian)*. Nauka, Leningrad, 1977.

[45] NI Kazimirov, RM Morozova, and VK Kulikova . *Organic Matter Pools and Flows in Pendula Birch stands of Middle Taiga. Nauka, Leningrad, Russia (in Russian)*. Nauka, Leningrad, 1978.

[46] Daniel Perruchoud, Fortunat Joos, Andreas Fischlin, Irka Hajdas, and Georges Bonani. Evaluating timescales of carbon turnover in temperate forest soils with radiocarbon data. *Global Biogeochemical Cycles*, 13:555–573, 1999.

[47] P Muukkonen and A Lehtonen. Needle and branch biomass turnover rates of Norway spruce (*Picea abies*). *Canadian Journal of Forest Research*, 34(12):2517–2527, 2004.

[48] P Muukkonen. Needle biomass turnover rates of Scots pine (*Pinus sylvestris* L.) derived from the needle-shed dynamics. *Trees - Structure and Function*, 19(3):273–279, 2005.

[49] M Starr, A Saarsalmi, T Hokkanen, P Merila, and HS Helmisaari. Models of litterfall production for Scots pine (*Pinus sylvestris* L.) in Finland using stand, site and climate factors. *Forest Ecology and Management*, 205:215, 2005.

[50] H Wachter. Zur Lebensdauer von Fichtennadeln in einigen Waldgebieten Nordrhein-Westfalens. *Der Forst- und Holzwirt*, 40:420–425, 1985.

[51] H Schmidt-Vogt. *Die Fichte*, volume II/2. Paul Parey, Hamburg, Berlin, 1989.

[52] Jari Liski, Daniel Perruchoud, and Timo Karjalainen. Increasing carbon stocks in the forest soils of western Europe. *Forest Ecology and Management*, 169:159–175, 2002. carbon.

[53] EA Davidson, K Savage, P Bolstad, DA Clark, PS Curtis, DS Ellsworth, PJ Hanson, BE Law, Y Luo, KS Pregitzer, JC Randolph, and D Zak. Belowground carbon allocation in forests estimated from litterfall and irga-based soil respiration measurements. *Agricultural and Forest Meteorology*, 345:1–13, 2003.

[54] Kristiina A. Vogt, Daniel J. Vogt, Peter A. Palmiotto, Paul Boon, Jennifer O'Hara, and Heidi Asbjornsen. Review of root dynamics in forest ecosystems grouped by climate, climatic forest type and species. *Plant and Soil*, 187:159–219, 1996.

[55] Knute J. Nadelhoffer and James W. Raich. Fine root production estimates and belowground carbon allocation in forest ecosystems. *Ecology*, 73(4):1139–1147, 1992.

[56] ZY Yuan and HYH Chen. Fine root biomass, production, turnover rates, and nutrient contents in boreal forest ecosystems in relation to species, climate, fertility, and stand age: Literature review and meta-analyses. *Critical Reviews in Plant Sciences*, 29:204–221, 2010.

[57] DL Godbold, HW Fritz, G Jentschke, H Meesenburg, and P Rademacher. Root turnover and root necromass accumulation of Norway spruce (*Picea abies*) are affected by soil acidity. *Tree Physiology*, 23:915–921, 2003.

[58] DL DeAngelis, R. H. Gardner, and H. H. Shugart. Productivity of forest ecosystems studied during the ibp: The woodlands data set. In D. E. Reichle, editor, *Dynamics of Forest Ecosystems*, number 23 in International Biological Programme, pages 567–672. Cambridge University Press, Cambridge, U.K., 1981.

[59] Aleksi Lehtonen. Carbon stocks and flows in forest ecosystems based on forest inventory data. Dissertationes Forestales 11, Finnish Forest Research Institute, 2005.

Chapter 14

Description of Case Study Areas for Deriving Management Strategies to Adapt Alpine Space Forests to Climate Change Risks

Robert Jandl, Frédéric Berger, Andrej Breznikar,
Giacomo Gerosa, Holger Veit, Gillian Cerbu and
Marc Hanewinkel

Additional information is available at the end of the chapter

1. Introduction

The Alpine Space is a heterogeneous region and practical forestry represents the cultural richness. In the transnational collaboration the intention was the selection of a number of representative regions for particular challenges of forestry in the Alpine Space. Forests are serving numerous societal demands and timber production is not always the prime priority [1].

Figure 1 shows the geographical distribution of the partners participating in the case studies. The case study areas are geographically widely distributed (Figure 2). In each case study area the scientists invoked a discussion with forest owners and stakeholders in forestry issues.

2. The case study areas

The forest stands of the case study area Rhône Alps (Drôme, Isère, Haute Savoie; France) and Val d'Aosta (Italy) represent the prevention of *natural hazards* (Figure 2). The area encompasses the Val d'Aosta, forests in the district of Vercors (France), Verbier (Switzerland), the region of Solčava-Luče (Slovenia), and the Montafon valley (Austria). In the test area an education program was developed enabling foresters to evaluate the current potential of forests of fulfilling their protective function. The areas are in difficult terrain. The eminent challenges are the continuous protection against rockfall and snow avalanches and the taking into account of the probable consequences of climate change on this protective function [2, 3].

Description of Case Study Areas for Deriving Management Strategies to Adapt
Alpine Space Forests to Climate Change Risks

227

Figure 1. Geographical distribution of the partner institutions of the Interreg project "MANFRED - Management strategies to adapt Alpine Space forests to climate change risks".

The transnational case study Slovenia-Austria covers a region with physico-geographical similarities, that has been divided by politics in the past (sites Solčava-Luče, Ossiacher Tauern, Dobrova in Figure 2). The regions have in common that *timber production* is the main target of forestry. Differences in the forest-management practice are visible. Slovenia has since long adopted the concept of continuous-cover forestry and is renown for its shelter-wood forests. In Austria, clear-cutting is still widely practiced within the regulations of the Austrian Forest Act. Presently, no severe challenges for practical forestry are ientified, but the consequences of climate change on forests are only vaguely known [4]. In the case study area an intensive exchange of views and experiences was conducted.

The case study area Valle Camonica within the province Lombardia (Italy) is expected to face a quick change in climatic conditions with adverse effects on forest productivity, an increase in the risk of *ozone damages* and *forest fires* (site Val Cominica, Lombardia in Figure 2). It is expected that the current forests cannot cope with the expected conditions and forestry will require a focus on tree species that are currently not dominantly managed [5]. The scientific work on the ozone exposure of forests was conducted jointly with the "Comunità Montana Valle Camonica - Parco dell'Adamello".

The focus of the transnational case study Montafon (Vorarlberg, Austria) - Baden Württemberg (Germany) - Bavaria (Germany) - Prättigau (Switzerland) is the increase in abiotic risks, most prominently the increasing risk of *storm damages* and *rock fall* (Baden-Württemberg, Bavaria, Vorarlberg in Figure 2). Evidence is given by the cumulation of storm damages in the recent past [6]. The geographical setting of the area includes mountains and the foothills of the Alps. A common consequence of storm damages is the disposition to *bark beetle* attacks, which is even attenuated by changing climate conditions. A part of the forests is dominated by secondary Norway spruce stands that are particularly vulnerable to biotic damages [7]. The intention of the case study work is providing tools for an up-to-date risk assessment for forester practitioners.

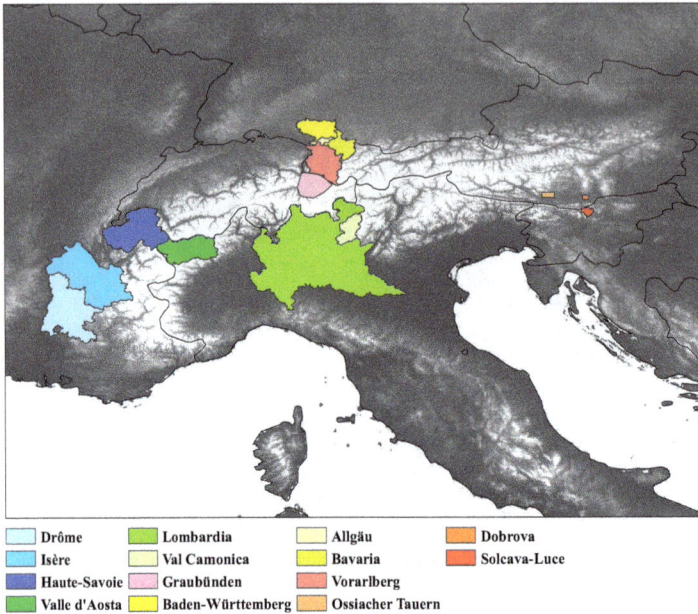

Figure 2. The test areas for the derivation of concepts of adaptive forest management. The sites Drôme, Isère, Haute-Savoie are in France, Valle d'Aosta, Valle Camonica, and Lombardia are in Italy, Baden-Württemberg, Bavaria, Allgäu are in Germany, Vorarlberg, Ossiacher Tauern, Dobrova are the Austrian sites, Solčava-Luče is located in Slovenia. Note that the site names are experimental codes and not political entities.

Acknowledgements

This chapter is an outcome of the Interreg project Manfred, conducted within the Alpine Space programme.

Author details

Robert Jandl[1], Frédéric Berger[2], Andrej Breznikar[3],
Giacomo Gerosa[4], Holger Veit[5],
Gillian Cerbu[5] and Marc Hanewinkel[6]

1 Forest Research Center (BFW), Vienna, Austria
2 IRSTEA, Grenoble, France
3 Zavod za gozdove Slovenije, Slovenia Forest Service, Ljubljana, Slovenia
4 DMF, Università Cattolica del Sacro Cuore, Brescia, Italy
5 Forstliche Versuchs- und Forschungsanstalt Baden-Württemberg (FVA), Department of Biometrics, Freiburg im Breisgau, Germany
6 Eidgenössische Forschungsanstalt für Wald, Schnee und Landschaft (WSL), Birmensdorf, Switzerland

References

[1] Martin F Price, Georg Gratzer, Lalisa Alemayehu Duguma, Thomas Kohler, Daniel Maselli, and Rosalaura Romeo. Mountain forests in a changing world - realizing values, addressing challenges. Technical report, FAO/MPS and SDC, Rome, 2011.

[2] Fred Berger, Franck Bourrier, Luuk Dorren, Charly Kleemayr, Bernhard Maier, Spela Planinsek, Christophe Bigot, Franck Bourrier, Oliver Jancke, David Toe, and Gillian Cerbu. *Eco-engineering and protection forests against rockfalls and snow avalanches*, chapter 12, In: Management Strategies to Adapt Alpine Space Forests to Climate Change Risks. InTech, 2013.

[3] Laurent Borgniet, David Toe, Frédéric Berger, Marta Galvagno, Cinzia Panigada, Roberto Colombo, Umberto Morra di Cella, Simone Gottardelli, Ivan Rollet, Mario Negro, Flavio Vertui, and Cédric Fermont. *Monitoring climatic change impacts on protection forests in Aosta Valley (Italy) and in Drôme (France) using medium and high resolution remote sensing and mateloscopes plots*, chapter 16, In: Management Strategies to Adapt Alpine Space Forests to Climate Change Risks. InTech, 2013.

[4] Robert Jandl, Andrej Breznikar, Marko Lekše, Christian Tomiczek, Silvio Schüler, Klaus Dolschak, and Hans Zöscher. *Case Study Carinthia / Slovenia – Productive Forests Affected by Climate Change*, chapter 17, In: Management Strategies to Adapt Alpine Space Forests to Climate Change Risks. InTech, 2013.

[5] Giacomo Gerosa, Angelo Finco, Stefano Oliveri, Riccardo Marzuoli, Alessandro Ducoli, Giambattista Sangalli, Bruna Comini, Paolo Nastasio, Giampaolo Cocca, and Elena Gagliazzi. *Case study Valle Camonica and the Adamello Park*, chapter 18, In: Management Strategies to Adapt Alpine Space Forests to Climate Change Risks. InTech, 2013.

[6] Barry Gardiner, Kristina Blennow, Jean-Michel Carnus, Peter Fleischer, Frederik Ingemarson, Guy Landmann, Marcus Lindner, Mariella Marzano, Bruce Nicoll, Christophe Orazio, Jean-Luc Peyron, Marie-Pierre Reviron, Mart-Jan Schelhaas, Andreas Schuck, Michaela Spielmann, and Tilo Usbeck. *Destructive Storms in European Forests: Past and Forthcoming Impacts*. EFI, Joensuu, 2010.

[7] Holger Veit, Bernhard Maier, Holger Griess, and Bion You. *Case study Oberschwaben / Allgäu / Vorarlberg – Risk Assessment of Abiotic and Biotic Hazards*, chapter 19, In: Management Strategies to Adapt Alpine Space Forests to Climate Change Risks. InTech, 2013.

Chapter 15

Provenance Trials in Alpine Range – Review and Perspectives for Applications in Climate Change

Stefan Kapeller, Silvio Schüler, Gerhard Huber,
Gregor Božič, Tom Wohlgemuth and
Raphael Klumpp

Additional information is available at the end of the chapter

1. Introduction

Effects of climate change on forest ecosystems and the mitigation of negative consequences to ecosystem functions and economic forest services are major challenges for current forest science and management [1-3]. Impact of climate change on forest ecosystems in Europe has been discussed controversially. It is generally agreed that increasing temperatures will lead to higher photosynthetic activity, faster growth and therefore higher forest net productivity, at least in large parts of central and northern Europe [4, 5]. However, increasing incidences of drought events and poor water supply are assumed to be limiting factors for tree growth in the future [6, 7], and the most productive coniferous species (e.g. Norway spruce) very likely cannot survive under future conditions in all parts of their natural range or cultivation area, respectively [8, 9]. This will have severe implications for forest management and the wood processing industry. For example, the financial effects of climate change due to biome shifts from productive coniferous forest to Mediterranean oak forests would result in a decrease of the expected value of European forest land between 14 and 50% [10].

Most tree species are distributed throughout large geographic ranges, where they experience a wide range of environmental conditions. The limits of these conditions are being considered to form the environmental niche of the species, for example see [11, 12]. However, the environmental variation within the species' ranges also creates manifold local adaptations to climate and site conditions. To identify and understand these local adaptations offers perspectives to maintain the existing tree species composition by using seed material that shows adaptation to future conditions.

Variation in climatic conditions is especially high throughout the Alpine region. Due to the multitude of environmental conditions within small distances and within a fragmented mountain terrain, Alpine space harbors high biological diversity [13]. At the same time, mountain ecosystems seem to be particularly vulnerable [14] and climate change is likely to induce severe habitat loss habitat for a majority of alpine plant species [15] and to threaten biodiversity in Alpine space [16, 17].

In order to mitigate the climate change impact on forestry in Alpine space and to adapt forest management accordingly, it is necessary to understand how forest tree species respond to changing climates. Provenance tests provide valuable data for the investigation of climate adaptation of forest species and intraspecific variation in climate response among tree populations.

The present study aims to review the utilization of provenance tests for revealing the intra-specific variation in climate response of trees and the development of adaptation measures for climate change. First, a general overview of climate adaptation of trees, affected adaptive traits and evolutionary and demographic processes is given and the application in climate change mitigation is discussed. Secondly, we describe general methodological issues of provenance tests and statistical approaches to reveal climate responses. Thirdly, we compile information about established provenance tests in the Alpine range and analyze their potential utilization for adaptation studies. Finally, we discuss abundant problems with climate response analyses and derive recommendations for the conception of future provenance trials.

2. Climate adaptation in trees

Studies on phenotypic traits demonstrated that forest tree populations exhibit substantial local adaptation, for example see [18]. This is thought to be mainly a result of diversifying selection caused by environmental heterogeneity [19]. Phenotypic traits that evolved through local adaptation are often referred to as adaptive genetic variation (e.g. [20]). Adaptive traits have been observed on the juvenile and the adult phase for many tree species. In the past, adaptive traits have been correlated mainly to the geographical origin of the analysed provenances, i.e. latitude or altitude. Recently, the availability of high resolution climate parameters allowed a direct relation of phenotypic traits to the climate parameters shaping it. Table 1 gives a short overview about typical adaptive traits that were found in the most important tree species of the Alpine space.

The evolutionary basis of local adaptation is natural selection causing changes of the mean phenotypic values to fit the climate optima of the respective environment [18]. In trees, selection is considered to be mainly active during seedling establishment, because trees possess long generation times and produce huge quantities of offspring [41]. Thus, mortality is strongest in the juvenile phases. The molecular basis for phenotypic variation is nucleotide variation at few or many individual loci (e.g. [42, 43]). The impact of selection on the traits underlying genes can be either through the fixation of single beneficial mutations [44] or on the basis of standing genetic variation on many genes where small changes of allele frequencies create substantial trait variation (e.g. [45]). Recent studies on quantitative trait loci and candidate genes suggest that for adaptive genetic variation the latter might be more common than previously thought [46]. Due to this complex trait architecture, selection on each gene is small, and thus high levels of standing genetic variation might be retained even in locally adapted populations [47].

Also, gene flow through steep environmental gradients may contribute substantially to genetic variation within populations and creates both, a genetic load and a higher adaptive potential in the subsequent populations [48]. However, even in cases of extensive gene flow, divergent selection acting on adaptive traits may result in a rapid increase of quantitative differentiation depending on the interpopulation allelic covariation of the underlying genes [49]. Further complexity might be added through epigenetic regulation of gene expression patterns [50, 51]. Overall, the contribution of nucleotide variation, allelic covariation, or gene expression patterns to phenotypic variation still needs to be deciphered.

Trait	Species	Literature
Bud break (S)	*Picea abies*	[21, 22]
	Abies alba	[23, 24]
	Larix decidua	[25]
	Pinus sylvestris	[26]
	Fagus sylvatica	[27]
Bud set (S)	*Picea abies*	[28, 29]
	Pinus sylvestris	[30]
Seedling height (S)	*Picea abies*	[28, 31]
	Pinus sylvestris	[30]
	Abies alba	[23, 24]
Biomass (S)	*Picea abies*	[31]
Tree height / Diameter at breast height (T)	*Picea abies*	[22, 32]
	Abies alba	[33]
	Larix decidua	[25, 34, 35]
	Pinus sylvestris	[36, 37]
	Fagus sylvatica	[38]
Frost resistance (S/T)	*Picea abies*	[39]
	Pinus sylvestris	[30]
Drought resistance	*Larix decidua*	[40]
	Abies alba	[23, 113]

Table 1. Overview of adaptive traits in alpine species, i.e. phenotypic traits of seedlings (S) or adult trees (T) that were found to be related to the environmental conditions of the seed provenance.

Besides selection and local adaptation, also neutral population genetic processes (i.e. drift, population expansion, migration, gene flow) are likely to affect the distribution of adaptive genetic variation. Overall, the genetic structure of temperate tree species has been strongly affected by historical processes: in particular, range contractions and expansions during quaternary climatic oscillations (ice ages) have shaped the current pattern of diversity (e.g. [52]). Thus, the admixture of adaptive variation from different postglacial lineages has been found to play a significant role in the current environmental clines of adaptive traits [53]. The decrease of genetic diversity from southern refugial populations to northern ones can be explained by bottleneck effects during colonisation or very rare events of long distance dispersal, e.g. [54, 55]. Although only few efforts have been made to test for geographical trends of quantitative genetic trait variation across species geographical range [56], recent comparisons of neutral vs. adaptive genetic variation show that quantitative differentiation due to adaptation is strong also at northern range margins, thus highlighting the role of local adaptation and adaptive plasticity [57, 58].

3. Application in climate change mitigation

The impact of adaptive traits on forest management, conservation biology and genetics, and on climate adaptation is tremendous: for example, in forest management adaptive traits are the basis for the delineation of breeding zones and provenance regions, for example see [59, 60]. In conservation biology adaptive traits are increasingly recognized as an ecologically important quality, thus need to be considered in restoration projects [61, 62]. For the genetic management of endangered populations, adaptive traits play a key role in developing translocation and genetic rescue schemes [63-65]. The high differentiation of adaptive traits and their strong correlation to environmental variables give them a paramount importance for developing adaptive forest management strategies to cope with climate change. Generally, two strategies can be used to test for local adaptation and to develop seed transfer schemes: First, seedling studies, where large numbers of seed provenances from different geographic origin (and therefore often also from different climatic origin) are planted under controlled conditions in a nursery of climate chambers, e.g. [28]. Seedling studies often reveal high correlations of phenotypic and phenological traits with environmental variables [66]. This correlation can be used as predictors for adaptation or maladaptation to current and future environmental conditions [67]. Secondly, provenance tests, where seed material from different origin is planted at several different test sites and being measured for at least 10-30 years, e.g. [68].

Seedling studies can be realised within few years. However, they are based only on one or very few test environments, and therefore their validity for predicting the long-term stability of provenances under a wide range of conditions is limited. On the other hand, seedling studies allow direct tests for extreme conditions, e.g. simulated drought or frost [69] and may therefore mimic effects that under natural conditions only take place on rare occasions. To observe the effect of such single extreme events in provenance tests is difficult, because they are usually controlled and measured in periods of one to five years. Therefore, they can only provide a surrogate for the general fitness of the tested genetic material under the given conditions. Fitness in provenance trials is measured mainly as survival, height or d.b.h., but with additional dendroclimatic measures also the intraspecific variation of climate-growth relationship can be assessed [70, 71].

True comparisons of seedling studies with subsequent provenance tests using the same genetic material are rare, but recent investigation with material from the same distribution area suggests, that seedling studies overestimate local adaptation and provenance test series show a more uniformly distributed genetic variation (for example, see [66, 72]).

4. Provenance trials history and test design

The creation and utilization of provenance tests for the analysis of intraspecific variation of trees goes back to the 18th century. Early common garden experiments were often established in a consistent test design and aimed at identifying well growing and suitable seed sources for reforestation. Langlet [73] credits the comparative cultures of pine established by H.L. Duhamel du Monceau between 1745 and 1755 as the first milestone of such genecological studies [74]. Although the experiments by Duhamel du Monceau rather aimed at gaining sufficient material for shipbuilding than scientific insights in intraspecific variation of pines [74], they attest an early awareness of local adaptations and the relationship between adaptive tree traits and different environmental conditions at the seed source origins.

Analyzing local adaptations and utilizing this information to find appropriate seed material has been a primary objective of provenance tests for more than 200 years [73, 75]. Later, provenance tests have also been used for seed transfer and utilization schemes and more recently, to develop adaptation measures for climate change, for examples, see [67, 68, 76-79]. Due to the wide range of objectives various test designs have been developed, all termed provenance tests or progeny trials. Morgenstern [75] classified all experimental setups with tree provenances according to the test environment in 1) growth cabinet and greenhouse experiments, 2) nursery experiments and 3) field experiments. While growth cabinets and nursery experiments are usually used for short term investigations, field experiments are established for periods of many years with iterated measurements. Actually, the very old provenance tests (tree age > 30 years) do often yield the most valuable results, since they enable reliable statements about the potential productivity of a specific stand.

The basic test design for provenance trials is a common garden test, where seedlings from different provenances are grown parallel under site conditions as similar as possible. However, the actual soil or environmental conditions for each plant are varying even within a mostly homogeneous test area (but see [80] on how to remove such effects). Therefore, trees are planted in multiple repetitions to compensate for any site heterogeneities. The most common layout plan is a randomized block design, which can be either complete, i.e. all provenances are planted throughout all blocks (repetitions) or incomplete, where the blocks do not contain the complete set of provenances. Moreover, in field trials over several sites, it can be distinguished between balanced and unbalanced designs, wherein the latter, the same provenances are not tested at all sites. For such unbalanced trials it has been recommended to use certain provenances as standard populations to which the others can be compared [74]. More details and guidelines for setting up field trials can be found in [74, 81, 82].

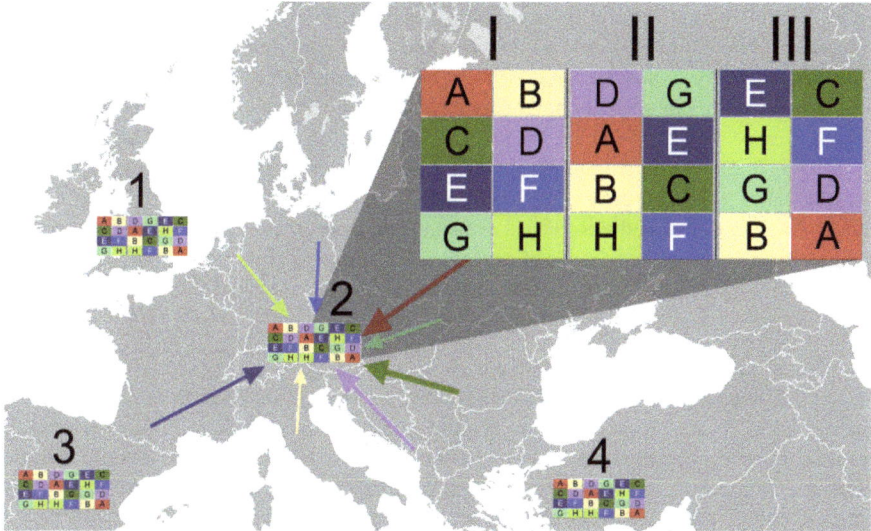

Figure 1. General scheme provenance test series. An exemplary trial series with 4 test sites (1-4) using a complete randomized block design is shown. Each provenance (A-H) is planted in (typically) three repetitions (block I-III) at each site. Each single plot (colored rectangles) consists of multiple individuals of the same provenance.

A scheme of the most common test design, the randomized complete block layout (RCB), is shown in figure 1. This test design has proved to be useful for many issues and environments and the basic setup did not changed during past decades. However, methods of data analyses did improve significantly. Powerful statistical tools are available due to increasing computer resources. Due to a growing amount of data derived from trials established in the past decades, it is very often worth utilizing this old data for further analyses employing modern statistical tools.

5. Climate response analyses with provenance test data

Ecological niches of species and populations are a fundamental concept in ecology. Usually the term refers to the 'Hutchinsonian multidimensional hypervolume', which is defined as the set of biotic and abiotic conditions in which a species is able to persist and maintain stable population sizes [83]. The 'fundamental niche' represents that portion of the environmental gradient within which existence is possible. By biotic interactions individuals of a species are excluded from a part of their fundamental niche, resulting in the realized niche that is observed in nature [84].

As a result of adaptation processes to a population's habitat, the fitness of the population has an optimum somewhere within the fundamental niche which decreases towards the edge areas. This linkage between fitness or fitness proxy parameters and climate variables has been studied for different tree species, e.g. *Pinus banksiana* [68], *Pinus sylvestris* [37], *Pinus contorta* [77], *Picea abies* [39]. To analyze the correlation between population specific quantitative traits and climate conditions several approaches have been developed. In its basic most common form, a climate response function is a univariate regression analysis with a single response variable (quantitative trait) and a single (climatic) predictor variable. Modelling approaches vary in regard to the chosen response and predictor variables and the applied mathematical model. In order to follow the ecological niche model, quadratic, Gaussian or Weibull models are preferred. If climate conditions of provenance origins and planting sites are both known it is possible to derive climate response functions and/or climate transfer functions.

Climate response functions depict the correlation between a quantitative trait of a specific tree population and distinct climate conditions at its planting sites. The respective genotype of a tree population translates to different phenotypes when exposed to variable growing environments (Fig. 2). The resulting pattern of the conditional trait response is also termed reaction norm. The specific objective of studies using climate response functions is to determine those climatic conditions, where the trait of interest is expressed in its most favorable shape. Such conditions may be defined as the climate optimum.

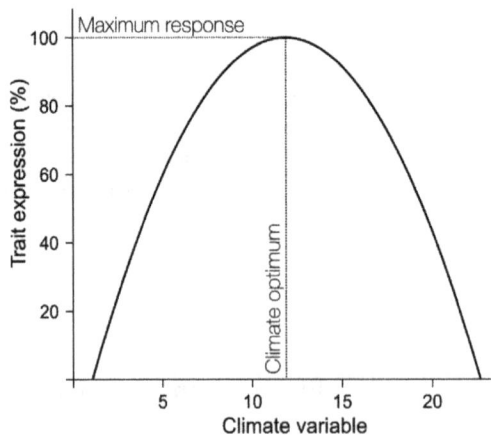

Figure 2. Schematized (quadratic) climate response function. Maximum response of the measured trait at a specific climate optimum is indicated.

Climate transfer functions include information about the climate of the provenance origins. Here, the quantitative traits are usually correlated to the climate transfer distance, i.e. the climatic difference between planting site and provenance origin. It is generally hypothesized that tree populations are best adapted to their local climate conditions. Therefore, the best growth performance is expected at small climate transfer distances, whereas at larger climatic distances, trees are maladapted and are thought to perform worse. Climate transfer functions can be derived for individual sites if data from sufficient populations from diverse climatic origins are available (Fig. 3). If the study comprises more than one planting site, data cannot directly be compared among sites with absolute units of measurements. Instead, relative performances to site means should be calculated. Thereby, site specific growth limitations (altitude, soil, climate) are eliminated and across site comparisons are possible.

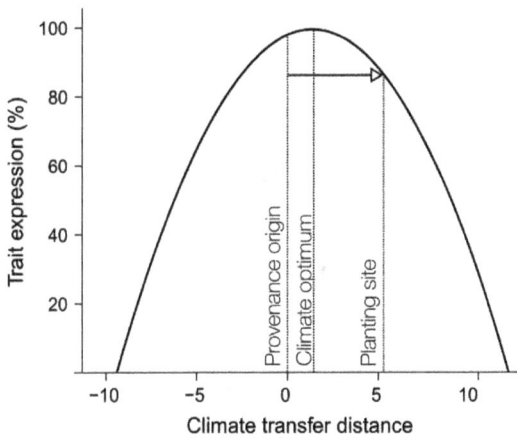

Figure 3. Schematized (quadratic) climate transfer function. The arrow indicates the transfer from the provenance origin to the planting site.

Both climate function types response functions and transfer functions can be utilized for combined analyses in universal response functions [85, 86, 88]. Universal response functions employ each site's transfer function to enhance response functions of all populations. In this way, all available information from any provenance at any test site is utilized to estimate a joined "universal" response of population to a given environmental factors. Such approach might compensate for unbalanced experimental setups with multiple test sites. Still, for statistical significant estimations, a large number of provenances and test sites are necessary. For example, reference [86] used results from a provenance test with lodgepole pine, comprising 140 populations at 62 sites, to calibrate a universal response function (Fig. 4). Test series spanning such a large climate range are rarely available.

However, incorporating information from both environmental effects of test site climate on phenotypes (accounted for by response functions) and among-population differentiation resulting from local adaptation to climate (accounted for by transfer functions) into a single "universal response function" reduces the necessary number of tested populations and test sites, while keeping the same predictive power [86]. Therefore, this integrative approach is promising for analyses of incomplete test designs that would be insufficient for establishing transfer or response functions [86]. Furthermore, combined with a sophisticated sampling strategy and choice of test sites, future provenance tests could be designed much more effectively. For example, both population and test site sample size could be by 65% without affecting the prediction accuracy of the universal response function [86].

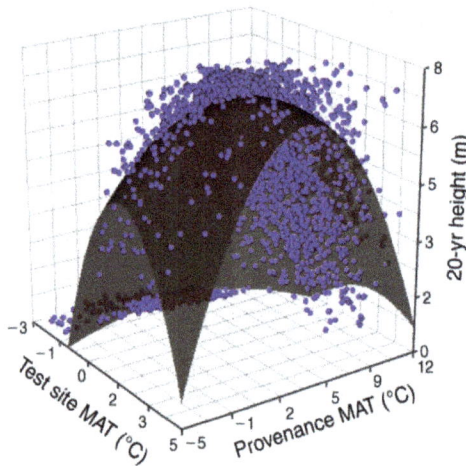

Figure 4. Universal response function including mean annual temperature (MAT) at test site and provenance origin (modified from Wang et al. [86]).

The mathematical model, which is fit to the obtained data, must be chosen carefully. Among the most commonly applied model functions are quadratic [37], Gaussian [87], Weibull [89] or Beta function [90]. In many cases more than one model would be appropriate from a statistical point of view. Several statistical tests and criteria may help in finding the most suitable model (goodness of fit). However, it is difficult to find the most appropriate approach, especially if only few data points are available or if upper and lower limits of the climatic niche are spurious.

Other problems arise for joint evaluations of provenance test series with multiple test sites, especially with incomplete designs, where different provenances have been tested at each site. This issue is addressed in reference [86] by utilizing universal response functions (see above). The classic approach to overcome this issue is the planting of a set of standard provenances at every site allowing for unbiased comparisons among sites. Though, the standard's value relative to other provenances need not be the same on all sites due to provenance x site interactions [32]. A method for joint analyses has been proposed in [32, 91], suggesting to present the results in standard deviations of the respective experimental mean at a given age [74]. However, this cannot effectively compensate inconsistent forest management activities or inconsistent data measuring practices. If such inhomogeneities and biases affect the validity of statistical analyses must be clarified case-to-case and should be well documented at any rate.

Aside from choosing an appropriate model function, the finding of the relevant climate parameter and predictor variables that drive selection is difficult. As local performance and productivity of tree populations are adapted to specific climate or environmental conditions, the selection forces driving this adaptation may vary from site to site. For example, whereas in warm regions tree productivity might possibly be limited by the extent of drought periods, in mountainous areas the limiting factor might rather be frost resistance and cold hardiness [92]. Especially in Alpine regions, where heterogeneous climates occur within small geographic ranges, a bunch of selection forces may play a role: for example, growth performance of Norway spruce was found to be positively related to temperatures in May and June at higher altitudes, and negatively related to temperatures of the same months at lower elevation [93]. In higher altitudes frost tolerance and stability is of paramount importance, in lower altitudes best exploitation of vegetation season has higher priority [92, 94]. However, most climate response functions concern only one or few climate parameters and do not reflect multiple effects, which may shape the actual population niche.

6. Minimum requirements for calculating response functions from provenance test data

Provenance experiments have been established with many different objectives, climate analyses being only one among them. Older trials were probably not established for the development of climate response functions or climate change adaptation measures. Such trials may be of limited or no use for the further analyses.

To use data of provenance experiments for the development of adaptation schemes, the field data have to fulfill few requirements: First, seed collections and the seed origins have to be well documented. Coordinates of provenances and test sites should be available as accurate as possible. Also, seed origins should be mostly autochthonous in order to find adaptations to the local climate. Second, provenances should be planted on a wide spectrum of test sites covering at least the natural range of climate conditions where the species occurs. If possible, few test sites should be beyond the natural range of the respective species. Third, for climate transfer and universal transfer functions it is also necessary to collect provenances from a wide climatic range in order to enable statistically valuable calibrations. Many provenance experiments do not meet all of these requirements sufficiently, which might result in low statistical power.

7. Provenance trials in Alpine space — Overview

From the oldest provenance trial with Norway spruce, established by Kienitz in 1881 (sowing 1878) at Gahrenberg, only descriptive observations are available [95]. In the Alpine space the altitude of population origins has early been recognized as a particular important factor. For

Requirement	CRF	CTF
Coordinates from provenance origins and test sites	X	X
Seed source material from presumably autochthonous stand	X	X
Seed collection involved multiple mother trees per population	X	X
Uniform forest management measures at test sites	X	X
Populations have been tested at sufficient test sites	X	
Wide climatic range of tested populations		X
Wide climatic range of test sites	X	
Sufficient climatic range of sites, where each populations has been tested	X	
Measurement data of the field trials should be available in the raw form	X	X

Table 2. Minimum requirements for climatic analyses of provenance test data. Right columns indicate if the respective criterion should be fulfilled to calibrate climate response functions (CRF) or climate transfer functions (CTF).

example, in Austria, Cieslar compared tree growth among Norway spruce offspring from different altitudinal ranges at three planting sites [96, 97] in 1899. Trees have been evaluated several times afterwards [96, 98]. Latest results at the tree age 58 indicate an assimilation and reduction of growth-differences among populations. Engler (see [99]) established a similar experiment in 1899 with five Suisse provenances of Norway spruce. The provenances were planted at 16 sites in altitudes from 380 m to 2150 m [99]. In both experiments, considerable differences in survival, growth and tree shape among provenances were found.

Table 3 gives an overview of provenance tests of Norway spruce that have been established in the Alpine space. Within the project MANFRED, they were reviewed and rated according to their potential application in climate adaptation studies. From the many trials that were established in the past 100 years, most trials did not fit. In some cases provenances were tested only under "colder" climate conditions, but not in warmer environments. Also, due to the long observation times, many trial data only exist in form of published tables or figures but not within their raw form. Provenance tests which might be used for climate adaptation analyses are listed in table 3 for *Picea abies* and table 4 for *Abies alba*.

The three IUFRO trial series (1938, 1964/68 and 1972) are especially interesting for potential climate related analyses, since both the test sites and the utilized provenances cover a wide climatic range. However, data are not fully available and despite many efforts, data exchange among involved institutions was not successful until now. Nevertheless, the national provenance tests mentioned in table 3 and 4 are also valuable sources for analyses and should be reanalyzed in relation to climate data. As an example we present some results from the Austrian provenance test of 1978 in chapter 4.

The IUFRO silver fir test series have been initiated by H. Mayer, who was impressed by reports on Silver fir dieback during the early sixties of the 20th century [112]. Aiming in a holistic

Name (initiator)	Country (coord.)	Year sown (plant.)	Sites	Prov.range	Design	Lit.	RF	TF
Cieslar	AT	1893 (1896)	1 (AT)	Alp. space	1 rep; ST harvest	-		(X)
Cieslar	AT	1899	3	Mostly AT (80 + 1)	1 rep; ST harvest	[96, 98] -		(X)
Engler	CH	1899	16	CH (5)	MT harvest	[99]	X	(X)
Holzer	AT	1978	44	AT (480) + Alp. space (60)	RCBD; 3 rep; MT harvest	[100]	X	X
Bouvarel	FR	1957	1 (FR)	FR (14) + AT (1)		[101]		
Fröhlich	DE	1962	5 (DE)	Nat. range (530)	Lattice square	[102]	(X)	X
Pavle	SI	1984 (1987)	1	SI (10)	3 rep			X
IUFRO 1938 (+39)	14 countries (Schmidt, DE)	1938 (1940-1944 (in))	25 (+2)	Nat. range (36)	MT harvest	[32]	X	X
IUFRO 64/68	13 countries (Langlet, SE)	1964 (1968)	21	Nat. range (1100)	RCBD; MT harvest	[102, 103]	X	X
IUFRO 1972	11 countries (Tyskiewicz / Matras)	1972	44	PL (20)	MT harvest		X	X

Table 3. Overview of Norway spruce trials in Alpine space. Information about the respective test series is given if available, including year of establishment, No. of involved sites, range of tested provenances, test design (RCBD = Randomized complete block design, rep = no. of repetitions, harvest of single (ST) or multiple (MT) trees). The potential of data for utilization in climate response functions (RF) and/or climate transfer functions (TF) is given. Brackets indicating limited use due to deficits in study design or data availability.

research design for testing genetic variation in general as well as drought resistance, he established a trial using 19 international provenances under summer warm climate conditions of the rim of the Viennese forest in Austria in 1970 [33]. The Austrian silver fir trial served as pilot study for the "1st International Silver fir provenance trial 1982" which was organized by Kramer [107]. Seeds from eleven provenances from central and southern Europe were sent to forestry institutes in Germany, Netherlands, Austria, Suisse, Slovenia, Poland and Slovakia to establish in total 27 trials [108]. Six of eleven provenances were tested at every test site as standards. In spite of detailed plans for a joint evaluation of all tests and a central database in Vienna [107], the data were recorded individually and published separately without any coordination. A further comprehensive test series was established in South Germany in 1982 using 42 south German and 17 provenances originating from the whole European Silver fir range, including the provenances of the 1st IUFRO trial [109]. Finally, the 2nd International Silver fir provenance test was established in 2005, in order to get better information on the variation of SE-European provenances [111].

Name (initiator)	Country (coord.)	Year sown (plant.)	Sites	Prov. range	Design	Lit.	RF	TF
Czecho-slovakian prov. test with Silver fir	SK	1965 (1970)	2 (SK)	27	RCBD , 3 rep.	[104]		X
Int. Silver fir prov. trial (IUFRO pilot study 1967)	H. Mayer (AT)	1967 (1970)	1 (AT)	19	RBD, 2 rep.	[33, 105]		X
South German Silver fir prov. trial	DE	1982/83 (1986-1989)	19 (GE)	DE (42) + Intern. (17)	3 rep.	[106]	X	X
1st Int. IUFRO Silver fir prov. test 1982	Kramer (DE)	1982 (1986)	27	Centr.+South. Europe (11)	RBD	[107, 108]	X	X
2nd Int. IUFRO Silver fir prov. test 2005	Eder (DE), Klumpp (AT)	2000 (2005)	5	Centr.+SE. Europe (17)	RCBD, 3 rep.	[111]	X	X

Table 4. Overview of *Abies alba* trials in Alpine space. Information about the respective test series is given if available, including year of establishment, No. of involved sites, range of testes provenances, test design (RCBD = Randomized complete block design, rep = no. of repetitions, harvest of single (ST) or multiple (MT) trees). The potential of data for utilization in climate response functions (RF) and/or climate transfer functions (TF) is given. Brackets indicating limited use due to deficits in study design or data availability.

Apart from the international series there have been several national provenance tests established. Earlier national experiments using seed provenances from the Alpine space had been started for example in Italy [110] and Slovakia 1965 [104].

8. Case study — Austrian provenance test with Norway spruce 1978

A comprehensive provenance test with Norway spruce has been established in 1978 at 44 sites across Austria. The trial included seeds from 480 Austrian provenances and 60 provenances from other countries [100]. At tree age 15, height has been recorded at 29 of these 44 trial sites. These data have been used in the present project to analyze the intraspecific variation in climate response [87]. In order to reveal more general insights, provenances with similar climatic conditions were aggregated in provenance clusters. The response of the nine resulting clusters to a heat-moisture index was calculated using a Gaussian distribution model.

The results hardly revealed any declines in potential tree growth of Norway spruce throughout its current distribution range in Austria. In fact, for most parts of Austria we found an expected increase of tree heights up to 45 percent until 2080. In general, provenances from currently warm and drought prone areas seem to be well adapted to respective climate conditions and may be appropriate candidates for extended utilization in future. However, the impact of a warming climate is different for individual provenance groups (Fig. 5). Thus, an optimized choice of seed material according to prospective future climate conditions has the potential for an additional increase of productivity up to 11 percent [87]. Although this study sets focus on provenances of the eastern Alps and the Bohemian Massif in Austria, some results of this most extensive provenance trial in the Alpine space can be generalized beyond the Austrian border, because firstly, although the provenances origin from a relatively small part of the natural range of Norway spruce, they cover the three main refugial lineages which build the basis of all natural Norway spruce populations in Central and Western Europe, and secondly, the provenance trial series has been established along a wide gradient of climate conditions ranging from 2.6°C to 9.2°C mean annual temperature and annual precipitation values from 535 mm to 2392 mm. Thus, it covers not only a large part of current Norway spruce habitats, but also extends into sites at the warm and dry edge of its distribution, making it highly suitable to analyze the potential response to a changing climate.

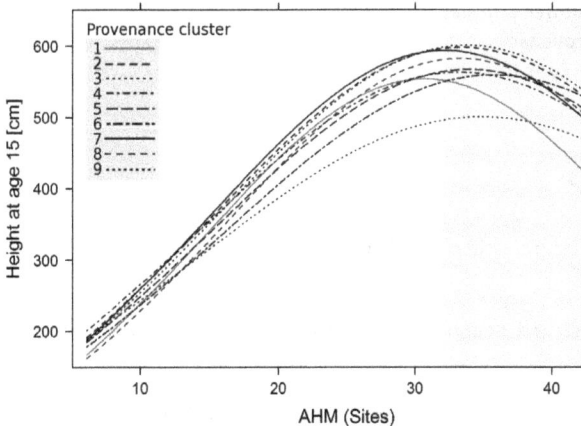

Figure 5. Response functions of nine provenance clusters of the Austrian Norway spruce provenance trial 1978. Gaussian functions of tree heights at age 15 fitted to heat-moisture index AHM (mean annual temperature + 10 / (Annual precipitation sum / 1000)) of planting sites [77]. Response functions are given for nine groups of provenances which originate from similar climates. Grey area denotes AHM-regions beyond current maximum (32 °C/mm) to future maximum (42.3 °C/mm) limits.

9. Conclusion

Many provenance tests have been established for tree species in the Alpine space within the last 200 years. These trials are based on a laborious work from its initial establishment to the often comprehensive and numerous measurements. Today, such data are strongly needed for a better understanding of the intraspecific variation of climate response of tree species and for the development of provenance transfer and deployment schemes in the light of climate change. Within the last 20 years, the theoretical and statistical basis for the analysis of provenance climate responses was strongly improved. Also, high-quality climate data are freely available now. Therefore, new statistical analyses including all available data could substantially improve our understanding of climate growth relationships and evolutionary processes that cause intraspecific variation in climate adaptation. However, raw data from historic provenance test are only partly electronically available, and if they are, then access is usually limited to staff of responsible national institutions. Hence, efforts have to be made to make data easy accessible. Moreover, since meaningful climate response analyses require a considerable climate range of tests sites and tested population, transnational exchange of data and joined analyses are strongly needed, in order to cover the natural range of the respective species.

Furthermore, test design of previous provenance trials was optimized to find most appropriate seed sources for specific test sites and not to investigate general climate growth relations. As a consequence, only few test sites are established at the edges of the natural climatic range of tree species. Those climatically extreme sites are most valuable to detect upper and lower limits of tree species and tree populations and to calibrate well-founded response functions. Therefore, for future provenance tests we recommend to include test sites at the edges of the natural range of the respective tree species or even beyond.

Author details

Stefan Kapeller[1], Silvio Schüler[1], Gerhard Huber[2], Gregor Božič[3], Tom Wohlgemuth[4] and Raphael Klumpp[5]

1 Federal Research and Training Centre for Forests, Natural Hazards and Landscape, Vienna, Austria

2 Bavarian Office for Forest Seeding and Planting, Teisendorf, Germany

3 Slovenian Forestry Institute, Ljubljana, Slovenia

4 Swiss Federal Institute for Forest, Snow and Landscape Research WSL, Birmensdorf, Switzerland

5 University of Natural Resources and Life Sciences, Vienna, Austria

References

[1] Prentice IC, Sykes MT, Cramer W. A simulation model for the transient effects of climate change on forest landscapes. Ecological Modelling 1993;65 51-70.

[2] Lindner M. Forest management strategies in the context of potential climate change. Forstwissenschaftliches Centralblatt 1999;118 1-13.

[3] Maracchi G, Sirotenko O, Bindi M. Impacts of present and future climate variability on agriculture and forestry in the temperate regions: Europe. Climatic Change 2005;70 117-135.

[4] Peltola H, Kilpeläinen A, Kellomäki S. Diameter growth of Scots pine (*Pinus sylvestris*) trees grown at elevated temperature and carbon dioxide concentration under boreal conditions. Tree Physiology 2002;22 963-972.

[5] Matala J, Ojansuu R, Peltola H, Raitio H, Kellomaki S. Modelling the response of tree growth to temperature and CO2 elevation as related to the fertility and current temperature sum of a site. Ecological Modelling 2006;199 39-52.

[6] Modrzynski J. Response of *Picea abies* populations from elevational transects in the Polish Sudety and Carpathian mountains to simulated drought stress. Forest Ecology and Management 2002;165 105-116.

[7] Bréda N, Huc R, Granier A, Dreyer E. Temperate forest trees and stands under severe drought: a review of ecophysiological responses, adaptation processes and long-term consequences. Annals of Forest Science 2006;3 625-644.

[8] Davis MB, Shaw RG. Range shifts and adaptive responses to Quaternary climate change. Science 2001;292 673-679.

[9] Pearman PB, D'Amen M, Graham CH, Thuiller W, Zimmermann NE. Within-taxon niche structure: niche conservatism, divergence and predicted effects of climate change. Ecography 2010;33 990-1003.

[10] Hanewinkel M, Cullmann DA, Schelhaas MJ, Nabuurs G-J, Zimmermann NE. Climate change may cause severe loss in the economic value of European forest land. Nature Climate Change 2012;3 203-207.

[11] Sykes MT, Prentice C, Cramer W. A Bioclimatic model for the potential distributions of North European tree species under present and future climates. Journal of Biogeography 1996;23 203-233.

[12] Thuiller W, Lavorel S, Sykes MT, Araújo MB. Using niche-based modelling to assess the impact of climate change on tree functional diversity in Europe. Diversity and Distributions 2006;12 49-60.

[13] Spehn EM, Koerner C. A global assessment of mountain biodiversity and its function. In: Huber UM. et al. (eds.) Global change and mountain regions: an overview of current knowledge. Springer; 2005. p393-400.

[14] Theurillat J, Guisan A. Potential impact of climate change on vegetation in the European Alps: a review. Climate Change 2001;50 77–109.

[15] Engler R, Randin CF, Thuiller W, Dullinger S, Zimmermann NE, Araújo MB, et al. 21st century climate change threatens mountain flora unequally across Europe. Global Change Biology 2011;17(7) 2330–2341.

[16] Thuiller W, Lavorel, S, Araújo MB, Sykes MT, Prentice IC. Climate change threats to plant diversity in Europe. Proceedings of the National Academy of Sciences of the United States of America 2005;102 8245–8250.

[17] Dirnböck T, Essl F, Rabitsch W. Disproportional risk for habitat loss of high-altitude endemic species under climate change. Global Change Biology 2011;17 990–996.

[18] Savolainen O, Pyhäjärvi T, Knürr T. Gene Flow and Local Adaptation in Trees. Annual Review of Ecology, Evolution, and Systematics 2007;38 595-619.

[19] Kremer A, Le Corre V, Petit RJ, Ducousso A. Historical and contemporary dynamics of adaptive differentiation in European oaks. In: DeWoody A, Bickham J, Michler C, Nichols K, Rhodes G, Woeste K (eds.) Molecular approaches in natural resource conservation. Cambridge University Press; 2010. p101-117.

[20] Holderegger R, Kamm U, Gugerli F. Adaptive vs. neutral genetic diversity: implications for landscape genetics. Landscape Ecology 2006;21 797-807.

[21] Lechner F, Holzer K, Tranquillini W. Über Austrieb und Zuwachs von Fichtenklonen in verschiedener Seehöhe. Silvae Genetica 1977;26 33-41.

[22] Hannerz M, Sonesson J, Ekberg I. Genetic correlations between growth and growth rhythm observed in a short-term test and performance in long-term field trials of Norway spruce. Canadian Journal of Forest Research 1999;29 768-778.

[23] Sagnard F, Barberot C, Fady B. Structure of Genetic diversity in *Abies alba* Mill.from southwestern Alps: multivariate analysis of adaptive and non-adaptive traits for conservation in France. Forest Ecology and Management 2002;157 175-189.

[24] Hansen JK, Larsen JB. European silver fir (Abies alba Mill.) provenances from Calabria, southern Italy: 15-year results from Danish provenance field trials. European Journal of Forest Research 2004;123 127-138.

[25] Schober R. Phänologie und Höhenwachstum der Lärche im Jahresverlauf in ihrer Abhängigkeit von Provenienz und Witterung. Beitrag zur Internationalen Ertragskundetagung 1966;433-483.

[26] Beuker E. Adaptation to climate changes of the timing of bud burst in populations of *Pinus sylvestris* L. and *Picea abies* (L.) Karst. Tree Physiology 1994;14 961-970.

[27] Wühlisch Gv, Krusche D, Muhs HJ. Variation in temperature sum requirement for flushing of beech provenances. Silvae Genetica 1995;44 343-346.

[28] Holzer K. Zur Identifizierung von Fichtenherkünften (*Picea abies* (L.) Karst.) Silvae Genetica 1975;24 169-175.

[29] Holzer K. Die Kulturkammertestung zur Erkennung des Erbwertes bei Fichte [*Picea abies* (L.) Karsten]. 2. Merkmale des Vegetationsablaufes. Centralblatt für das gesamte Forstwesen 1978;95 30-51.

[30] Hurme P, Repo T, Savolainen O, Pääkkönen T. Climatic adaptation of bud set and frost hardiness in Scots pine (*Pinus sylvestris*). Canadian Journal of Forest Research 1997;27 716-723.

[31] Holzer K. Die Kulturkammertestung zur Erkennung des Erbwertes bei Fichte [*Picea abies* (L.) Karsten]. 3. Quantitative Merkmale. Centralblatt für das gesamte Forstwesen 1979;96 129-144.

[32] Giertych M. Summary results of the IUFRO 1938 Norway spruce (*Picea abies* (L.) Karst.) provenance experiment. Height growth. Silvae Genetica 1976;25 154–164.

[33] Mayer H, Reimoser F, Kral F. Ergebnisse des internationalen Tannenherkunftsversuches Wien 1967-1978, Morphologie und Wuchsverhalten der Provenienzen. In: Mayer H (ed.) 3. Tannen-Symposium. Vienna: Österreichischer Agrarverlag; 1980.

[34] Schober R. Neue Ergebnisse des II. Internationalen Lärchenprovenienzversuches von 1985/59 nach Aufnahmen von Teilversuchen in 11 europäischen Ländern und den U.S.A. Schriften aus der Forstlichen Fakultät der Universität Göttingen und der Niedersächsischen Forstlichen Versuchsanstalt 1985;83 165.

[35] Giertych M. Summary of results on European larch (*Larix decidua* Mill.) height growth in the IUFRO 1944 provenance experiment. Silvae Genetica 1979;28 244-256.

[36] Shutyaev AM, Giertych M. Height Growth Variation in a Comprehensive Eurasian Provenance Experiment of (*Pinus sylvestris* L.) Silvae Genetica 1997;46 332-349.

[37] Rehfeldt GE, Tchebakova NM, Parfenova YI, Wykoff WR, Kuzmina NA, Milyutin LI. Intraspecific responses to climate in *Pinus sylvestris*. Global Change Biology 2002;8 912-929.

[38] Mátyás C, Božic G, Gömöry D, Ivankovic M, Rasztovits E. Transfer analysis of provenance trials reveals macroclimatic adaptedness of European Beech (*Fagus sylvatica* L.) Acta Silvatica et Lignaria Hungarica 2009;5 47-62.

[39] Gömöry D, Foffová E, Kmet J, Longauer R, Romšáková I. Norway Spruce (*Picea abies* [L.] Karst.) Provenance Variation in Autumn Cold Hardiness: Adaptation or Acclimation? Acta Biologica Cracoviensia s. Botanica 2010;52 42-49.

[40] Kral F. Vergleichende Transpirationsstudien an Herkünften der europäischen Lärche. Centralblatt für das gesamte Forstwesen 1962;79 222-238.

[41] Petit RJ, Hampe A. Some evolutionary consequences of being a tree. Annual Review of Ecology, Evolution, and Systematics 2006;37 187–214.

[42] Hurme P, Sillanpää M, Arjas E, Repo T, Savolainen O. Genetic basis of climatic adaptation in Scots pine by Bayesian quantitative locus analysis. Genetics 2000;156 1309–1322.

[43] Scotti-Saintagne C, Bodenes C, Barreneche T, Bertocchi E, Plomion C, Kremer A. Detection of quantitative trait loci controlling budburst and height growth in *Quercus robur*. Theoretical and Applied Genetics 2004;109 1648–1659.

[44] Kopp M, Hermisson J. The Genetic Basis of Phenotypic Adaptation I: Fixation of Beneficial Mutations in the Moving Optimum Model. Genetics 2009;182 233–249.

[45] Hermisson J, Pennings PS. Soft sweeps – molecular population genetics of adaptation from standing genetic variation. Genetics 2005;169 2335–2352.

[46] Le Corre V, Kremer A. The genetic differentiation at quantitative trait loci under local adaptation. Molecular Ecology 2012;21 1548–1566.

[47] Aitken S, Yeaman S, Holliday J, Wang T, Curtis-McLane S. Adaptation, migration or extirpation: climate change outcomes for tree populations. Evolutionary Applications 2008;1 95-111.

[48] Yeaman S, Jarvis A. Regional heterogeneity and gene flow maintain variance in a quantitative trait within populations of lodgepole pine. Proceedings Royal Society B 2006;273 1587-1593.

[49] Kremer A, Le Corre V. Decoupling of differentiation between traits and their underlying genes in response to divergent selection. Heredity 2011;108 375-85.

[50] Bird A. Perceptions of epigenetics. Nature 2007;447 396–398.

[51] Kvaalen H, Johnsen Ø. Timing of bud set in *Picea abies* is regulated by a memory of temperature during zygotic and somatic embryogenesis. New Phytologist 2008;177 49–59.

[52] Hewitt GM. The genetic legacy of the Quarternary ice ages. Nature 2000;405 907-913.

[53] De Carvalho D, Ingvarsson PK, Joseph J, Suter L, Sedivy C, Macaya-Sanz D, Cottrell J, Heinze B, Schanzer I, Lexer C. Admixture facilitates adaptation from standing variation in the European aspen (*Populus tremula* L.), a widespread forest tree. Molecular Ecology 2010;19 1638-1650.

[54] Hewitt GM. Post-glacial distribution and species substructure: lessons from pollen, insects and hybrid zones. In: Lees DR, Edwards D. (eds.) Evolutionary Patterns and Processes. London: Linnean Society Symposium Series, Academic Press; 1993. p97-123.

[55] Bialozyt R, Ziegenhagen B, Petit RJ. Contrasting effects of long distance seed dispersal on genetic diversity during range expansion. Journal of Evolutionary Biology 2006;19 12-20.

[56] Eckert CG, Samis KE, Lougheed SC. Genetic variation across species' geographical ranges: the central–marginal hypothesis and beyond. Molecular Ecology 2008;17 1170-1188.

[57] Mimura M, Aitken SN. Adaptive gradients and isolation-by-distance with postglacial migration in *Picea sitchensis*. Heredity 2007;99 224–232.

[58] Mimura M, Aitken SN. Local adaptation at the range peripheries of Sitka spruce. Journal of Evolutionary Biology 2010;23 249-258.

[59] Crowe KA, Parker WH. Provisional breeding zone determination modelled as a maximal covering location problem. Canadian Journal Forest Research 2005;35 1173–1182.

[60] Hamann A, Gylander T, Chen PY. Developing seed zones and transfer guidelines with multivariate regression trees. Tree Genetics & Genomes 2011;7 399-408.

[61] Bischoff A, Vonlanthen B, Steingera T, Müller-Schärer H. Seed provenance matters — effects on germination of four plant species used for ecological restoration. Basic and Applied Ecology 2006;7 347-359.

[62] O'Brien EK, Mazanec RA, Krauss SL. Provenance variation of ecologically important traits of forest trees: implications for restoration. Journal of Applied Ecology 2007;44 583-593.

[63] Tallmon DA, Luikart G, Waples RS. The alluring simplicity and complex reality of genetic rescue. Trends in Ecology & Evolution 2004;19 489-496.

[64] Hedrick PW. 'Genetic restoration:' a more comprehensive perspective than 'genetic rescue.' Trends in Ecology & Evolution 2005; 20 109.

[65] Bouzat JL, Johnson JA, Töpfer JE, Simpson SA, Esker TL, Westemeier RL. Beyond the beneficial effects of translocations as an effective tool for the genetic restoration of isolated populations. Conservation Genetics 2009;10 191-201.

[66] St.Clair JB, Mandel NL, Vance-Borland KW. Genecology of Douglas fir in Western Oregon and Washington. Annals of Botany 2005;96 1199-1214.

[67] St.Clair, JB, Howe GT. Genetic maladaptation of coastal Douglas-fir seedlings to future climates. Global Change Biology 2007;13 1441–1454.

[68] Matyas C. Modeling climate change effects with provenance test data. Tree physiology 1994;14 797–804.

[69] St.Clair JB. Genetic variation in fall cold hardiness in coastal Douglas-fir in western Oregon and Washington. Canadian Journal of Botany 2006;84 1110-1121.

[70] Grabner M, Karanitsch-Ackerl S, Schueler S. The influence of drought on density of Norway spruce wood. In: Kúdela, J., Lagaňa, R. (eds.) Wood Structure and Properties '10. Zvolen: Arbora Publishers; 2010. p27–33.

[71] McLane SC, Daniels LD, Aitken SN. Climate impacts on lodgepole pine (*Pinus contorta*) radial growth in a provenance experiment. Forest Ecology and Management 2011;262 115-123.

[72] Krakowski J, Stoehr MU. Coastal Douglas-fir provenance variation: patterns and predictions for British Columbia seed transfer. Annals of Forest Science 2009;66 1-10.

[73] Langlet O. Two hundred years genecology. Taxon 1971;20 653–721.

[74] König AO. Provenance research: evaluating the spatial pattern of genetic variation. In: Geburek T, Turok J. (eds.): Conservation and Management of Forest Genetic Resources in Europe. Zvolen: Arbora Publishers; 2005. p275-333.

[75] Morgenstern E. Geographic variation in forest trees: genetic basis and application of knowledge in silviculture. Vancouver: UBC Press; 1996.

[76] Rehfeldt GE, Ying CC, Spittlehouse DL, Hamilton DA. Genetic responses to climate in *Pinus contorta*: Niche breadth, climate change, and reforestation. Ecological monographs 1999;69 375–407.

[77] Wang T, Hamann A, Yanchuk A, O'Neill GA, Aitken SN. Use of response functions in selecting lodgepole pine populations for future climates. Global Change Biology 2006;12 2404–2416.

[78] McLachlan JS, Hellmann JJ, Schwartz MW. A framework for debate of assisted migration in an era of climate change. Conservation Biology 2007;21 297–302.

[79] Ukrainetz NK, O'Neill GA, Jaquish B. Comparison of fixed and focal point seed transfer systems for reforestation and assisted migration: a case study for interior spruce in British Columbia. Canadian Journal of Forest Research 2011;41 1452–1464.

[80] Zas R. Iterative kriging for removing spatial autocorrelation in analysis of forest genetic trials. Tree Genetics & Genomes 2006;2 177-185.

[81] Burley J, Wood PJ. A manual of species and provenance research with particular reference to the tropics. University of Oxford: CFI Tropical Forestry Paper No. 10; 1976.

[82] Cochran WG, Cox GM. Experimental designs. New York: Wiley; 1957.

[83] Hutchinson GE. Concluding Remarks. Cold Spring Harbor Symp. Quantitative Biology 1958;2 415–427.

[84] Guisan A, Thuiller W. Predicting species distribution: offering more than simple habitat models. Ecology Letters 2005;8 993–1009.

[85] O'Neill GA, Hamann A, Wang T. Accounting for population variation improves estimates of the impact of climate change on species' growth and distribution. Journal of Applied Ecology 2008;45 1040–1049.

[86] Wang T, O'Neill GA, Aitken SN. Integrating environmental and genetic effects to predict responses of tree populations to climate. Ecological Applications 2010;20 153–163.

[87] Kapeller S, Lexer MJ, Geburek T, Schüler S. Intraspecific variation in climate response of Norway spruce in the eastern Alpine range: Selecting appropriate provenances for future climate. Forest Ecology and Management 2012;271 46–57.

[88] O'Neill GA, Nigh G. Linking population genetics and tree height growth models to predict impacts of climate change on forest production. Global Change Biology 2011;17 3208–3217.

[89] Rehfeldt GE, Ferguson DE, Crookston NL. Quantifying the abundance of co-occurring conifers along inland northwest (USA) climate gradients. Ecology 2008;89 2127–2139.

[90] Austin MP, Nicholls AO, Doherty MD. Determining species response functions to an environmental gradient by means of a beta-function. Journal of Vegetation 1994;5 215–228.

[91] Giertych M. Report of the IUFRO 1938 and 1939 provenance experiments on Norway spruce (*Picea abies* L. Karst.). Kórnik: Polish Academy of Sciences; 1984.

[92] Aitken SN, Hannerz M. Genecology and gene resource management strategies for conifer cold hardiness. In: Bigras FJ, Columbo SJ. (eds.) Conifer cold hardiness. Dordrecht: Kluwer Academic Publishers; 2001. p23–53.

[93] Mäkinen H, Nojd P, Kahle HP, Neumann U, Tveite B, Mielikainen K, Rohle H, Spiecker H. Radial growth variation of Norway spruce (*Picea abies*) across latitudinal and altitudinal gradients in central and northern Europe. Forest Ecology and Management 2002;171 243–259.

[94] Howe GT, Aitken SN, Neale DB, Jermstad KD, Wheeler NC, Chen THH. From genotype to phenotype: unraveling the complexities of cold adaptation in forest trees. Canadian Journal of Botany 2003;81 1247–1266.

[95] Schmidt-Vogt H. Die Fichte. Band II/1. Hamburg: Parey; 1986.

[96] Melzer H. Der Fichten-Herkunftsversuch in Loimannshagen, Ergebnisse einer Aufnahme im Jahre 1936. Centralblatt für das gesamte Forstwesen 1937;63 225-232.

[97] Günzl L. Internationale Fichten-Provenienzversuche der IUFRO 1938 und 1964/68 sowie Versuche mit österreichischen Herkünften. Allgemeine Forstzeitung 1979;90 182-190.

[98] Günzl L. Ergebnisse aus einer Fichtenprovenienzforschung. Forstliche Bundesversuchsanstalt Wien, Informationsdienst 120; 1969.

[99] Nägeli W. Einfluss der Herkunft des Samens auf die Eigenschaften forstlicher Holz-
      gewächse; Die Fichte. Eidgenössische Anstalt für das Forstliche Versuchswesen, Mit-
      teilungen 1931;17 150-237.

[100] Nather J, Holzer K. Über die Bedeutung und die Anlage von Kontrollflächen zur
      Prüfung von anerkanntem Fichtenpflanzgut. Forstliche Bundesversuchsanstalt Wien,
      Informationsdienst 181; 1979.

[101] Bouvarel P. Observation sur la date de l'aoutement de quelques provenances franca-
      ises d'Epicea. Annales Nationale des Eauxet Forets de la Station de Recherche et Ex-
      perience 1961;18 99-129.

[102] Liesebach M, Rau H, König AO. Fichtenherkunftsversuch von 1962 und IUFRO-Fich-
      tenherkunftsversuch von 1972 - Ergebnisse von mehr als 30-jähriger Beobachtung in
      Deutschland. Beiträge aus der Nordwestdeutschen Forstlichen Versuchsanstalt 5;
      2010.

[103] Krutzsch P. The IUFRO 1964/1968 Provenance Test with Norway Spruce (*Picea abies*
      (L.) Karst.). Silvae Genetica 1974;23 58–62.

[104] Korpel S, Paule L. Ergebnisse eines Provenienzversuches mit tschechoslowakischen
      und polnischen Herkünften der Weißtanne (*Abies alba* Mill.). Silvae Genetica 1984;33
      (6) 177-182.

[105] Kral F. Untersuchungen zur physiologischen Charakterisierung von Tannenprove-
      nienzen. In: Mayer H (ed.) 3. Tannen-Symposium. Vienna: Österreichischer Agrar-
      verlag; 1980.

[106] Rütz WF, Franke A, Stimm B. Der Süddeutsche Weisstannen (*Abies alba* Mill.)-Prove-
      nienzversuch. Jugendentwicklung auf den Versuchsflächen. Allgemeine Forst- und
      Jagdzeitung 1998;169 (6/7) 116–126.

[107] Kramer W. Vorschlag für einen internationalen Herkunftsversuch von Weißtanne
      (*Abies alba* Mill.). In: Mayer H (ed.) 3. Tannen-Symposium. Vienna: Österreichischer
      Agrarverlag; 1980.

[108] Larsen JB. Provenienzen und Versuchsteilnehmer des IUFRO-Weißtannen-prove-
      nienzversuches von 1982. Schriften aus der Forstlichen Fakultät der Universität Göt-
      tingen und der Niedersächsischen Forstlichen Versuchsanstalt 1985;80. 239-241.

[109] Wolf H, Rütz WF, Franke A. Der süddeutsche Weißtannenprovenienzversuch: Er-
      gebnisse der Baumschulphase und Anlage der Versuchsflächen. In: Wolf H. (ed.)
      Weißtannenherkünfte - Neue Resultate zur Provenienzforschung bei *Abies alba* Mill.:
      Contributiones Biologiae Arborum; 1994. p107–130.

[110] Pavari A. Esperienze e indaginisu le provenienze e razzedell'abetebianco (*Abies alba*
      Mill.). Firenze: Pubblicazioni della Regio Stazione Sperimentale di Selvicoltura 8;
      1951.

[111] Tabel U. Stand der Vorbereitungen zum 2. IUFRO-Weißtannen-Herkunftsversuch. Proceedings of the 9th International European Silver Fir Symposium, Skopje / Macedonia. IUFRO WP 1.05-16; 2000.

[112] Mayer H. Zur waldbaulich-genetischen Beurteilung der Tanne. Bericht über ein „Tannen-Symposium" der Arbeitsgruppe Gebirgswaldbau IUFRO Sektion 23, Vienna, 1969. In: Mayer H (ed.) 3. Tannen-Symposium. Vienna: Österreichischer Agrarverlag; 1980. p169-179.

[113] Kral F. Untersuchungen zur physiologischen Charakterisierung von Tannenprovenienzen. In: Mayer H (ed.) 3. Tannen-Symposium. Vienna: Österreichischer Agrarverlag; 1980. p139-157.

Chapter 16

Monitoring Climatic Change Impacts on Protection Forests in Aosta Valley (Italy) and in Drôme (France) Using Medium and High Resolution Remote Sensing and Mateloscopes Plots

Laurent Borgniet, David Toe, Frédéric Berger,
Marta Galvagno, Cinzia Panigada,
Roberto Colombo, Umberto Morra di Cella,
Simone Gottardelli, Ivan Rollet, Mario Negro,
Flavio Vertui and Cédric Fermont

Additional information is available at the end of the chapter

1. Introduction

Land management in relation to climate change has become a major issue since many studies have confirmed significant changes at local to global scales over recent decades.

At the Alps scale, observations showed an increase in average temperature of 0.9 ° C over the period 1901-2000, and of 0.6 ° C during the last 30 years [1]. In addition, daily maximum temperatures have risen faster: between 0.9 ° C and 1.1 ° C over the same period. The majority of observations converge to an increase in the number of hot summer days and a decrease in the number of frost days. Thus, the years 1994, 2000, 2002 and 2003 were the warmest years since the 16th century [2] [3] [4]. In terms of projections, an increase of 5.5 to 6 ° C in summer temperature and a decrease in summer rainfall of 15 to 20% are expected in the next 100 years (this last prediction is very variable according to altitude)[5]. Some scenarios announce more extreme precipitation events [6] [7].

The impacts of these changes on vegetations' dynamics are extensively studied by the scientific community. These impacts also concern policy makers and forest managers who are trying to find ways to prepare forests to face with the predicted climate changes.

The first consequences of global change on forest habitats have already been demonstrated and reveal two contradictory aspects:

• a positive evolution of species' productivity during the last century;

a probable movement of the spatial distribution of forest species' habitats northward (in our hemisphere) and/or in altitude. For mountainous French forest, the upslope migration of forest species has been estimated to 65 m during the last century. If the projected warming up will be effective, then the corresponding latitudinal shift of forest species' habitats will vary from 250 to 500 km from their current positions.More locally, foresters observe diebacks more or less localized, without knowing their root causes and if they should attribute them to an epiphenomenon related to extreme weather events or to the initiation of a sustainable evolution of climate conditions.

In recent years climate-induced forest stress and dieback have been apparently increasing in Europe [8][9][10]. The two key drivers of climate-induced forest dieback are aridity increasing (e.g., drought intensity and duration) and temperatures warming, resulting in physiological stress that can exceed mortality thresholds for particular tree species. The combined effect of lack of precipitation over a certain period with other climatic anomalies, such as high temperature, high wind and low relative humidity over a particular area may result in reduced green vegetation cover [11]. When drought conditions end, the following vegetation recovery process may last for longer periods of time. The recent availability of reliable satellite imagery covering wide regions over long periods of time has progressively strengthened the role of remote sensing in environmental studies [12] [13]. The use of vegetation index obtained from satellite imagery allows overcoming the limit of discrete point data provided by conventional drought monitoring tools. In recent years there has been a lot of studies dealing with drought events over different regions of Europe or even covering the entire European continent [14] [15] [16] [17] [18] [19]. Summer droughts are the most important in terms of human perception, but water shortages during the remaining seasons may also have significant impacts on vegetation. The spatial distribution and severity of dieback and mortality is not routinely captured and areas affected by dieback often go unreported. Efficient and accurate mapping of disturbances over large geographical areas are possible with satellite remote sensing. This method offers the possibility of time- series analysis given the large quantity of archived data spanning many years.

Recently, forest dieback have been observed in the Southern Alps [20] [21] affecting protection function against rockfalls hazards. High mortality of Scots Pine (*Pinus silvestris* L.) have been observed and studied in large areas of Europe since the early 1980's. The Scots Pine decline is a complex phenomenon that may develop in response a variety of triggering agents (e.g. high temperature and drought) and ecological characteristics related to the stand location. The process can affect single plants or the entire stand and the effects, visible on the crown, may appear both as a discoloration process and as a progressive thinning and loss of needles [22]. In Aosta Valley acute Scots Pine decline events occurred starting from 2005 in large sub-alpine areas located within 650 and 1000 m a.s.l and specifically in North exposed slopes. Several studies on this subject have been conducted in previous years [23] [24], but the causes for these diffusive deaths are not well clarified. It has been argued that the decline process is triggered by several concurrent factors (e.g. drought, land use changes, parasites). Since these forests have a part in protective functions against natural hazards, monitoring the status and the evolution of the decline process is of primary importance to lead decisions on current and future management strategies.

According to these elements, foresters and land use managers can legitimately wonder about the consequences of climate change on forest stands in short, medium and long term. The future of forests in the Alps is a major issue at the the European scale, which requires a high anticipation from managers, policy makers and scientists for understanding phenomena and their evolutions, defining management strategies for risks prevention and mitigation.

In order to help forest managers to cope with climate changes' consequences in protection forests, one of the objectives of the project MANFRED has been to test models, tools and technologies which can be used for diagnoses and for building up forest management guidelines. The contribution of remote sensing techniques has been tested in the Aosta Valley. The contributions of specific training plots (marteloscopes), rockfall protection eco-engineering works and a model for mapping probable rockfall risk below a forest screen have been tested, in the two French case studies (in the Drôme administrative area) for the achievement of a black pine forest new magagement plan. This chapter presents the mains results obtained.

2. The Aosta Valley case study

The Aosta Valley case study had two main objectives:

• to test remote sensing techniques for mapping the degree and the evolution of Scots Pine decline

• to localize and quantify dieback sectors impacts comparing ground measurements and satellite remote sensed data.

2.1. Data

The response of vegetation was assessed with:

• Normalized Difference Vegetation Index (NDVI) as derived from two types of sensors: a medium resolution (MODIS) and a high resolution (Landsat)

• Very high resolution images analysis and field data.

2.1.1. Modis

MODIS is an optical multi-spectral instrument onboard TERRA satellite that performs an almost complete cover of the Earth surface in 8 spectral bands, on a bi-daily basis.

2.1.2. NDVI

Satellite vegetation index (VI) products are commonly used in a wide variety of terrestrial science applications in order to monitor and characterize the Earth's vegetation cover from space. VIs are optical measures of vegetation canopy "greenness", a composite property of leaf chlorophyll, leaf area, canopy cover, and canopy architecture. Although VIs are not intrinsic to physical quantities, they are widely used as proxies in the assessment of many biophysical and biochemical variables. VI time series data records have played an important role in measuring and characterizing land surface responses to climate variability and change. In our study, NDVI data were extracted from the so-called 16 days Vegetation Index Products (MOD13Q1) with 250m resolution of the MOLT database [25].

The technique of compositing data for 16 days considerably reduces noise in the surface reflection signal including QA data sets with statistical data indicating the quality of the VI product. It makes use of a filter for data dependent on cloud and viewing geometry [26]. This VI compositing algorithm includes the maximum value composite (MVC) and a constraint on view angle – maximum value composite (CV-MVC). The objective of this compositing methodology is to determine a single value per pixel from all the data retained by the filter. This value is assumed to be representative for each pixel over the 16-day period of interest. As a result, the quality of MODIS NDVI data is significantly enhanced [27]. The NDVI, a normalized ratio of the near infrared (NIR) and red spectral light reflected (red) by the land surface back up into space is linked to the presence, density and condition of vegetation:

$$\text{NDVI} = \left[\left(\text{NIR} - \text{Red} \right) / \left(\text{NIR+Red} \right) \right] \tag{1}$$

The raw NDVI values are fractional real numbers that range between -1 to +1 units, but when used to define the land cover classes, the values are normally constrained from + 0.1 units in rocks and up to + 0.8 units in dense forest.

First we analyze the intra and inter annual variations of NDVI for protection forests over the period 2000-2011. The temporal evolutions of the VI is bound to the vegetation development phases, the established profiles allow to characterize the specific phenological cycle of the considered stands and to localize keys phenological stages in particular the spring development. Temporal evolution can be defined from « phenologic metrics ». Over a year (Fig. 1), the NDVI time-series shows a typical pattern with three phases: an increase corresponding with the start of vegetation growth (peak),a plateau period of high photosynthetic activity, a sharp decrease indicating vegetation senescence.

The temporal evolution of monthly values of NDVI, spatially averaged over protection forests in the Aoste Valley between 2000 and 2010, is presented in Fig. 2. The general linear trend of the vegetation was plotted to identify trends towards degradation or vegetation recovery.

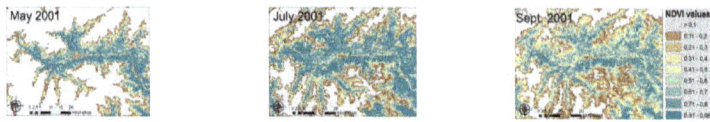

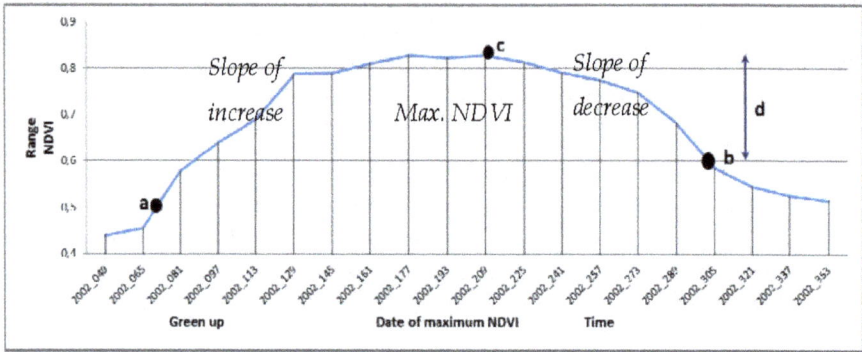

Figure 1. Annual cycle of NDVI in Aoste valley protection forest (year 2002). Points (a) and (c) mark start and end of season, point (c) displays the largest value and (d) displays the seasonal amplitude.

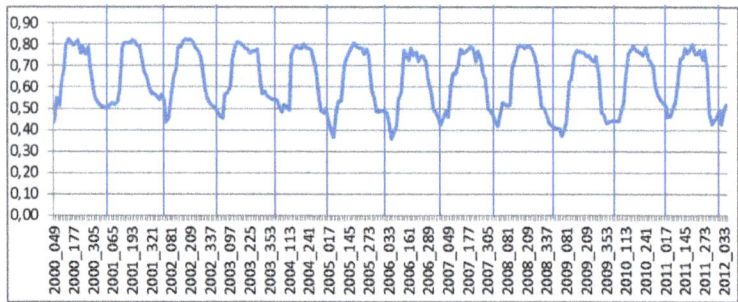

Figure 2. Time-series NDVI data for representative pixels of Aosta valley protection forests observed on a biweekly basis from 2000 to 2012.

2.1.3. VCI

16 days-NDVI composites (250 m spatial resolution) for 2000–2011 were used to calculate the Vegetation Condition Index (VCI) developed to control for local differences in ecosystem productivity [28]. The VCI is a pixel-wise normalization of NDVI accumulated over a long period that is useful for making relative assessments of changes in the NDVI signal by filtering

out the contribution of local geographic resources to the spatial variability of NDVI. The VCI is computed as:

$$VCI = 100(NDVI - NDVImin) / (NDVImax - NDVImin) \qquad (2)$$

where NDVI, NDVImax and NDVImin are the smoothed bi-weekly NDVI, multi-year maximum NDVI and multi-year minimum NDVI, respectively, for each grid cell. VCI changes from 0 to 100, corresponding to changes in vegetation condition from extremely unfavorable to optimal. Individual years can then be compared and assessed against the 'normal' conditions. The VCI smoothes out non-uniformity in the MODIS data and it indicates how weather conditions have influenced the relative vigor of the vegetation with respect to the ecologically defined limits.

2.1.4. SG Index

We used an annual index of vegetation vigour able to characterize quantitatively the vegetation activity during the period of spring increase: SG phenologic metric. This index corresponding to the accumulation Spring NDVI that can be likened to the net primary production. SG is obtained by making the sum of NDVI during the green up period.

$$SG \ = \ \Sigma \ NDVI \ from \ 15^{th} \ april \ to \ 15^{th} \ june \qquad (3)$$

2.1.5. Landsat

Many factors affect NDVI variations within a pixel: plant architectural arrangement, interactions with canopy cover, height, composition of species, vegetation vigour, leaf properties and vegetation stress are some factors that can significantly affect the remotely sensed information. In order to improve the accuracy of the spatial analysis, Landsat data were selected. Spectral and temporal resolution was found to be adequate for vegetation stress detection and a single scene covered a large enough area [29]. 30 Landsat 5 TM and Landsat 7 ETM+ images (Path 195 and Row 28) were provided courtesy of the USGS EROS Data Center. The full scenes (standard L1G product) are available at http://glovis.usgs.gov.

Atmospheric correction of satellite measurements is critically important in remote sensing especially when using multiple images. Most of the radiation detected by a satellite sensor is backscattered from the atmosphere. Therefore, removing atmospheric effects is important [30]. The images were geometrically and radiometrically corrected. First, all scenes have been calibrated into radiance in units of [μW/(cm2*sr*nm)] using sensor calibration coefficients [31]. Later, an atmospheric correction of the images was carried out to minimize the effects of the different atmospheric conditions on the images that should compensate for the "skylight" (atmospheric aerosol scattering) to produce spectra that more truly depict surface reflectance.

The FLAASH (Fast Line-of-sight Atmospheric Analysis of Spectral Hypercubes) software package in ENVI (by Research Systems) provides an accurate means of compensating for atmospheric effects. It incorporates MODTRAN 4 radiative transfer code with all MODTRAN atmosphere and aerosol types to calculate a unique solution for each image [32]. FLAASH includes also a correction for the adjacency effect, provides an option to compute a scene-average visibility (aerosol/haze amount). The method is based on observations [33] of a nearly fixed ratio between the reflectance for such pixels at 660 nm and 2100 nm (FLAASH User's Guide). It includes also the average elevation of the study area, scene center coordinates, sensor type, flight date and time, and information about aerosol distribution, visibility, and water vapor conditions [34]. In this study, model parameters describing a mid-latitude summer atmosphere and rural aerosols together with automatic aerosol retrieval were used in FLAASH to correct the Landsat images.

2.2. Aerial images and field validation

Airborne and satellite remote sensing is a valuable tool for a synoptic view of the terrestrial ecosystems in order to monitor the environmental processes. In particular, in the case of forest decline, the spatial distribution of environmental parameters (such as canopy greenness or species distribution) and their associated temporal dynamics are detectable by remote sensing, allowing understanding, monitoring and predicting the evolution of the process [35] [36].

In detail, the analysis and the numerical classification of digital aerial photographs at high spatial resolution allows understanding the intensity of the decline process. Moreover the availability of several acquisitions is useful to monitor the decline process evolution over time. Aerial images acquired in 2011 (20 cm spatial resolution) were used in order to semi-automatically map the presence and distribution of Scots Pine and to identify the state of canopy decline. The map obtained in 2011 was then compared with digital aerial photographs acquired in 2006 (50 cm spatial resolution) in order to evaluate the evolution of the decline process, in terms of potential reversibility. Finally, the analysis computed on digital aerial photographs have been validated using direct field observations carried out in summer 2011 in six sample areas of the central valley.

The method for the processing of aerial images involved the computation of spectral indices from the digital numbers (DN) values of the images in the red (R), green (G) and blue (B) channels. These indices together with the values of the original DN were used to make a classification of the different images. In detail, 12 spectral classes were identified which were then assigned to 5 information classes : 1) shadows, 2) deciduous tree species, 3) not damaged Scots Pine, 4) damaged Scots Pine, 5) other surfaces (e.g. roads). The two classes of interest (3 and 4) were used to analyze differences between the 2006 and 2011 images.

2.3. Results

2.3.1. Annual cycle of NDVI using Modis data

In order to smooth Modis time-series, we experimented different processing methods in Timesat program [37] (Fig.3). The performance of different processing methods have been evaluated in a recent study by Hird and McDermid [38] „concluding that they are highly competitive and that they preserve the signal integrity". Three different processing methods are experimented: the local Savitzky-Golay filter [39], the asymmetric Gaussian and double logistic model functions that are well suited for describing the shape of the time-series in overlapping intervals around maxima and minima. In our case, fits to the asymmetric Gaussians appeared to be the better choice.

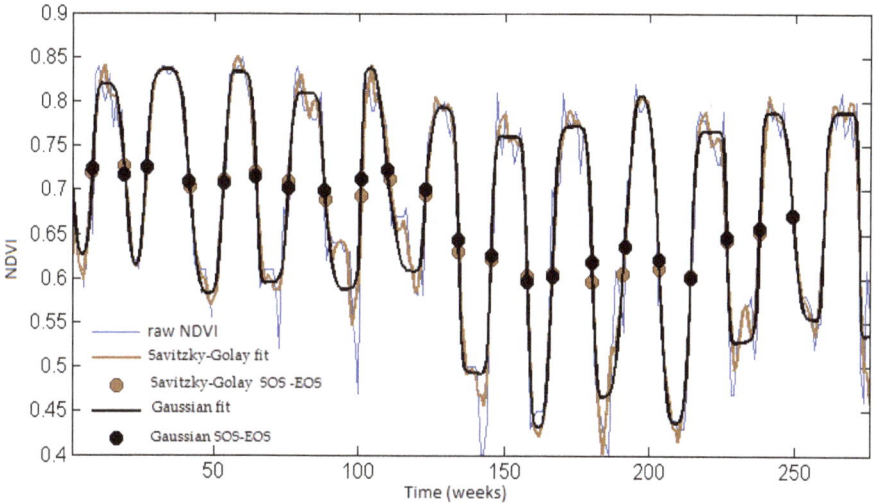

Figure 3. Smoothed NDVI time-series for decade 2000-2010 in Aoste valley protection forests. The blue thin line represent the original NDVI data, the brown thin line shows the Savitsky-Golay fitted function, the black solid line displays the Gaussian fit.

The curve exhibits a consistency both in amplitude and in length of growing season and low peaks values appears in 2006 and 2009.

Some seasonality parameters computed using TIMESAT are shown in Fig.4 and Fig.5:

a. Start of season defined from the fitted function currently set to 10% of the distance between the left base level and the maximum ;

b. End of season show a decay of 15 days for SOS and 10 days for EOS with a trend to occur later over the decade 2000-2010.

c. point with the largest NDVI value during the season

d. Seasonal amplitude during the decade

SOS shows a heterogeneous pattern with high variability from day 105 in 2000 to day 120 in 2010. EOS data indicates a later onset of autumnal phenological events, but these shifts are less pronounced.

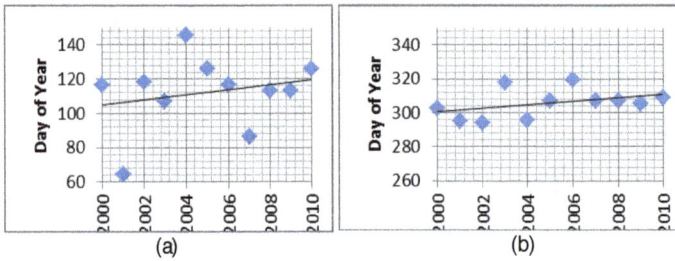

Figure 4. SOS (a) and EOS (b) estimated from fitted functions (Timesat) from 2000 to 2010 in Aosta Valley protection forests.

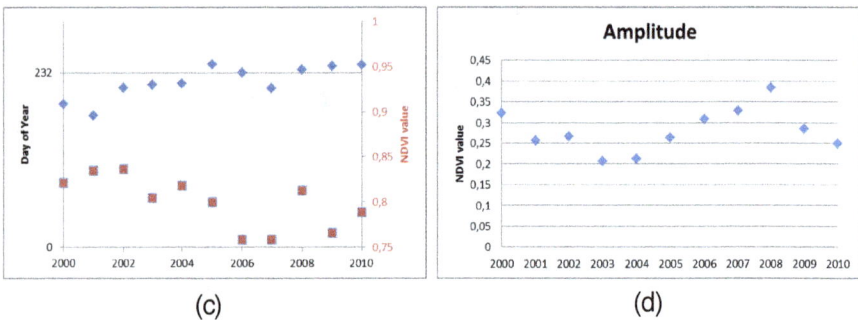

Figure 5. NDVI peaks values (c) and seasonal amplitude (d) estimated from fitted function (Timesat) over the decade 2000-2010 in Aosta valley protection forests.

NDVI peak points over the season with the largest value calculated along the decade begin at day 190 in 2000 (mid June) and day 243 in 2010 (mid august). The mean at Day 232 is situated in 2006 (end of July). We observe a progressive delay of NDVI peak days during the decade, with very low variation of NDVI peak value (around 0, 8).

The seasonal amplitude, reflecting the "biomass production "vary from 0, 21 in 2003 and 2004, to 0,39 in 2008.

2.3.2. Vegetation condition index

The VCI index indicates percent change of the difference between the current NDVI index and historical NDVI time series minimum with respect to the NDVI dynamic range (Fig.6). VCI was calculated by seasons, aggregating values from april to june (spring) and from july to September (summer).

The trend for spring values shows a continuous decrease from 2002 but staying over 70%. Summer vegetation conditions show a decrease over the period 2000-2011 with two marked inflexions (under 60%) for the years 2006 and 2009.

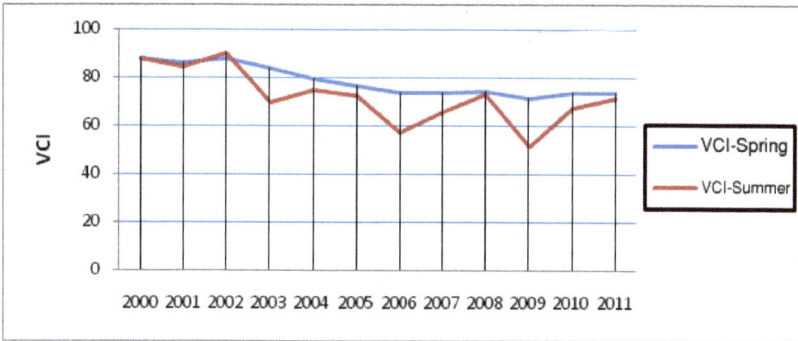

Figure 6. Vegetation Condition Index in Aoste protection forests (2000-2012).

2.3.3. SG index

SG Index values allow characterizing each year and indicating annual fluctuations.

We can highlight and quantify abnormal variations of the vegetation activity over the period of observation. The series of annual SG calculated from 2000 till 2012 can be characterized by a trend which reports the variation of the vegetation activity between two extremes: Downward trend if physiological disturbance or upward if the vegetation increases. A strong fall of vegetation activity, over the period of observation, is to be put in connection with important disturbances which would have occurred within forest stands (Decay or sanitary cutting).

Fig. 7 shows a valuable example of annual SG for all protection forests between 2000 and 2006. The line of trend of the series is characterized by a value of slope (α) which reports the importance of the fall of activity over seven years (2000-2006). The principle to use the slope on series of temporal images was already implemented in the case of analysis of trend to measure variations of behavior of the vegetation [32]. The slope (α) of the trend line, calculated for every pixel, establishes our indicator of "change" or disturbance observed over 2000-2006. This one can be mapped for all the protection forest areas.

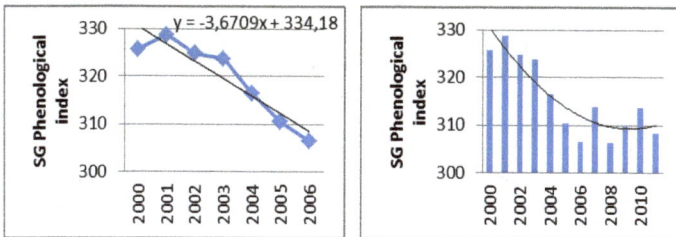

Figure 7. SG index (sum of spring NDVI values) indicating the vigor of vegetation. SG values for all forest protection pixel; slope of trend line (α) =-3.67

These monitoring tools are able to specifically detect periods of varying drought conditions. Findings indicate a clear trend of decrease starting from summer 2002 and reaching the lowest values at summer 2006.

Further results at finer resolution using Landsat images should confirm and refine these results.

2.3.4. Monitoring impacts at high resolution using Landsat sensors

At the high resolution of Landsat sensor (30m), the objective was to determine the patterns of vegetation variability and change between 1999 and 2011. We took into account seasonal variations in the distributions of different vegetation types. Forests show contrasting seasonal variations in vegetation activity with a peak activity occurring in summer [39].

Landsat 7 ETM+ SLC-off data refers to all Landsat 7 images collected after May 31, 2003, when the Scan Line Corrector (SLC) failed. A hardware component failure left wedge-shaped spaces of missing data on either side of images. The sensor still acquires approximately 75 percent of the data for any given scene (240*240km). The gaps in data form alternating wedges that increase in width from the center to the edge of a scene. In our case study, over the 30 Landsat TM5 and ETM+, we calculate a loss of information between 4 and 6% due to SLC off on protection forest areas (Fig. 8).

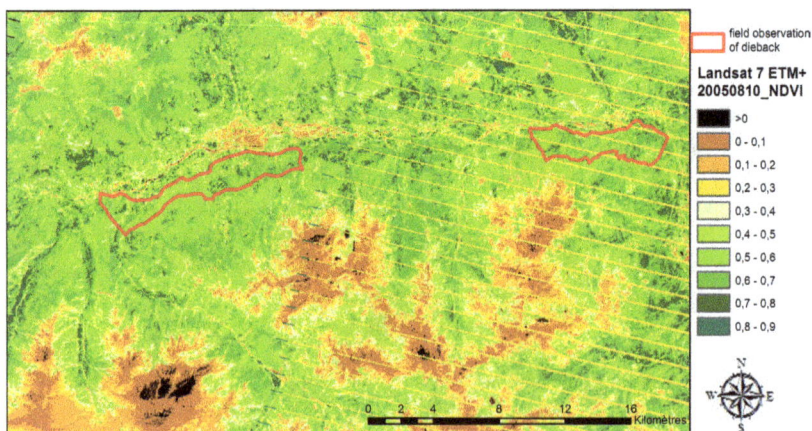

Figure 8. NDVI in Aoste valley 2005/08/10 Landsat 7 ETM+ (SLC off) and localisation of dieback fields observations.

The temporal evolution of monthly values of NDVI (end of july- first days of august), spatially averaged over Aoste region south of river between 1999 and 2011, is presented in Fig. 9.

Time series of monthly NDVI are also shown for three different types of forests (Fig.10 and 11). Information about the forest type associated to each pixel was obtained overlaying the images with the official forestry database: the Regione Autonoma Valle d'Aosta made available

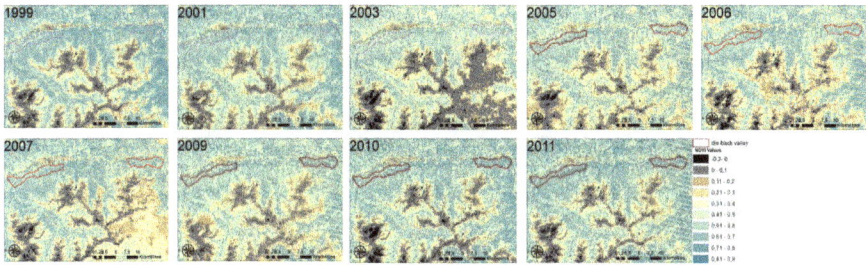

Figure 9. NDVI evolution in Aoste valley 1999-2011 Landsat sensors.

a map of protection forests from the „dipartimento Agro.Selvi.Ter. dell'Università degli Studi di Torino". This map covers the entire regional territory (arround 3.000 km²).

The comparison between the NDVI values obtained from MODIS and Landsat images has shown a good consistency of the temporal dynamics but a systematic error that can be read as bias (MODIS NDVI over estimation). We took NDVI values from end of july-beginning of august images as it was determined as the peak date during the decade using Modis data.

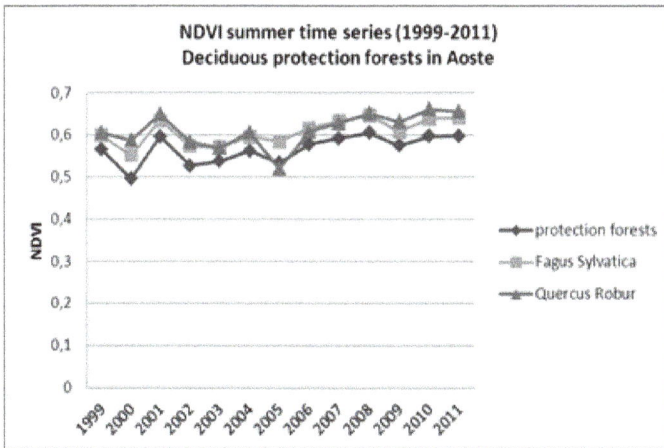

Figure 10. NDVI summer time series (1999-2011) - Deciduous protection forest

Monitoring Time series of summer NDVI values in the area of protection forest in Aosta valley allowed detection of good (2001, 2008) and bad years (2000, 2002, 2005, 2009) relatively to a general increasing trend. Spatial characterization of forest patch (900m2) showed a decrease of NDVI during bad years where trees are affected by drought. At regional scale we achieve a precise localization of die-back and ability to qualify stand types and to separate behavior of coniferous and deciduous forests.

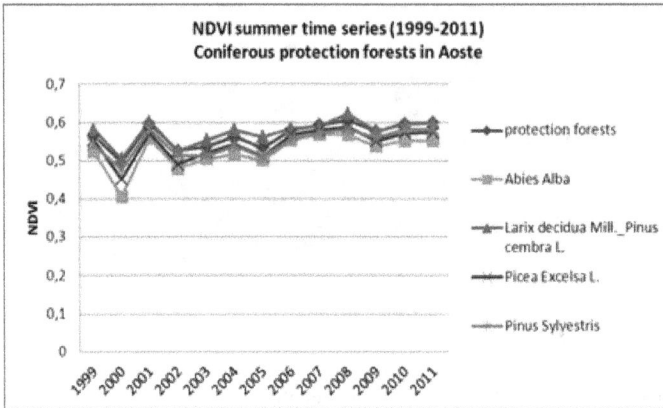

Figure 11. NDVI summer time series (1999-2011) - Coniferous protection forest

2.4. High resolution analysis on ortho-images and field validation

Results of this analysis showed good agreement between the classified images and field data collected (coefficient of determination R2 = 0.74) in the sample areas (fig. 12).

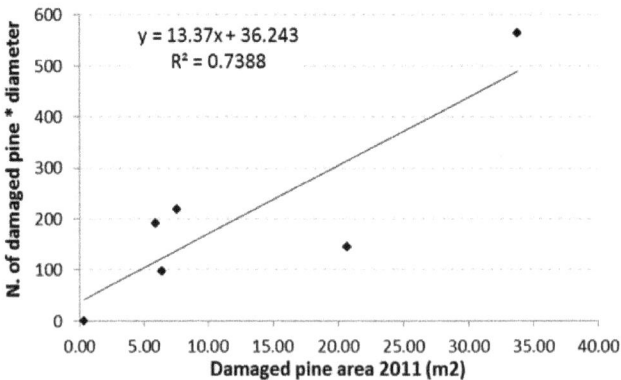

Figure 12. Relationship between the area affected by Scots Pine decline, as classified by the digital aerial photograph processing, and the number of damaged Pine trees observed in the field. The average diameter of trees measured in the sample area has been multiplied by the number of Pine individuals in order to better relate the field observations to the mapped area.

Analysis on images collected in 2011 allowed to distinguish healthy Pine from damaged Pine. The field observations evidenced that damaged pine mapped from 2011 aerial photographs are trees characterized by severe crown defoliation (i.e. needle loss > 60%) or dead trees. On the contrary crown discoloration was rare (discoloration % < 10) in all the monitored areas (percentage of canopy discoloration < 10 %). Images collected in 2006 allowed to distinguish only the damaged Pine, characterized by a brownish colour typical of severe discoloration, while the healthy Pine was not well distinguished from other tree species, probably due to the worse spatial and radiometric image resolution.

Finally, results showed that the area interested by Scots Pine decline in 2006 was higher compared to 2011 as it can be appreciated by Fig. 13 and 14.

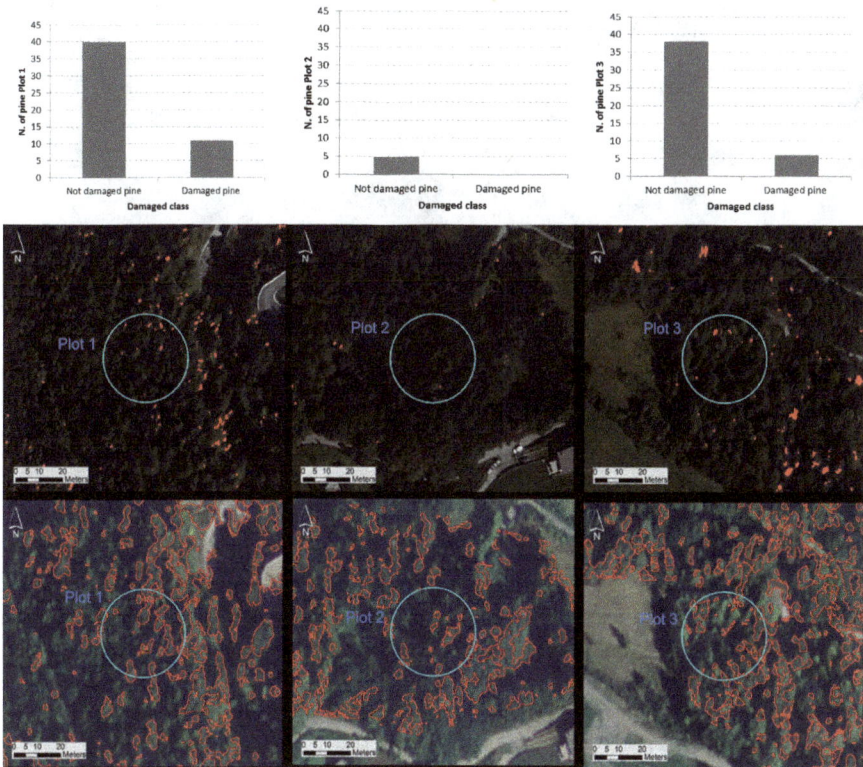

Figure 13. Plots represent the number of pine sampled on field belonging to "not damage pine" (needle loss < 60 % and discoloration < 10 %) and "damaged pine" (needle loss > 60 % or dead trees and discoloration < 10%) classes. The red area depicted on the digital photographs represent damaged pine mapped in 2011 (images above) and 2006 (images below) in plot 1, 2 and 3.

This observation lead to the hypothesis that the stands which experienced the decline observed since 2005 (i.e. severe discoloration) are likely going towards a recovering process. This

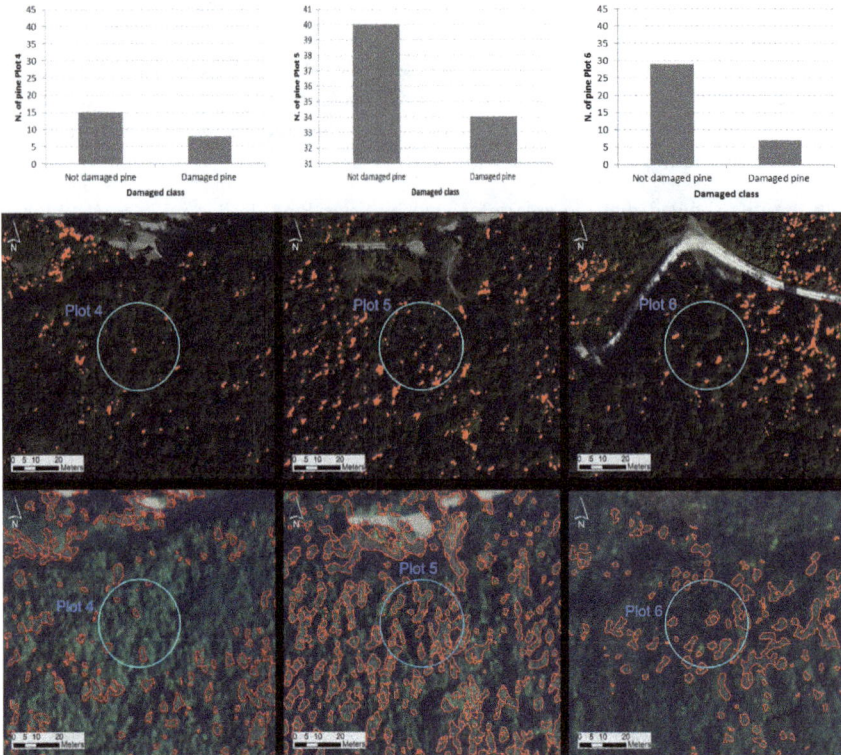

Figure 14. Plots represent the number of pine sampled on field belonging to "healthy pine" (needle loss < 60 %) and "damaged pine" (needle loss > 60 % and dead trees) classes. The red area depicted on the digital photographs represent damaged pine mapped in 2011 (images above) and 2006 (images below) in plot 4, 5 and 6.

hypothesis is supported by field observations conducted in summer 2011: the average discoloration was low (< 10 %) in all the study plots and the average needle loss was moderate (> 25-60 %) or low (< 10 %) meaning that all the plots investigated belongs to the category of forest damage absent or very low [41] [42].

2.5. Spatial analysis of protection loss against rockfall

First we used an energy line model in order to map rockfall hazards in Aosta Valley based on topographic criteria: RockForLIN has been developed by the French research centre IRSTEA and used to develop a method for protection forest mapping using Geographic Information Systems (GIS) [43].

First, all potential release points have to be mapped. 2D GIS models have been developed to localize them depending on topographic conditions: a simple slope threshold is applied to the

slope surface raster (computed from the raster Digital Elevation Model [DEM]), according to
the equation:

$$\alpha = 55 \times RES^{-0.075}$$ (4)

Where RES is the DEM resolution. All cells with values higher than the threshold α are
qualified as potential release zones for rock falls.

Then, from each of the identified potential release points, 2D GIS models simulate the probable
run out envelops. RockForLIN is based on the Energy Line principle, which allows relating
rockfalls run out envelops to slope angles. The maximal spread of a block is determined by
intersecting the ground and an imaginary line drawn from its release point with angle β.
Different values for β are used: 32, 35 and 38°. Areas between 32 and 35° have a low but not
null probability to be reached by rockfalls ; between 35 and 38°, a intermediate probability ;
and higher than 38°, a high probability.

Second we overlayed NDVI data at the two resolutions with Modis (250m) and Landsat (30m)
images, representing forest pixels as points (Fig.15 and 16).

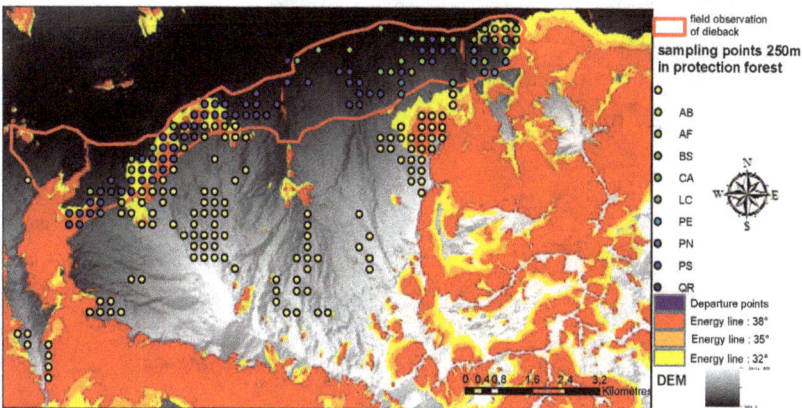

Figure 15. Rockfall mapping in Aoste valley using energy line model created with 5m resolution DEM. Sampling
points representing Modis pixels (250m) in the west sector.

The GIS environment alowed us to map the "loss" of protection forest capacity against rockfall,
taking into account the values of NDVI decrease at two periods 2001-2005 and 2001-2009 (fig.
17 and Fig.18).

The areas affected by die-back are overlaid with the field plots inside the red polygon defined
on the ortho-images. Many affected zones overlay the rockfall hazard mapping calculated with
the energy line model.

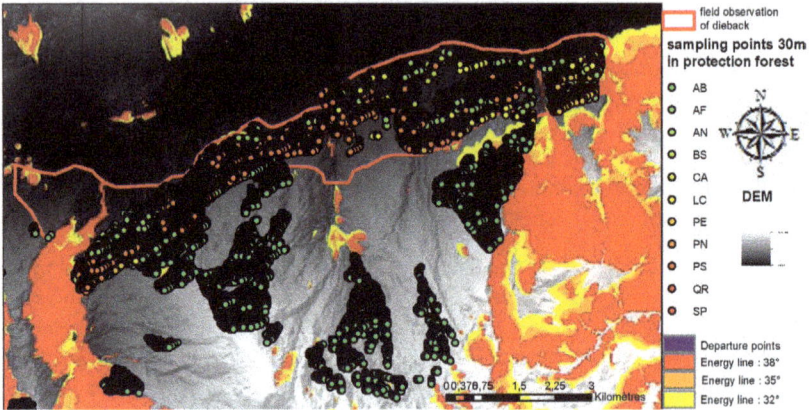

Figure 16. Rockfall mapping in Aoste valley using energy line model created with 5m resolution DEM. West sector. Sampling points from Landsat 30m (1999-2011).

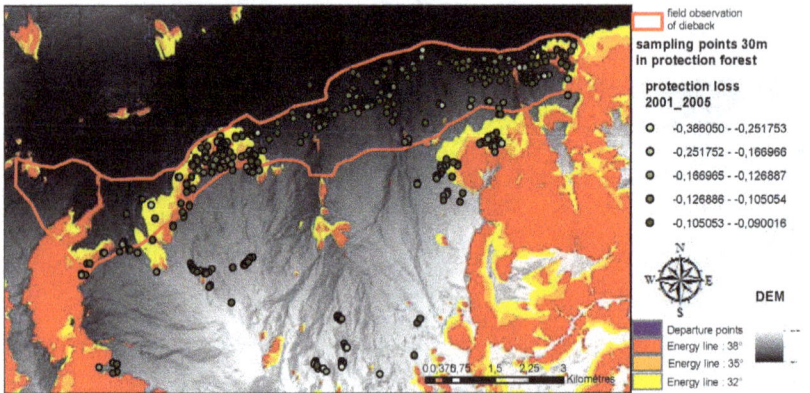

Figure 17. Localization of NDVI loss in protection forest between 2001 and 2005. West sector

2.6. Conclusions

This study takes place in the long series of environmental and risk assessment with a particular focus on spatial methods multi-sensors multiresolution. Satellite-based assessment of die-back have been successfully used and validated. An integrated remote sensing methods has been tested for detection and localization of die-backs in protection forests. It was coupled with a hazard characterization in order to monitor cross-impacts. Method appear to be consistent for detection of phenologic parameters giving indications on forest health at Modis resolution (250m). The use of satellites for vegetation monitoring provides several key advantages over other methods. It can provide near real-time data over large areas at a relatively high spatial

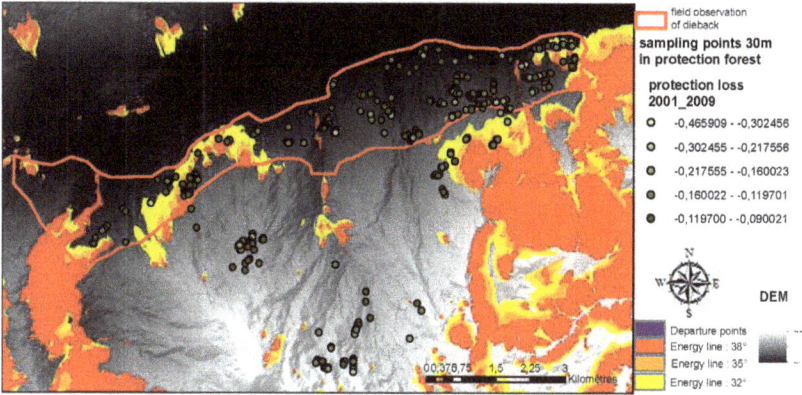

Figure 18. Localization of NDVI loss in protection forest between 2001 and 2009. West sector

resolution. It is often difficult to calculate station-based drought indices in a timely fashion because the required data is usually not available in real-time.This means that satellite-based drought/vegetation indices may able to detect consequences of droughts earlier and more accurately than other methods. The processes identified and localized in this study(die-off and rockfall) would be a first step of diagnosis that allows to develop management strategies able to adapt protection forest to changing environmental conditions. The energy line model is an empirical model which principal bias is the detection of the block departure zone directly linked to the DEM resolution. More calibration work would be necessary which implies more in situ experimentations and field validation of the areas affected by rockfall.

3. The Val Drôme case study and its new rockfall black pine protection forest management plan

One of the MANFRED French case study sites is located in the administrative department of la Drôme. More precisely, it is located in the district of le Diois (administrative district around the town of Die). Within this case study a sylvicultural training plot (also named martlesocope) has been implemented. The main thematic of this case study is an Austrian black pine rockfall protection forest management. The results of the 5 training sessions held on the marteloscope, have been used in connexion with rockfalls propagation modelling results for designing the new forest management plan of this district.

3.1. The historic context of natural risks protection forests in the Diois

Historically, the protection function of the forest was introduced by the "RTM" (Restoration of Mountain Greunds) policy to fight against natural hazards in mountain areas and the devastating floods of the XIXth century. This policy was applied in mountain forests especially in Diois, a land next to the town of Die, in the French administrative area of Drôme.

During the Little Ice Age (1450-1850), the Diois district was concerned by an important torrential activity while the watersheds were weakened due to an agropastoral overexploitation and a low afforestation (in the 1850s, forests covered only 30 % of the area in Diois, now 70 %). The torrential activity caused changes in Alpine rivers: beds were very high and large so the risk of flooding and the erosion significantly increased. Due to the lack of forest cover, soil materials were washed away by rain water of heavy thunderstorms, which are frequent events in Diois. Soils were strongly degraded.

Figure 19. An example of the appearance of soil cobber in the early twentieth century in Menglon in the Diois

In this difficult context for local people, the repetition of large floods in the 1840s and 1850s had a decisive impact on public opinion. Decision-makers established different policies: in 1860, a first law has been stated for reforestation of degraded land, in 1882 a law on restoration and conservation of mountain grounds has been voted. All degraded grounds have been identified and included in perimeters of "restoration". These restoration perimeters generally include a large number of villages in the watershed of the torrential river but they only content degraded grounds. The watersheds of the Drôme have been divided into six areas: Basse-Drôme, Haute-Drôme, Roanne, Bez, Eygues, Oule. Then, the state purchased the land perimeter and the Waters and Forests Administration undertook significant work.

The first "RTM" missions done by the Administration of Waters and Forests have been:

• From 1863 to 1867: torrential correction and weeding are common. They increase the soil stability and the proportion of successful planting and seedling.

• From 1887 to 1914: reforestation begins, principally by planting because seeding is too uncertain in these difficult environmental conditions. The main species used are the Austrian Pine and Scots.

Due to these actions, 18 500 ha of Austrian black Pine have been planted in the French administrative area of Drôme between 1860 and 1930.

Figure 20. Water and Forests Engineers in Die, early twentieth century

Figure 21. The effectiveness of the RTM laws, the reforestation of the Marignac and le But St Genis districts between 1900 and 2012.

Nowadays the French Forest Commission (ONF in French) continues the actions by promoting natural regeneration of Austrian Black Pines and by promoting the return of native species such as Beech or Fir, with the aim to protect fragile mountain lands on 40,000 ha of RTM forests in the Drôme. About the rivers, in addition to overseeing the maintenance and up keeping of the torrential works, the ONF provides new missions, for instance, a new management of sediments in the upper watershed as the mobilization of sediment stocks in stable sectors.

3.2. "The rockfall marteloscope" in the Haut Diois

As part of the RTM actions, forest managers must adapt their silviculture in order to sustainably and effectively reduce the consequences of natural risks in mountains by optimizing the protective actions of mountains forests. In mountain area, the predominant natural risk is the one generated by rockfalls. All the aspects of human socio-economic activities are concerned: houses, industrial areas, roads, railways, energy transportation lanes… Forest stands can considerably reduce the impact of rockfalls by stabilizing mountainsides, anchoring the rocks, slowing boulders and so imitating boulders kinematics. A forest can play an important role in rockfalls mitigation and its efficiency largely depends on the silviculture used. Thus, to maintain the protective function of these forests, an adapted silviculture anticipating the consequences of climate changes must be applied to them.

In this context, the marteloscope of the Haut Diois (and also all the other ones implemented during the MANFRED project but not presented in this chapter) aims to:

• educate and train foresters to consider this problem of rockfalls into their forest management;

• test virtually different kinds of sylviculture in order to select the most adapted one to the local problems and conditions.

3.2.1. General data

The marteloscope is located in the French administrative area of Drôme (26), in the Haut Diois, France. Its surface area is 1 ha. It settles on the parcel 2 of the «forêt domaniale du Val de Drôme". This forest is classified as RTM forest, with a status of protection forest. The marteloscope is on the South-West facing slopes in the upper valley of the Drôme, at an average altitude of 700m. The average slope of the versant is about 40%.

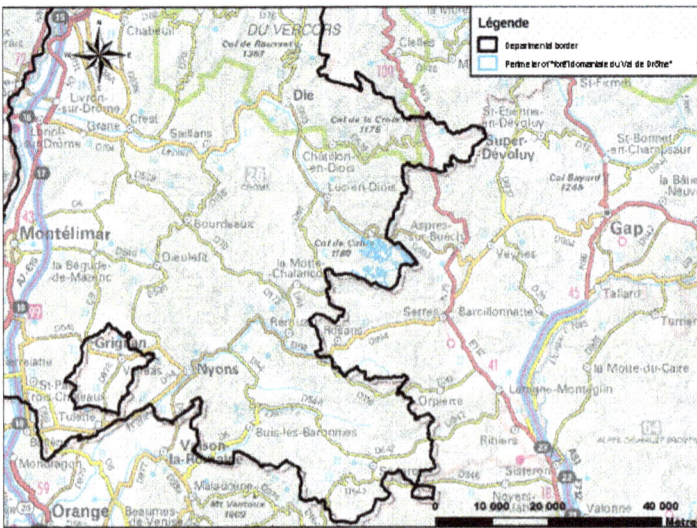

Figure 22. The state forest of Val Drôme and its localization in the "départment de la drôme"

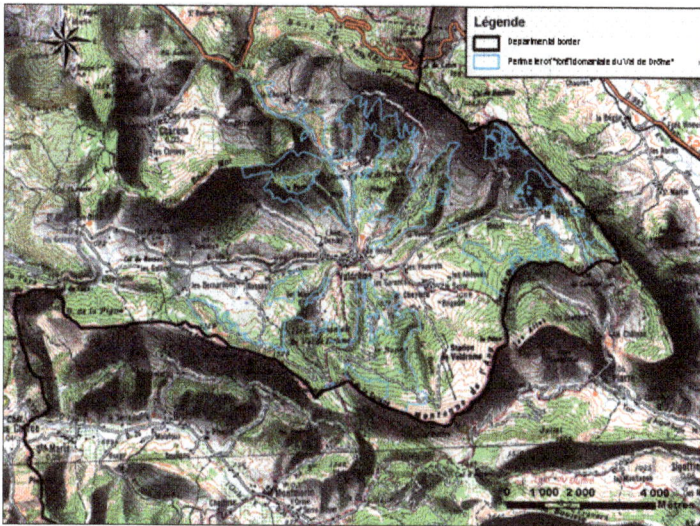

Figure 23. The state forest of Val Drôme and the localization of the marteloscope.

The climate is mountainous, with a Mediterranean influence. The annual average temperature is 10-11 ° C and the annual precipitation is of 1000 mm. The marteloscope is on a warm slope (South-West facing, annual average temperature is high) with enough water for the vegetation but with a significant risk of summer drought. Late frozen are common. Seasonal and annual variations are very important. But generally, summers are short, very hot and stormy, while the winters are long and cold.

The soil comes from massrocks. It is deep, not compact and composed of many limestone disorganized elements. Some places are less deep with a marly limestone cracked rock in surface, which gives rendosols or rendzines. The marteloscope is located on a soil of good quality for a forest soil with high slopes, well drained and offering a good prospection for the root systems.

3.2.2. Marteloscope's stand description and historical overview on the forest management

The stand of the marteloscope is composed at 100% of Austian Black Pines, planted in 1902 in the framework of the application of the RTM laws. This forest is managed into even-aged forest (" futaie régulière " in French). The distribution of tree by classes of diameters is represented by the following histogram. This graph shows that all classes of diametres are present; the average diameter is 22 cm. It is a dense forest stand, with 1026 stems/ha. The marteloscope is belonging to the parcel number 2 of the state forest (Forêt domaniale in French) of Val Drôme.

Figure 24. A general overview on the slope on which the Val Drôme marteloscope is located.

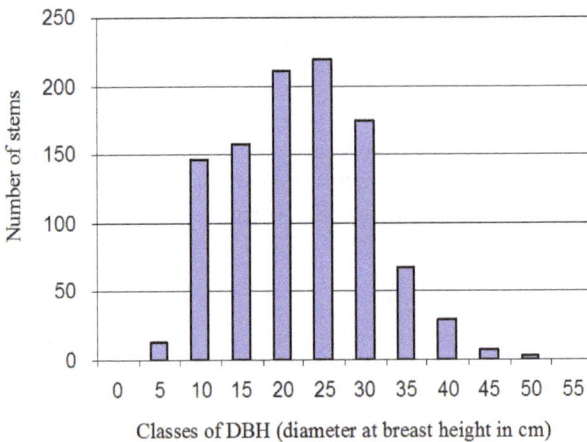

Figure 25. The DBH distribution of the martelsocope's Austrian black pine stand.

The parcel 2 of the forest « Val de Drôme » is bounded by 10 m high cliff upstream, and by a departmental road D106 (previously D306) downstream (cf. Figure 23). This road allows users from Valence, Die, or other towns to reach Valdrôme and its ski resort or to reach Baronies at the South and East. It is usually quite busy. Moreover, traffic is still increasing during winter and summer with the arrival of tourists (Valdrôme ski resort, hiking, biking...). However, this parcel is located on a steep slope (between 35-40%) and the cliff represents a potential starting zone of rockfalls. Indeed, the rocks became dislodged because of climatic constraints (alternating freezing and thawing, erosion...). Many impacts on the trees have been identified in the marteloscope (cf. Figure 26). The average size of the block is 60cm × 60cm × 60cm. Some trees have big impacts. Two ravines, located on either side of the marteloscope in the direction of the slope, are natural ways often taken by the falling blocks. In this context, rockfalls are a real risk for the user of the departmental road below this forest screen. Up to now no catastrophic events occurred but each year at least one rock is reaching this road.

Figure 26. Photo of the marteloscope's stand.

All the trees have been mapped in x, y, z using a theodolithe. The figures below present a 2D and a 3D maps of this plot.

According to the forest management plan for the period 1975-1994, the parcel 2 is a part of the series in «even-aged forest " whose vocation is Austrian Black Pine forest, with a production target, like most of the Val de Drôme forest (70% surface) at that time. Therefore, improvement cuts were made: a first thinning was conducted in 1986 with 3726 trees cut which represents a total volume of 706 m³ and a density of 190 stems/ha on this 19.6 ha parcel. Thinning should be "fairly light", even possibly restrict the natural regeneration. In addition, the final cutting was prohibited by this plan. Then, a second thinning was conducted in 2006 (rotation of 20 years) with 760 trees cut, for a total volume of 276 m³. This cutting is located on the downstream portion of the parcel to limit potential damage caused by rockfalls. Few trees have been cut in order to respect the protection forest function and their branches were cut and arranged along the path for retaining the falling rocks.

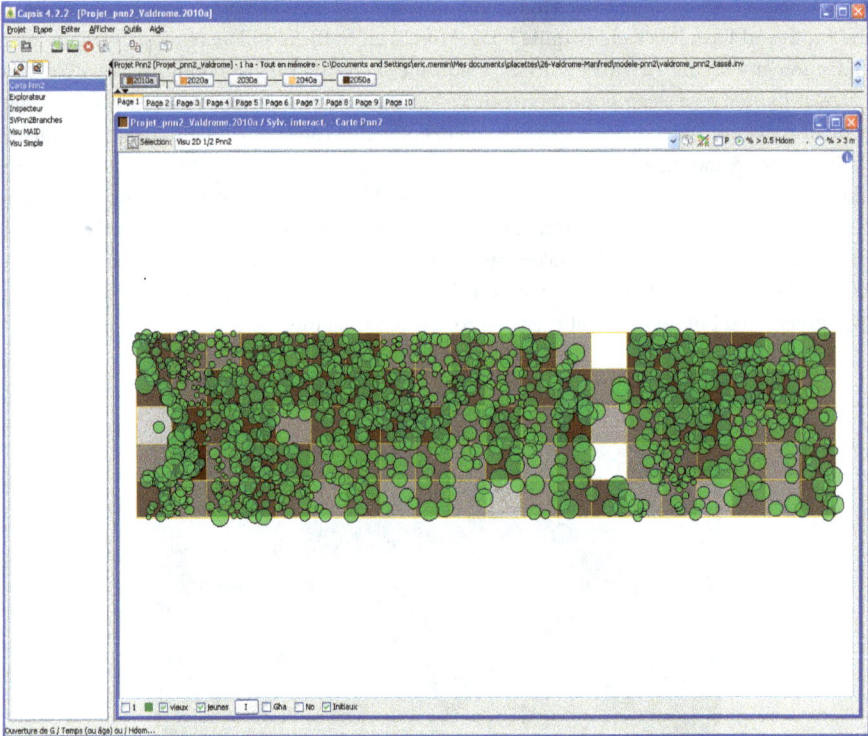

Figure 27. The trees' position map of the marteloscope of Val Drôme. A green circle represents the position of a tree and it is proportional to the tree's diameter.

The current forest management plan of this parcel, according to the high risk of rockfalls, avoids logging operation. But silvicultural interventions are now necessaries for improving and helping natural regeneration dynamics.

3.3. Rockfalls risks assessment using the modelling tools: Rockfor^{LIN},Rockyfor3D and Rockfor^{NET}

The rockfalls risk assessment in this case study has been done using tree scale modelling tools. The first one used has been the model Rockfor^{LIN}. The advantage of this empirical model is that only the Digital Terrain Elevation model is required to perform the analysis. This model has been built up using the energy line principle and the notion of mobility index [46]. This model allows having a first overview on the probable maximal envelop of rockfalls propagation areas with and without taking into account the forest stands. The map below presents the results obtained with this model without taking into account the presence of forest stands.

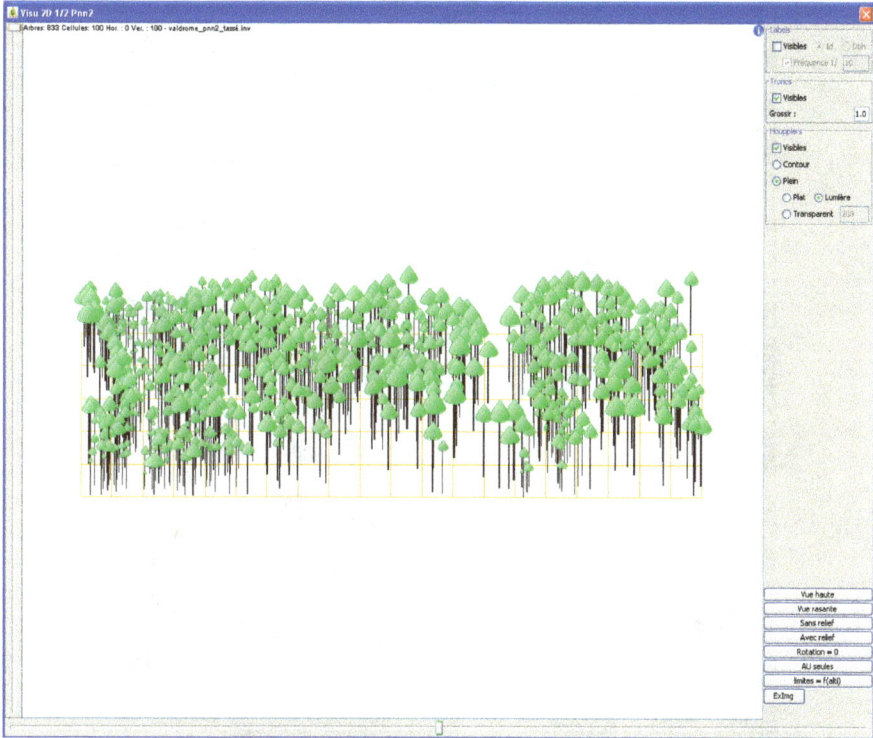

Figure 28. A 3D view of the marteloscope's stand of Val drôme.

According to the figure 27, the rockfall risk on the road below the martelsocopes will be very high if the forest stands will disappear one day.

To confirm this first analysis, the 3D trajectories simulation model Rockyfor3D has been used. To use this model, the following data are needed:

- the Digital Terrain Model with the highest accuracy and resolution available. For this case study the LiDAR data (with 5 points/m^2) has been acquired

- the map of the geomorphological units : it classifies areas according to specific soils' characteristics (topographical and geological parameters, soil roughness...)

- the map of the current deposit rocks: all rocks landed in the marteloscope are spatially represented. From this map an average volume of the largest block is defined

- impacts data : the impacts on trees due to rockfall are analyzed (location, height, dating by analysis of tree rings)

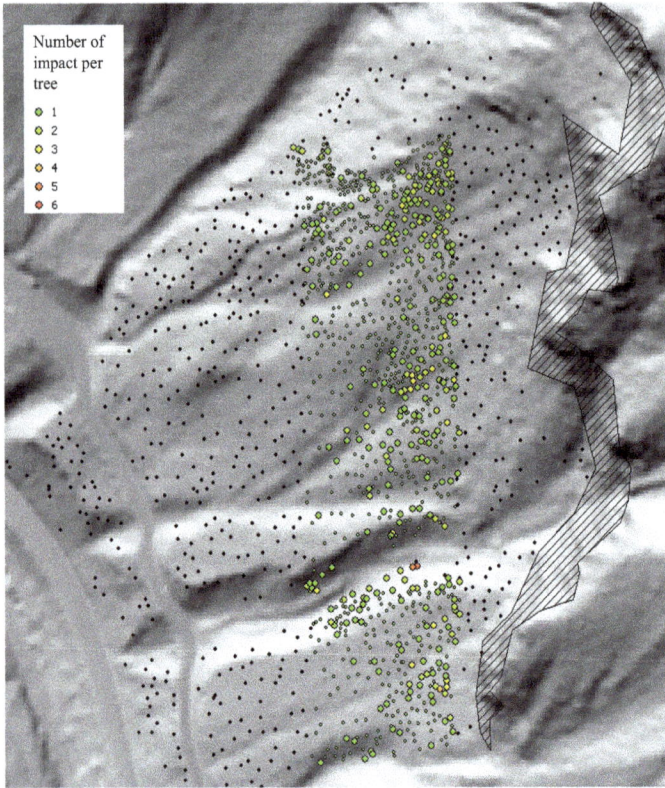

Figure 29. Map of the number of rockfall impact per marteloscopes's tree.

- the map of release areas : a linear symbol drawn thanks to the DEM and an informatics program shows the cliffs where rockfalls come from

- the map of forest stands according to the the main dendrometrical parameters : the mean DBH with its standard deviation, the stem density and the species distribution.

All these input data are used by Rockyfor-3D for simulating rockfalls trajectories. The user has to select a number of simulations. 500 simulations for each release points is a good compromise for having results in both reliable and fast.

The software allows:

- a visualization of all the simulated trajectories

- a display in each points of the DTM of the main statistics of these trajectories : passing frequencies,..

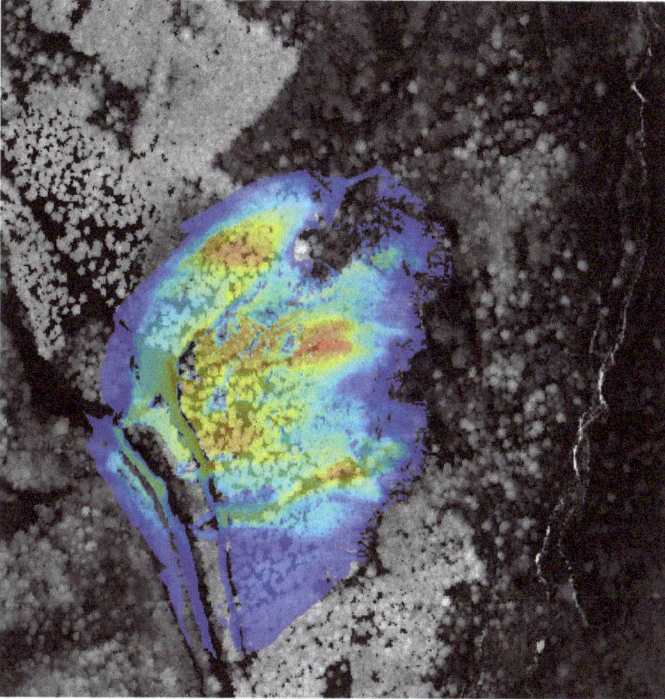

Figure 30. Map of the maximal envelop of probable rockfalls propagation areas obtained with the empirical model Rockfor^{LIN}. The difference of colour represents the rockfall passing frequency: Blue = low frequency, Red = high frequency. The results are expressed on the LiDAR Digital Surface Model.

- an estimation of the kinematic parameters in each point of the DTM.

With all these calculated output data the user can provide the map of the endangered areas, It is also possible for the forest manager to test if his silviculture has effectively strengthen the protection function by comparing before and after tree marking, all trajectories and their frequency and the value of the kinematic parameters of boulders in each point of the DTM. The maps below present the results (passing frequencies) obtained with Rockyfor3D for 3 different rock volumes and taking into account the role played by the current forest stand.

The results obtained with Rockyfor3D confirm the first ones obtained with Rockfor^{LIN} interm of potential endangered areas. But Rockyfor3D allows the forest manager to go furthermore in the protection forest assessment. For rocks having a volume less than 0.5m^3 the current forest stand offers an efficient protection : none rocks is able to reach the road. For rocks of 0.5m^3 the current forests stand is only able to limit the number of boulders able to reach the road. This action of clipping can be expressed using the concept of Probable Residual Rockfal Hazard (PRH). The PRH represents the percentage of rocks which are able to surpasse the forested area of a slope. What has also to be noticed is that the results obtain with Rockyfor3D fit perfectly with the map of the observed rockfall impacts on trees.

According to these results, it appears that the upslope part of the forest has a very important role for all the categories of rocks' volumes. This forested strip of 50m is one of the key components of the protective action of the current forest stand.

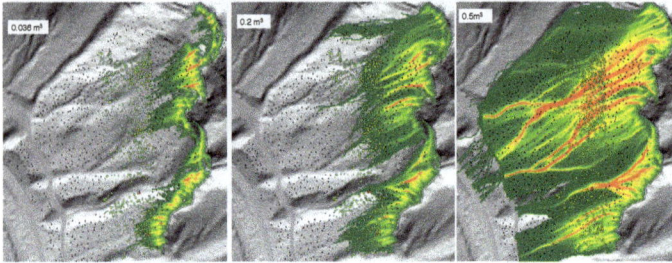

Figure 31. Maps of rockfalls propagation areas, for boulders of 0.036, 0.2 and 0.5 m³, obtained with Rockyfor3D and taking into account the role played by the current forest stand. The difference of color represents the rockfall passing frequency: Green = low frequency, Red = high frequency. The results are expressed on the LiDAR Digital Terrain Model.

The last modeling tool which has been used is Rockfor^NET. This tool is specifically dedicated to the Probable Residual Rockfall Hazard assessment. The PRH is a synthetic indicator of the level of protection that a forested area can provided. This probabilistic tool also provides theoretical stand density intervals and Mean Stem Diameter value in order to have a PRH less than 1%. This tool is described and freely usable via the web site: www.rockfor.net. The input parameters are presented in the figure 29.

Before making a silvicultural treatment in a rockfalls area, this one can be tested in term of consequences on the rockfall risk using Rockfor^NET. This tool is usable for checking: if the PRH has been increased after the realization of the planned silvicultral treatment and if this increasing is acceptable or not, if the theoretical stand density and average DBH are realistic or not, and if not then the question of using civil engineering works has to be asked. So this tool has to be used as expertise support to evaluate, via the quantification of the PRH, the protective role played by current forest stands and to adapt in consequences the forest management.

Rockfor^NET has been used in the Val Drôme case study for displaying the current PRH before any silvicultural operations (cf. figure 30). The results obtained with Rockfor^NET are also fitting with the ones of Rockyfor3D and Rocfor^LIN.

3.4. The new forest management plan elaborated using Rockfor^NET

According to the results of the workpackage 4 of the Manfred project, the probable consequences of climate changes in the region, in which the Val Drôme case study is located, are a

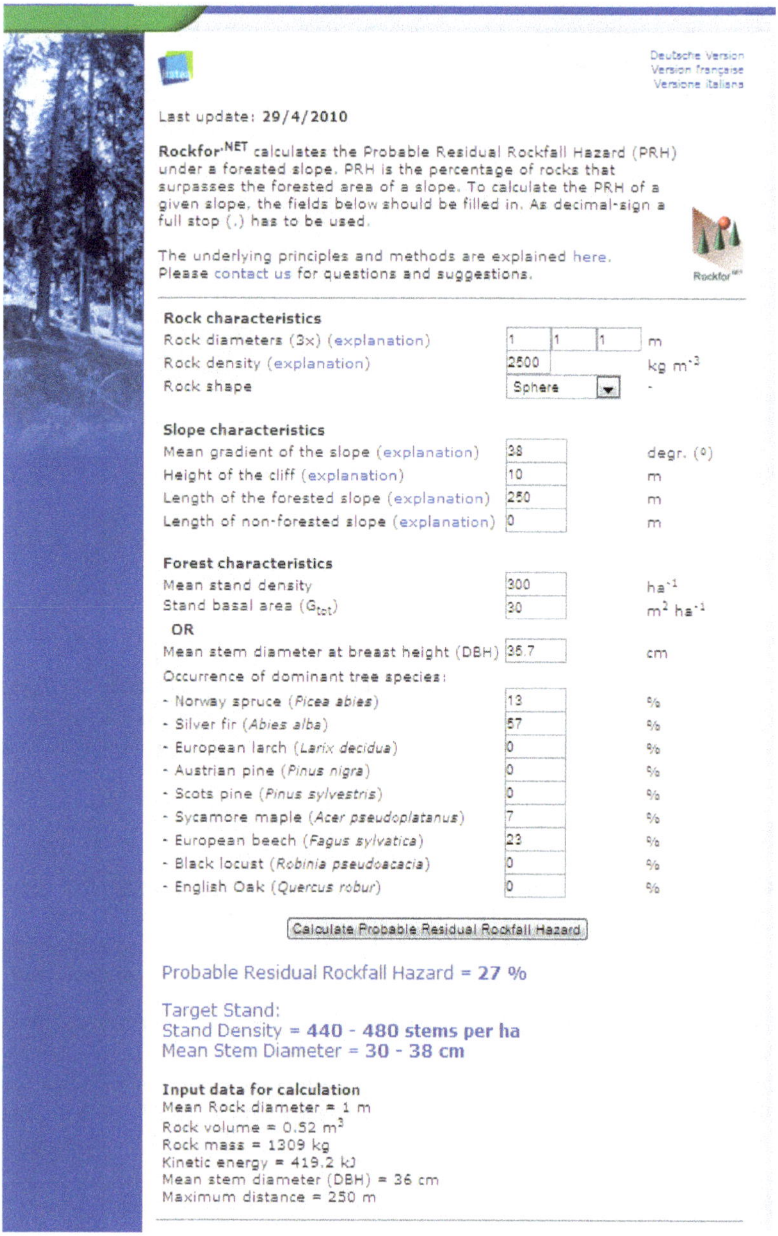

Figure 32. The tool Rockfor[NET] freely usable via the web site www.rockfor.net

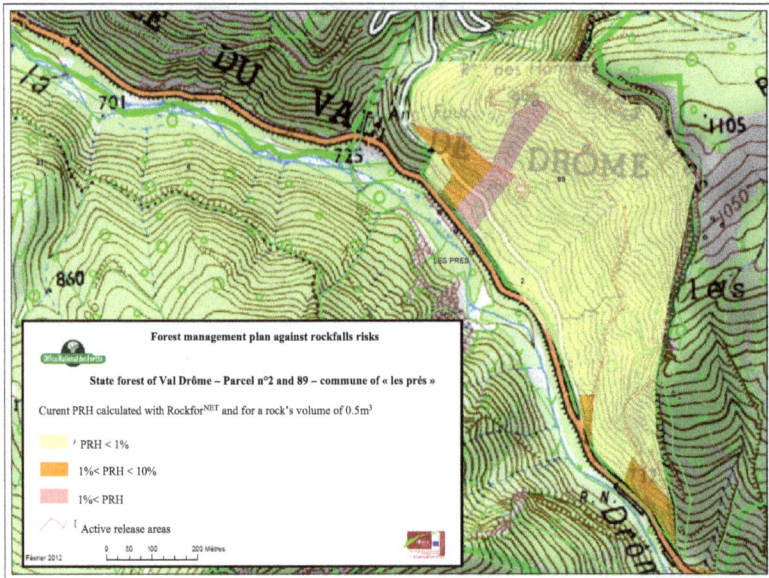

Figure 33. Map of the current PRH obtain with Rockfor^{NET}

migration up slope of the broadleaves species and the maintenance in the next 100 years of Austrian black pine in this area. These two situations are currently observed in this case study: broadleaves trees are settled in the understory of the present Austrian black pine stand and in open areas the Austrian black pines are naturally regenerating.

So, local forest managers have decided to: shift from an even-aged structure to an uneven-aged one, promote the current settlement of broadleaves trees in the understory and to develop the natural regeneration of Austrian black pine by opening or promoting regeneration gaps. Due to the important protective effect of the 50m width forest strip just below the release areas, it has been decided to ban any logging activities in this zone. In this forest strip the natural regeneration is the result of the "cutting" activities of the falling rocks on mature trees.

Rockfor^{NET} has been used for the dimensioning and spacing of the regeneration gaps to be opened. It has been decided to make the regeneration gaps without any complementary eco-engineering works, only in the areas where the PRH after the opening will be still of less than 1%. The other driver for the implementation of regeneration gaps has been to promote the current and viable regeneration patches. According to these drivers the final planning is the following one:

- The maximal length of a regeneration gap in the direct down slope direction could not exceed 30m

- The shape of the gaps will be a circular one

- The distance between to gaps (from centre to centre) will be at minima of 140m

- The spatial distribution of the gaps will follow up a quinconce pattern on the slope

- The gaps radius will be increase of 10 m each 20 meters if and only if the regeneration development will be efficient and viable

- All the slashes due to the silvicultural interventions will be used for increasing the roughness of the slope. They have to be positioned diagonally to the slope on the upslope part of each gap.

If regeneration gaps have to be implemented in areas where the current PRH is higher than 1%, then specific eco-engineering works have to be used. This specific eco-engineering works are:

- Trees felled diagonally to the direct slope and anchored on other trees or stumps (cf. Figure 31).

- Obligation to cutting trees leaving high stumps: at minima the height of the stump has to be equal to 1.30m, measured upslope (cf. Figure 31).

- To use Treenet or Fasnet works in thalwegs in order to promote the settlement of natural regeneration and the development of vegetation works (plantation of broadleaves trees).

Treenet (cf. figure 32) are eco-engineering works based on the use of mature trees as support poles of rockfall nets, and Fasnet structures are based on the use forest exploitation products as trunks and branches for making fascines (cf. figure 34) which are also anchored on living trees. These two eco-engineering have been experimentally tested within the Manfred project and are able to dissipate kinetic energy less than 100 kJ.

Figure 34. Examples high stumps and felled trees diagonally to the direct slope and anchored on high stumps in the case study Val Drôme

Figure 35. Examples of uses of Treenet eco-engineering works in the case study Val Drôme

Figure 36. Example of use of a Fasnet eco-engineering work in the case study Val Drôme

The figure 34 presents the map of the future PRH after the implementation of the new forest management plan elaborated using the support expertise tool Rockfor^{NET}.

In parallel to these eco-engineering technics, all the unstable deposit rocks on the slope have been anchored using metallic cable (Cf. Figure 35).

Monitoring Climatic Change Impacts on Protection Forests in Aosta Valley (Italy) and in Drôme (France) Using Medium and High Resolution Remote Sensing and Mateloscopes Plots

287

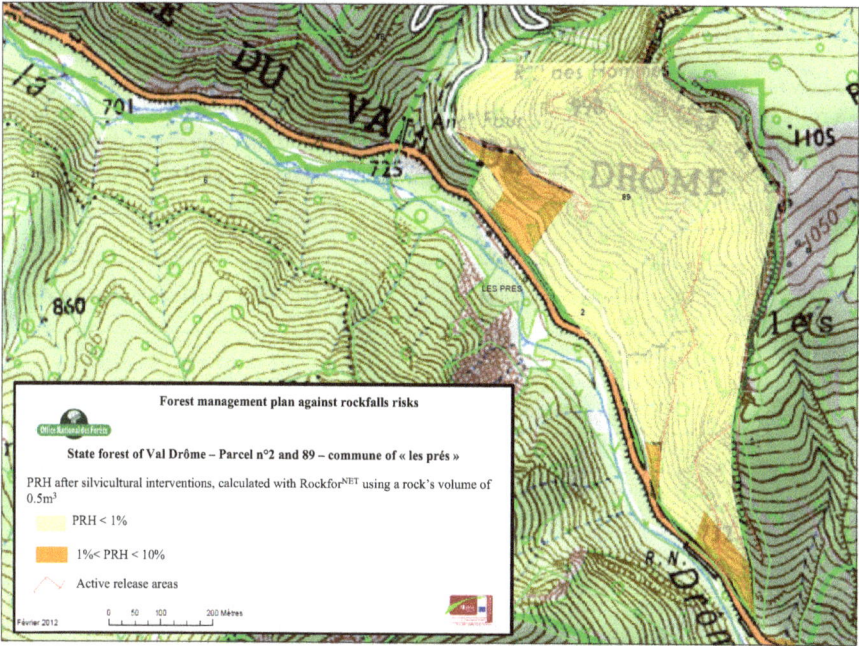

Figure 37. Map of the future PRH after the implementation of the new forest management plan and using Treenet and Fasnet eco-engineering works.

Figure 38. Example of anchorage of an unstable deposit rock in the Val Drôme case study.

3.5. Technical and financial report of the implementation of the new forest management plan of the Val Drôme case study

The new forest management plan of the Val Drome case study is dedicated to the regeneration of a rockfall protection Austrian black pine forest which covers a surface of 36 ha. The forest interventions have been targeted on a forested strip of 6 ha and they have been defined using the expertise support tool Rockfor[NET]. The following table gives the main technical items and their associated costs.

Type of intervention	Details of the intervention	Duration man.day	Staff costs	Consumable costs	Total costs
Forest management plan elaboration	Diagnoses on the site Definition of the interventions Public presentations	7	3000 €	0 €	3000 €
Project management	Technical preparation of the site	6	3000 €	0 €	3000 €
	Administrative preparation of the site	1	500 €	0 €	500 €
	Monitoring of the works	3	1500 €	0 €	1500 €
Logging, revegetation and eco-engineering works	Regeneration gaps opening	24	9120 €	996 €	10216€
	Revegetation of one thalweg (plantation of 300 trees and associated digging in works)	8	2240 €	600 €	2840 €
	2 Treenets structures (2 * 25m) 1 Fasnet structure (1*25m)	16	4480 €	2000 €	6480 €
	Total	65	23840€	3696€	27536€

Table 1. Technical and financial report of the implementation of the new forest management plan in the Val Drôme case study.

For this site the elaboration and realization of an adapted to climate changes forest management plan dedicated to the optimization of the protection role of a Austrian black pine forest has cost 765 euros per ha. For only the elaboration phase of the management plan the cost is 84 Euros per ha. The cost of the eco-engineering works is 86.4 Euros per linear meter. For the same condition, a protection strategy based only on civil engineering works will have cost about 1000 Euros per linear meter of rockfall defense structure. For this site the total cost of a pure civil engineering strategy is estimated to 75000 Euros.

3.6. Conclusions

The set of modelling tools Rockfor[LIN], Rockyfor3D and Rockfor[NET] is perfectly adapted to the building up of protection forest management plan. At minima forest managers have to use Rockfor[LIN] and Rockfor[NET]. The first model is able to give an accurate evaluation of the maximal probable envelop of rockfalls propagation zone. The second one helps managers to 1) provide a quick and efficient analysis of the protective effect of forest stands and 2) display the results via the mapping of the values of the Probable Residual rockfall Hazard. The use of these expertise supporting tools is optimized if accurate and high resolution geographical data are available like LiDAR ones.

The probable consequences of climate changes trends on forest stands dynamics has to be taken into account for producing a long term efficient forest management plan. In the field of protection forest management, the main objective is to promote the tree species both adapted in term of climatic conditions and risk mitigation. In the case of rockfalls protection, broad-leaves tree species are the most adapted ones. That's the reason why, forest managers have to promote in coniferous stands the natural settlement of broadleaves trees.

The costs associated to the implementation of an adaptive forest management in rockfall protection forests, based on a technical approach coupling eco-engineering technics and regeneration by gaps, are 3 times cheaper than the ones associated to a pure civil engineering protection strategy. The magnitude of the difference of these two categories of costs is: the cost of 1 linear meter of a classical rockfall net (not ones like Treenet) is equal to one at three times the cost management of 1 ha of a protection forest. If forest stands are able to offer an efficient protective function then their management is much more sustainable and cost effective for the society then protection strategies only based on civil engineering technics. But if some subventions exist for civil engineering building up, this is not the case for protection forest management. For subventionning protection forests maintenance, which have to be considered as natural protection works, the most efficient solution is to fix a subvention per hectare and for a certain period. According to the results obtained within the Manfred project the minimal amount of the subvention should be of 150 Euros/ha/year for a period of 5 years. As the objective is to maintain protection forest, any benefits coming from the sale of timbers have to be subtracted from the subvention. This subventionning will be fully effective if a map of protection forests has been firstly done.

4. Conclusions

Forest managers have now a set of efficient tools for expertising gravitational risk protection forests. Depending on the available data, this expertise can be provided using the last development in remote sensing technic for assessing the evolution of dieback in protection forest and their consequences in term of natural risk activity. The minimal set of data to be used is at least an accurate digital terrain model and a forest stands description map. For rockfalls protection forest assessment the main modelling tools are usable or available via the web sites : www.rockfor.net and www.ecorisq.org. The assessment processes presented in this chapter are easy to reproduce in any other sites.

The valorization of protection forest stands as real natural protective works, needs to firstly produce on large scale the map of localization of protection forests and secondly to develop an adapted financial support for covering the specific costs associated to protection forest adaptive management plan. The MANFRED project offers, via the models developed and the case study examples, adapted and efficient solutions to these questions.

Acknowledgements

This work has been financed by the the European project MANFRED - INTERREG Alpine Space programme. We want also to thank the Regione Autonoma Valle d'Aosta for the topographic and LiDAR data, and the observers of the Val Drôme case study for their inputs.

Author details

Laurent Borgniet[1], David Toe[1], Frédéric Berger[1], Marta Galvagno[2], Cinzia Panigada[3], Roberto Colombo[3], Umberto Morra di Cella[2], Simone Gottardelli[3], Ivan Rollet[4], Mario Negro[5], Flavio Vertui[4] and Cédric Fermont[6]

1 Institut national de recherche en sciences et technologies pour l'environnement et l'agriculture, Irstea, Grenoble, France

2 Agenzia Regionale per la Protezione dell'Ambiente della Valle d'Aosta, ARPA, Aosta, Italy

3 Remote Sensing of Environmental Dynamics Laboratory, DISAT, Università degli Studi di Milano Bicocca, Milano, Italy

4 Corpo Forestale della Valle d'Aosta, Dip. Risorse Naturali e Corpo Forestale, Regione Autonoma Valle d'Aosta, Aosta, Italy

5 Forestazione e Sentieristica, Dip. Risorse Naturali e Corpo Forestale, Regione Autonoma Valle d'Aosta, Aosta, Italy

6 Office National des Forêts, Département de la Drôme, France

References

[1] Hansen, J., Sato, M., Ruedy, R., Lo, K., Lea, D.W., Medina-Elizade, M., 2006. « Global temperature change », PNAS, vol. 103, n°. 39, p. 14288-14293.

[2] Durand, Y., Laternser, M., Giraud, G., Etchevers, P., Lesaffre, B., Mérindol, L., 2009. « Reanalysis of 44 Yr of Climate in the French Alps (1958–2002): Methodology, Model Validation, Climatology, and Trends for Air Temperature and Precipitation », Journal of Applied Meteorology and Climatology, vol. 48, n°. 3, p. 429-449.

Monitoring Climatic Change Impacts on Protection Forests in Aosta Valley (Italy) and in Drôme
(France) Using Medium and High Resolution Remote Sensing and Mateloscopes Plots

291

[3] Smiatek, G., Kunstmann, H., Knoche, R., Marx, A., 2009. « Precipitation and temperature statistics in high-resolution regional climate models: Evaluation for the European Alps », *J. Geophys. Res.*, vol. 114, n°. D19, p. D19107.

[4] Beniston, M., 2012. « Exploring the behaviour of atmospheric temperatures under
dry conditions in Europe: evolution since the mid-20th century and projections for
the end of the 21st century », *International Journal of Climatology*, in press.

[5] Beniston, M., 2003. « Climatic Change in Mountain Regions: A Review of Possible
Impacts », *Climatic Change*, vol. 59, n°. 1, p. 5-31.

[6] Lenderink, G., van Meijgaard, E., 2008. « Increase in hourly precipitation extremes
beyond expectations from temperature changes », *Nature Geoscience*, vol. 1, n°. 8, p.
511-514.

[7] Gimmi, U., Wohlgemuth, T., Rigling, A., Hoffmann, C., Bürgi, M., 2010. « Land-use
and climate change effects in forest compositional trajectories in a dry Central-Alpine
valley », *Annals of Forest Science*, vol. 67, n°. 7, p. 701-701.

[8] Leonelli, G., Pelfini, M., Morra di Cella, U., Garavaglia, V., 2010. Climate Warming
and the Recent Treeline Shift in the European Alps: The Role of Geomorphological
Factors in High-Altitude Sites. *AMBIO* 40, 264–273.

[9] Seidl, R., Rammer, W., Lexer, M., 2010. Climate change vulnerability of sustainable
forest management in the Eastern Alps. *Climatic Change* 1–30.

[10] Lindner, M., Maroschek, M., Netherer, S., Kremer, A., Barbati, A., Garcia-Gonzalo, J.,
Seidl, R., Delzon, S., Corona, P., Kolström, M., Lexer, M.J., Marchetti, M., 2010. Climate change impacts, adaptive capacity, and vulnerability of European forest ecosystems. *Forest Ecology and Management* 259, 698–709.

[11] Gouveia, C., Trigo, R.M., DaCamara, C.C., 2009. Drought and vegetation stress monitoring in Portugal using satellite data. *Nat. Hazards Earth Syst. Sci.* 9, 185–195.

[12] Liu, W.T., Kogan, F.N., 1996. Monitoring regional drought using the Vegetation Condition Index. *International Journal of Remote Sensing* 17, 2761–2782.

[13] Kogan, F.N., Zhu, X., 2001. Evolution of long-term errors in NDVI time series:
1985-1999. *Advances in Space Research* 28, 149–153.

[14] Lloyd-Hughes, B., Saunders, M.A., 2002. A drought climatology for Europe. *International Journal of Climatology* 22, 1571–1592.

[15] Lehner, B., Döll, P., Alcamo, J., Henrichs, T., Kaspar, F., 2006. Estimating the Impact
of Global Change on Flood and Drought Risks in Europe: A Continental, Integrated
Analysis. *Climatic Change* 75, 273–299.

[16] Reichstein, M., Ciais, P., Papale, D., Valentini, R., Running, S., Viovy, N., Cramer, W.,
Granier, A., Ogée, J., Allard, V., Aubinet, M., Bernhofer, C., Buchmann, N., Carrara,
A., Grünwald, T., Heimann, M., Heinesch, B., Knohl, A., Kutsch, W., Loustau, D.,

Manca, G., Matteucci, G., Miglietta, F., Ourcival, J. m., Pilegaard, K., Pumpanen, J., Rambal, S., Schaphoff, S., Seufert, G., Soussana, J.-F., Sanz, M.-J., Vesala, T., Zhao, M., 2007. Reduction of ecosystem productivity and respiration during the European summer 2003 climate anomaly: a joint flux tower, remote sensing and modelling analysis. *Global Change Biology* 13, 634–651.

[17] Jung, M., Verstraete, M., Gobron, N., Reichstein, M., Papale, D., Bondeau, A., Robusteli, M., Pinty, B., 2008. Diagnostic assessment of European gross primary production. *Global Change Biology* 14, 2349–2364.

[18] Morán-Tejeda, E., Ceglar, A., Medved-Cvikl, B., Vicente-Serrano, S.M., López-Moreno, J.I., González-Hidalgo, J.C., Revuelto, J., Lorenzo-Lacruz, J., Camarero, J., Pasho, E., 2012. Assessing the capability of multi-scale drought datasets to quantify drought severity and to identify drought impacts: an example in the Ebro Basin. *International Journal of Climatology*, in press. DOI 10.1002/joc.3555.

[19] Bontemps, J.-D., Herve, J.-C., Duplat, P., Dhôte, J.-F., 2012. Shifts in the height-related competitiveness of tree species following recent climate warming and implications for tree community composition: the case of common beech and sessile oak as predominant broadleaved species in Europe. *Oikos* 121, 1287–1299.

[20] Giuggiola, A., Kuster, T. & Saha, S., 2010. Drought-induced mortality of Scots pines at the southern limits of its distribution in Europe: causes and consequences. *iForest - Biogeosciences and Forestry* 3, 95–97.

[21] Vacchiano, G., Garbarino, M., Borgogno Mondino, E. & Motta, R., 2012. Evidences of drought stress as a predisposing factor to Scots pine decline in Valle d'Aosta (Italy). *European Journal of Forest Research* 131, 989–1000.

[22] Vertui F. and Tagliaferro, 1998. Scots Pine (Pinus sylvestris, L.) die-back by unknown causes in the Aosta Valley, Italy. *Chemosphere*, 36: 1061-1065

[23] Rebetez M. and Dobbertin M., 2004. Climate change may already threaten Scots pine stands in the Swiss Alps. *Theoretical and Applied Climatology*, 79: 1–9

[24] Bigler C., Braker O.U., Bugmann H, Dobbertin M. and Rigling A., 2006. Drought as an inciting mortality factor in Scots Pine stands of the Valais, Switzerland. *Ecosystems*, 9: 330–343.

[25] NASA MOLT database: ftp://e4ftl01u.ecs.nasa.gov/MOLT/MOD13Q1.005.

[26] Huete, A.R., Justice, C., van Leeuwen, W.J.D., 1999. MODIS vegetation index algorithm theoretical basis document,version3. http://www.modis.gsfc.nasa.gov/data/atbd/land_atbd.html.

[27] Huete,A., K. Didan, T. Miura, E.P. Rodriguez, X. Gao and L.G. Ferreira., 2002. Overview of the radiometric and biophysical performance of the MODIS vegetation indices. *Remote Sensing of Environment* 83:195–213.

[28] Kogan, F. N., 1995. Droughts of the Late 1980s in the United States as Derived from NOAA Polar-Orbiting Satellite Data. *Bulletin of the American Meteorological Society* 76, 655–668.

[29] Miura, T., Huete, A. & Yoshioka, H., 2006. An empirical investigation of cross-sensor relationships of NDVI and red/near-infrared reflectance using EO-1 Hyperion data. *Remote Sensing of Environment* 100, 223–236.

[30] Kutser, T., Pierson, D. C., Kallio, K. Y., Reinart, A. & Sobek, S., 2005. Mapping lake CDOM by satellite remote sensing. *Remote Sensing of Environment* 94, 535–540.

[31] Vicente-Serrano, S. M., Pérez-Cabello, F. & Lasanta, T., 2008. Assessment of radiometric correction techniques in analyzing vegetation variability and change using time series of Landsat images. *Remote Sensing of Environment* 112, 3916–3934.

[32] Kaufman, Y. J., Wald, A.E., Remer, L.A., Bo-Cai G., Rong-Rong L., 1997. The MODIS 2.1- mu;m channel-correlation with visible reflectance for use in remote sensing of aerosol. *IEEE Transactions on Geoscience and Remote Sensing* 35, 1286 –1298.

[33] Matthew, M. W., Adler-Golden, S.M., Berk, A., Felde, G., Anderson, G.P., Gorodetzky, D., Paswaters, S., Shippert, M., 2002. Atmospheric correction of spectral imagery: evaluation of the FLAASH algorithm with AVIRIS data. *Applied Imagery Pattern Recognition Workshop,. Proceedings. 31st* 157 – 163.doi:10.1109/AIPR.2002.1182270.

[34] Yuan, J. & Niu, Z., 2008. Evaluation of atmospheric correction using FLAASH. *International Workshop on Earth Observation and Remote Sensing Applications, 2008. EORSA 2008* 1 –6 (2008).doi:10.1109/EORSA..4620341.

[35] Panigada C., Rossini M., Busetto L., Meroni M., Fava F., Colombo R., 2010. Chlorophyll concentration mapping with MIVIS data to assess crown discoloration in the Ticino Park oak forest. *International Journal of Remote Sensing*, 31: 3307-3332.

[36] Rossini, M., Panigada, C., Meroni, M., Colombo, R., 2006. Assessment of oak forest condition based on leaf biochemical variables and chlorophyll fluorescence. *Tree Physiology*, 26: 1487- 1496.

[37] Jönsson, P., Eklundh, L., 2004. TIMESAT—a program for analyzing time-series of satellite sensor data. *Computers & Geosciences*, 8, 833-845.

[38] Hird, J. and G.J. McDermid, 2009. Noise reduction of NDVI time series: An empirical comparison of selected techniques. *Remote Sensing of Environment*, 113(1): 248-258.

[39] Jönsson, P., Eklundh, L., 2002. Seasonality extraction from satellite sensor data. In *Frontiers of Remote Sensing Information Processing*, edited by Chen, C.H. World Scientific Publishing. pp 487-500.

[40] Walther, G-R., Post, E., Convey, P., Menzel, A., Parmesan, C., Beebee, T. J. C., Fromentin, J-M., Hoegh-Guldberg O. & Bairlein F., 2002. Ecological responses to recent climate change. *Nature* 416, 389-395.

[41] Gonthier, P., Nicolotti, G., Linzer, R., Guglielmo, F., Garbelotto, M., 2007. Invasion of European pine stands by a North American forest pathogen and its hybridization with a native interfertile taxon. *Molecular Ecology* 16 (7) 1389–1400.

[42] UN-ECE, 2004. Manual on methodologies and criteria for harmonized sampling, assessment, monitoring and analysis of the effects of air pollution on forests, part II - International Co-operative Programme on Assessment and Monitoring of the effects of air pollution on forests, pp.77.

[43] Rammer, W., Brauner, M., Dorren, L. K. A., Berger, F., Lexer, M. J., 2010. « Evaluation of a 3-D rockfall module within a forest patch model », *Nat. Hazards Earth Syst. Sci.*, vol. 10, n°. 4, p. 699-711.

Chapter 17

Case study Oberschwaben / Allgäu / Vorarlberg / Prättigau – Risk Assessment of Abiotic and Biotic Hazards

Holger Veit, Holger Grieß, Bernhard Maier and
Peter Brang

Additional information is available at the end of the chapter

1. Introduction

1.1. The present condition

The transnational case study of the MANFRED Project includes regions of southern Germany, western Austria and eastern Switzerland (Oberschwaben / Allgäu / Montafon / Prättigau) and is represented by the forest administrative district of Ravensburg in Baden-Württemberg, the forest administration unit of Kempten in Bavaria, the mountain forest enterprise Stand Montafon in the Montafon Valley of Vorarlberg and the region of Prättigau as a part of Grisons (see figure 1). According to the encompassed landscapes it comprises a wide range of forest management aspects.

Timber production is carried out throughout the whole case study area from the lower foothills on the border of Lake Constance up to the steep regions of high the mountain ranges where Norway spruce (*Picea abies* (L.) Karst.) is by far the most abundant tree species. In the past violent storm events and bark beetle mass outbreaks have raised forest management problems [1] especially for secondary growth forests dominated by Norway spruce in the test areas of Ravensburg and Kempten. Particularly in the Montafon valley the protection of human settlements and traffic infrastructure from rock fall and landslides is the most important forest function [2, 3].

Within this case study information on management practices dealing with protection forest issues has been exchanged and discussed with stakeholders for advice and colleagues from German speaking countries to be able to cope with the challenges of climate change and

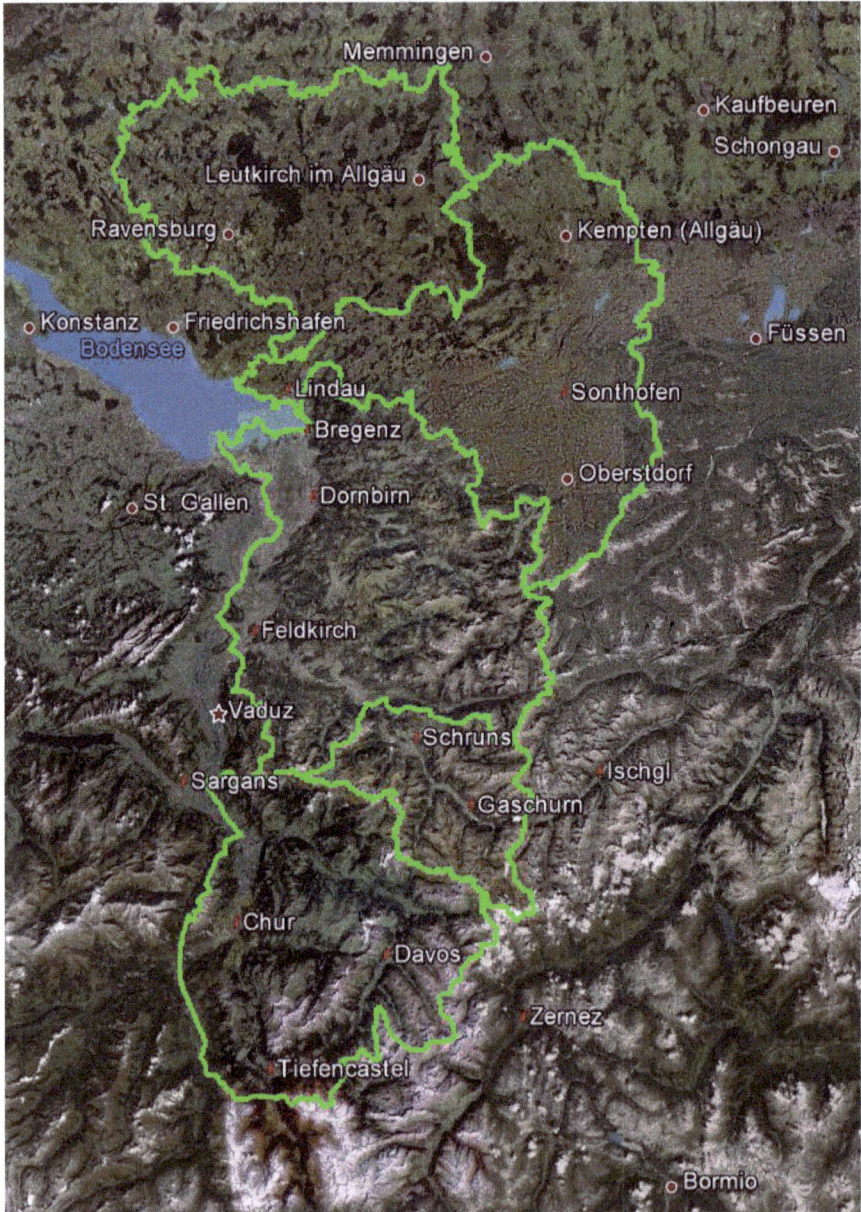

Figure 1. The case study Oberschwaben / Allgäu / Montafon / Prättigau in the cross border region of Germany, Austria and Switzerland (North-South ~ 150 km, West-East ~ 55 km)

address existing knowledge gaps. Know-how and specific already made up guidelines were shared through training courses and made available for professional practitioners, decision makers and scientists.

2. Oberschwaben / Ravensburg

2.1. Geographical setting

Ravensburg is a district in the far south-east of Baden-Württemberg, Germany. Neighbouring districts are Bodensee, Sigmaringen and Biberach, the Bavarian urban district Memmingen and the districts Unterallgäu, Oberallgäu and Lindau. The landscape in the district are the hills of Upper Swabia which rise from 458 metres above sea level in the valley of the river Danube to a maximum of 833 metres above sea level in the south-west and drop again to 395 meters above sea level at Lake Constance (figure 2). Upper Swabia is a region in Germany in the federal states of Baden-Württemberg and Bavaria. The name refers to the area between the Swabian Jura, Lake Constance and the Lech. It is situated in the central south of Germany consisting of the south-east of Baden-Württemberg and the south-west Bavarian Swabia region and is part of the Iller-Lech-Plateau, also known as the Upper Swabian Plain. Its landscape was formed by retreating glaciers after the Riss glaciations, leaving behind a large number of shallows which quickly filled up with water. This led to the large quantity of lakes in the area. The European watershed also passes through the region, with some rivers draining into the river Danube (ultimately flowing into the Black Sea) and others emptying into Lake Constance (ultimately ending in the North Sea).

2.2. Climate

The climate of the Ravensburg district and the Upper Swabian region as a whole is strongly influenced by the alpine range and Lake Constance expanding for 500 km² on the borders of Germany, Switzerland and Austria. Temperatures are balanced due to this tremendous water reservoir. Mean annual temperature is 9.1°C and annual precipitation sums up to 1000 mm. Winter frost periods are normally diluted unless the lake freezes itself. Especially in autumn and winter foggy periods may last for several weeks at a stretch. Resembling the weather regimes of other pre alpine foothills foehn winds are well known to speed up to gale force.

2.3. Forest types, function and distribution

Most of the forest in the district (48000 ha) is owned privately (67%), whereas the state of Baden-Württemberg is in charge of 23% of the area. Another 10%, to a lower extent by local munici-palities and private forest enterprises. The Altdorfer Wald which is the largest coherent forest of Upper Swabia with a size 82 km², is situated in the district of Ravensburg. Its elevation ranges from 450 m in the Schussen valley to 776,6 m a. s. l. in the southern part of the ridge.

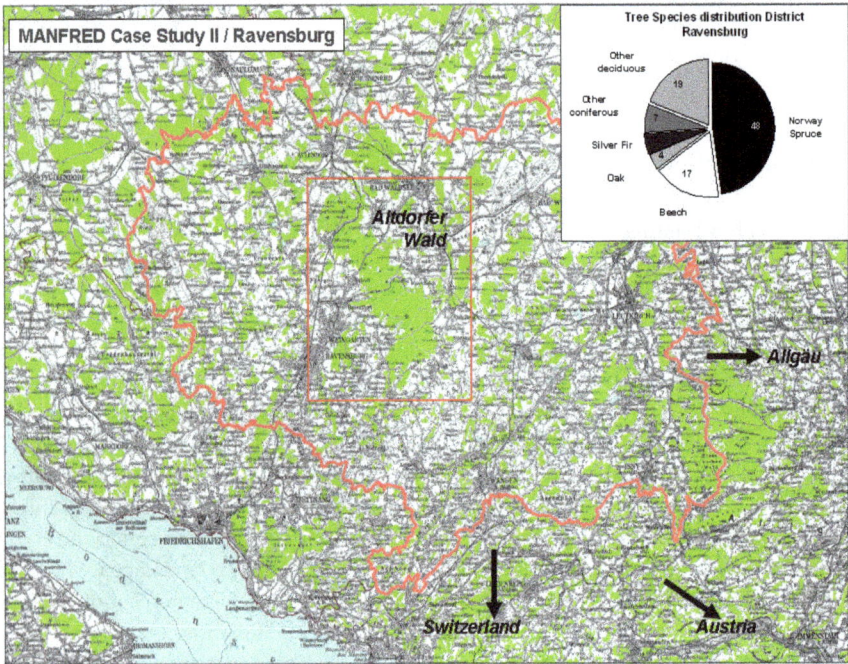

Figure 2. Case study subarea Ravensburg and tree species distribution

Mixed and pure Norway Spruce forests dominate throughout the district whereas European Beech and other deciduous tree species play a minor role. Forests are interspersed with bogs and a few open grasslands. Besides management of water protection and nature conservation areas half a million cubic meters of timber are sustainably harvested every year providing valuable resources for the regional sawmill, paper and plywood industry. Unfortunately many of the even aged Norway spruce dominated forests have been subject to wind throw and following bark beetle mass outbreak events in the past [1]. According to current climate projections there is a strong need for alternative silvicultural management options to preserve their growth and economical productivity.

3. Allgäu / Kempten

3.1. Geographical setting

The case study area Allgäu / Kempten (figure 3) is located in the south-western part of Bavaria (47.27° and 47.864° latitude and 9.563° and 10.513° longitude) between the foothills of the Alps close to Kempten and Füssen in the north and the high Alps in the south bordering the frontier to Austria. It includes the administrative districts Lindau in the West and Oberallgäu in the East. Altitudes range from 400m on Lake Constance up to the 2649 m high alpine summit of Hochfrottspitze.

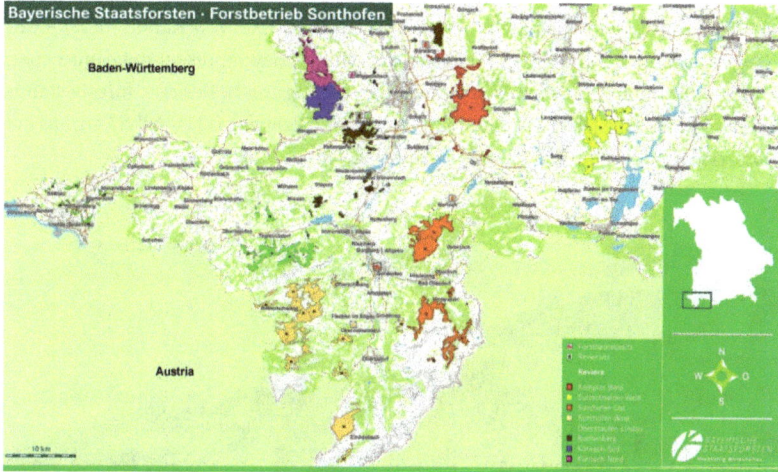

Figure 3. The case study area Allgäu / Kempten, Forest Management unit Sonthofen (BaySF)

The administrative district of Lindau is separated in 18 municipalities with approx. 76.000 inhabitants and a total area of 32.330 ha. The administrative district of Oberallgäu is the southernmost administrative district of the Federal Republic of Germany with an area of 1.527,55 km² and approximately 150.000 inhabitants. Oberallgäu is separated in 28 municipalities.

3.2. Climate

In the western lowlands (400 – 500 m a.s.l.) along Lake Constance, annual temperatures average around 8 - 9°C. In the Western Allgäu Uplands (500 – 700 m a.s.l.) the temperature mean is 7 - 8°C. In both areas mean annual precipitation is 1.400 – 1.600 mm. Within the administrative district of Oberallgäu mean annual temperatures range from 6.6°C in the northern parts to 5.8°C in the south. The mean annual precipitation in northern areas sums up to 1.000 – 1.800 mm per year whereas the south receives 1.700 – 2.400 mm annually.

3.3. Forest functions

The case study area comprises 18.000 ha of forests of which about a third (5.500 ha) are declared protection forests securing settlements and infrastructural facilities from avalanches, rockfall, landslides and flooding. They develop under geologic conditions characterized by limestone, flysch and tertiaries with sandy clayey soil or bogs. Besides their essential protective function, forests are crucial for tourism and recreation activities (200 km of pedestrian walkways, 100 km of bicycle tracks) as well as for landscape and nature conservation.

3.4. Forest types and distribution

The most common tree species are Norway Spruce (*Picea abies* (L.) Karst.) (68%), European Beech (*Fagus sylvatica* L.) and Silver Fir (*Abies alba* Mill.). They represent the climax species of typical mixed mountain forests in southern Germany (figure 4). Besides, forests stands also include Common maple (*Acer campestre* L.), Elm (*Ulmus campestris* L.), Ash (*Fraxinus excelsior* L.) and European mountain ash (*Sorbus aucuparia* L.) on a small scale.

Figure 4. Mixed Forest on molasses conglomerate at Rottachberg in Upper Allgäu

The abundance of European Beech decreases with increasing altitude. Norway Spruces and Silver Firs take their place (*Galio-Abietum* and *Homogyne-Piceetum*). Above 1200 m a. s. l. spruce forests (*Honogyne-Piceetum, Asplenio-Piceetum* and *Adenostyloglabrae-Piceetum*) naturally dominate the vegetation. Above 1400 m a. s. l. single firs and beeches are able to survive and eventually only spruces cope with the climatic conditions of the high mountainous areas. From 1600 m a.s. l. on only knee timber and Green alders (*Alnus viridis* L.) are growing. These species are crucial for the function of protection forests.

4. Vorarlberg / Montafon

4.1. Geographical setting

The Montafon is a 39 km long valley located in the southern part of the Vorarlberg province, the westernmost federal state of Austria (see Figure 5). It extends from the mountain ranges of Rätikon and Silvretta peaking in the 3312 m high Piz Buin in the South and the Verwall mountains in the North. It stretches from the district capital of Bludenz in the West to the Bielerhöhe in the East which is the water divide of the Rhenish and Danubian stream systems and connects the Montafon to the Paznaun valley in Tyrol. The river Ill drains water from the mountain ranges to the river Rhine. The Valley hosts 10 municipalities associated in the administrative collectivity of the Stand Montafon. Forestry, Tourism and Hydropower serve as the backbone of local economy.

4.2. Climate

The Montafon is characterized by transitional climatic elements from the sub continental dry inner alpine valleys to the more cool humid areas of the alpine rims. Precipitation ranges from below 1000 mm to more than 1900 mm with a distinct summer maximum. Winters are snowy and milder than adjacent alpine areas to the East and South. Maximum annual temperatures range from about 5 to 13.3°C on the valley floor. Minimum temperatures are between -2.0 and 3.7°C. Foehn winds are quite abundant in the Montafon.

4.3. Forest functions

The Stand Montafon Forstfonds is an alpine forestry enterprise that administers and manages about 70 % of the forested area (8400 hectares). The forests predominantly grow on steep terrain at 1200 m above sea level and higher where 90% of all the forests have a protective function (figure 5). They offer essential protection against avalanches and landslides to the villages and infrastructural facilities in the valley. The most important objectives are the maintenance of the forests safeguarding the inhabited areas of the valley floor and ensuring the expected forest functionalities by managing them in a multifunctional sustainable way. Consequently the Stand Montafon is specialized in mountain forest silviculture. Harvesting is frequently carried out by means of cable cranes, in order to protect forest soil and remaining trees. Besides the essential protective function, forests in the valley do also serve for timber production and play an important role in tourism and recreation as well as landscape and nature conservation [2]. Very often management decisions consider various aspects of forest functions and need support by up-to-date and site-related silvicultural methods. Therefore the Stand Montafon has been cooperating with different Austrian, Swiss and Dutch research institutes in order to investigate certain aspects of the forests and to evaluate existing forest management techniques.

Due to the high elevation of the Stands forest property about 96 % of the Montafon valley is naturally dominated by Norway Spruce (*Picea abies* (L.) Karst.), complemented by 3% Silver Fir (*Abies alba* Mill.). The remainder of species are European Beech, Scots Pine and European Larch and further Maple, Mountain ash and Swiss stone pine.

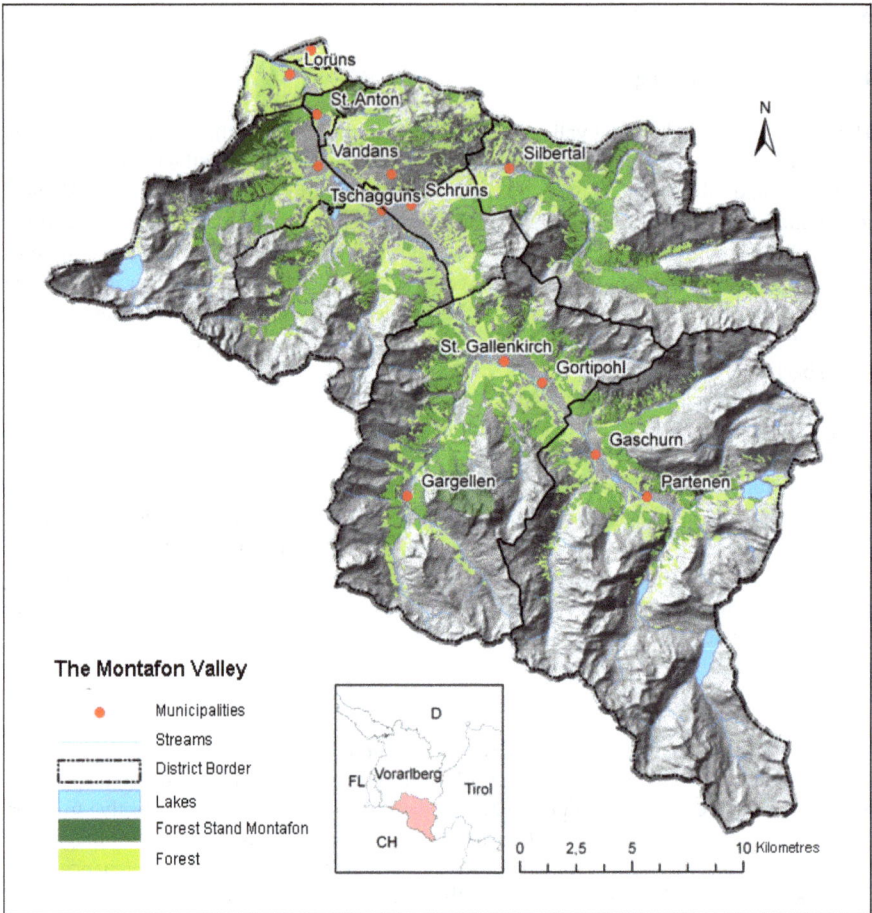

Figure 5. Stand Montafon, forested areas

5. Prättigau, project partner: WSL, Birmensdorf

5.1. The expected climate

Climate Modelling for the case study was carried out by WSL for different climatic parameters and summarized for specific time periods characterizing the past and the future climate according to the IPCC A1B SRES scenario. Comparing the climate scenarios for the case study sub areas indicated almost overall similar trends. Therefore the presentation of climate data was limited to a single set of figures showing precipitation and temperature development assuming representativeness for the case study Oberschwaben / Allgäu / Vorarlberg / Prättigau.

5.2. Precipitation

The development for four 30-year mean time slices of daily precipitation (1971-2000, 2001-2030, 2031-2060 and 2071-2100) is presented in figure 6. Generally overall precipitation seems to decrease up to the end of the century. It can be assumed that summer precipitation will decrease significantly whereas early spring may be slightly wetter than in the past. The availability of water during the vegetation period will decrease.

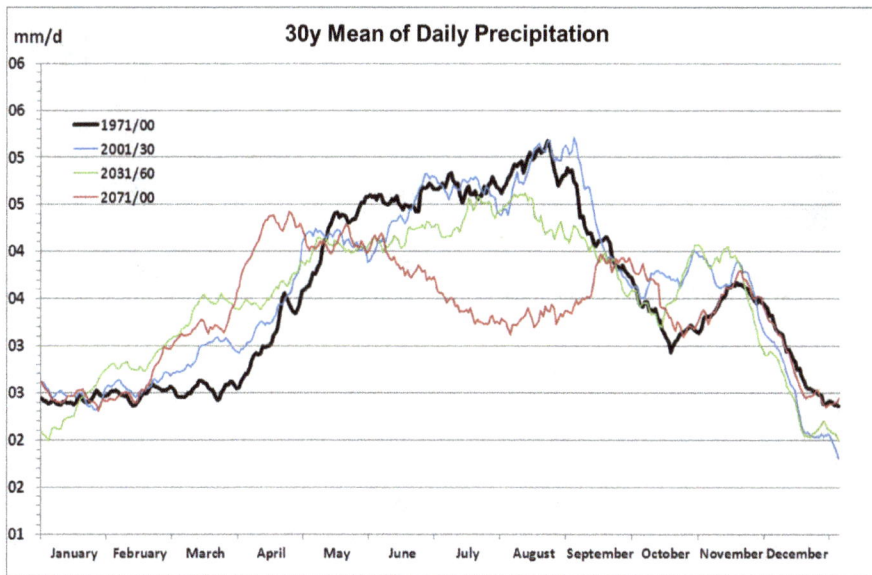

Figure 6. Development of Precipitation (30-year Mean of Daily Sum)

The modelling results presented in figure 6 are in line with the data presented in figure 7. Up to the first half of this century there will be hardly any change in the number of dry days in the vegetation period. This will change dramatically as the region will face 30% increase of dry days until 2100 favouring drought tolerant tree species.

5.3. Temperature

According to the A1B climate scenario of the IPCC there will be a steady temperature increase up to the end of this century (figure 8). Within the second and especially the last third of the modeling period temperatures will increase in all months of the year and stronger than in the periods before. Particularly the summer and winter months seem be subject to these alterations.

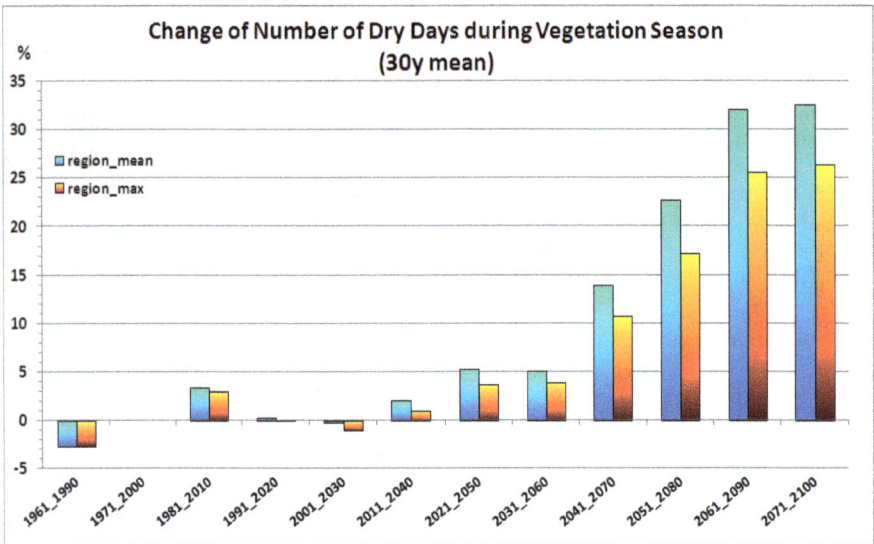

Figure 7. Change of the Number of Dry Days During the Vegetation Period

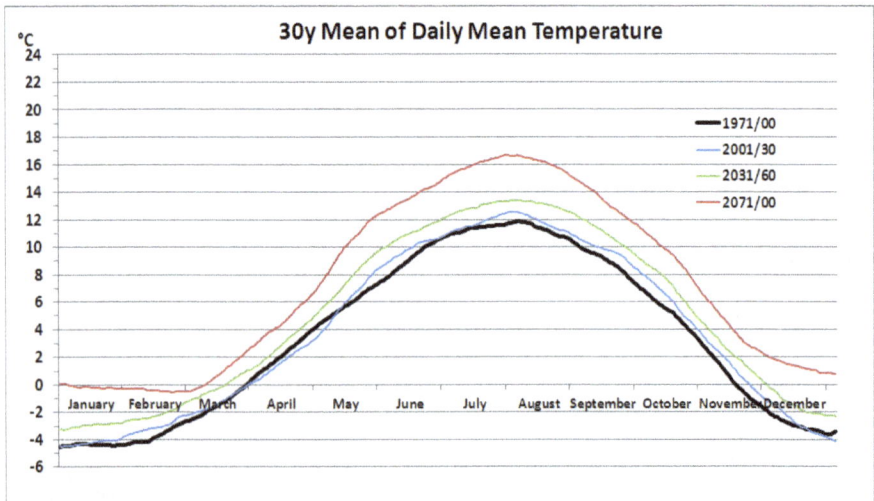

Figure 8. Development of the Daily Mean Temperature

Also the number of temperature dependant heat waves (figure 9) will more than double in the course of the present century becoming more and more noticeable from mid-century on.

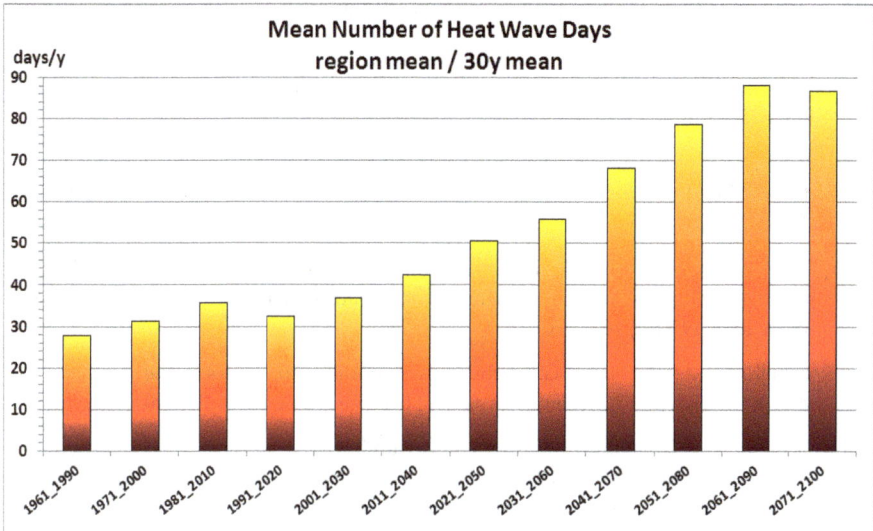

Figure 9. Maximum Length of Heat Waves

Author details

Holger Veit[1], Holger Grieß[2], Bernhard Maier[3] and Peter Brang[1]

1 Department of Biometrics, Forest Research Institute of Baden Württemberg, Freiburg, Germany

2 Department of Forest Management, Bavarian Forest Institute, Freising, Germany

3 Stand Montafon – Forstfonds, Schruns, Austria

References

[1] Schröter H.; Becker T.; Weigerstorfer D.; Veit H. Waldschutzprobleme nach "Lothar". In: Ministerium für Ernährung und Ländlichen Raum Baden-Württemberg (Hrsg.) Orkan "Lothar" – Bewältigung der Sturmschäden in den Wäldern Baden-Württem-

bergs, Dokumentation, Analyse, Konsequenzen. Schriftenreihe LFV Baden-Württemberg 2004; Bd. 83 341-365.

[2] Dönz-Breuss M., Maier B. & Malin H. Management for forest biodiversity in Austria - the view of a local forest enterprise. Ecological Bulletin 2004, 51.

[3] Brang P., Schöneberger W., Bachofen H., Zingg A. & Wehrli A. Schutzwalddynamik unter Störungen und Eingriffen: Auf dem Weg zu einer systemischen Sicht. Forum für Wissen 2004, 55-66.

[4] Brang, P. Resistance and Elasticity: promising concepts for the management of protection forests in the European Alps. Forest Ecology and Management 2001; 145 107-117.

[5] Dorren, L. K. A., Berger, F., Imeson, A.C., Maier, B. & Rey, F. Integrity, stability and management of protection forests in the European Alps. Forest Ecology and Management 2004; 195(1-2) 165-176.

[6] Schelhaas, M.J., Nabuurs, G.J., Schuck, A. Natural disturbances in the European forests in the 19th and 20th centuries. Global Change Biology 2003; 9 1620–1633.

[7] Schmidt M., Hanewinkel M., Kändler G., Kublin E & U. Kohnle. An inventory-based approach for modeling single-tree storm damage — experiences with the winter storm of 1999 in south western Germany. Canadian Journal of Forest Research, 2010, 40(8): 1636-1652.

[8] Solomon S., D. Qin, M. Manning, Z. Chen, M. Marquis, K.B. Averyt, M. Tignor and H.L. Miller, editors. Contribution of Working Group I to the Fourth Assessment Report of the Intergovernmental Panel on Climate Change, Summary for Policymakers. New York, USA: Cambridge University Press; 2007.

[9] Coumou D., Rahmstorf S. A Decade of Weather Extremes. Nature Climate Change 2012; DOI: 10.1038/NCLIMATE1452

[10] Hanewinkel M., Breidenbach J., Neeff T., Kublin E. Seventy-seven years of natural disturbances in a mountain forest area – the influence of storm, snow and insect damage analyzed with a long-term time series. Canadian Journal of Forest Research 2008;38 (8) 2249-2261. DOI 10.1139/X08-070*

[11] Schelhaas M-J., Nabuurs G.J. and Schuck A. Natural disturbances in the european forests in the 19th and 20th centuries. Global Change Biology 2003; 9 1620-1633.

[12] Lindner M, Maroschek M, Netherer S, Kremer A, Barbati A, Garcia-Gonzalo J, Seidl R, Delzon S, Corona P, Kolström M, Lexer M J, Marchetti M. Climate change impacts, adaptive caoacity, and vulnerability of European forest ecosystems, Forest Ecology and Management 2010; 259, 698-709.

[13] Forster B. et. al. Erfahrungen im Umgang mit Buchdrucker-Massenvermehrungen (Ips typographus L.) nach Sturmereignissen in der Schweiz. Schweizerische Zeitschrift für Forstwesen 2003;154 (11) 431–436.

[14] Meier F., Gall R. & B. Forster: Ursachen und Verlauf der Buchdrucker-Epidemien (Ips typographus L.) in der Schweiz von 1984 bis 1999. Schweizerische Zeitschrift für Forstwesen 2003; 154 11 437–441.

[15] Steyrer G., Krehan H. Borkenkäferkalamität 2010: Schäden weiterhin sehr hoch. BFW aktuell 2011; 52 4-5.

[16] Becker T., Schröter H. Ausbreitung von rindenbrütenden Borkenkäfern nach Sturmschäden. Allgemeine Forstzeitung 2000; 55 280-282.

[17] Wermelinger B. Ecology and management of the spruce bark beetle Ips typographus – a review of recent research. Forest Ecology and Management 2004; 202 67-82.

[18] Seidl R., Baier P., Rammer W., Schopf A., Lexer M. J. Modeling tree mortality by bark beetle infestation in Norway spruce forests. Ecological Modeling 2007; 206(3-4) 383-399.

[19] Seidl R., Rammer W., Jäger D., Lexer M. J. Impact of bark beetle (Ips typographus L.) on timber production and carbon sequestration in different management strategies under climate change. Forest Ecology and Management 2008; 256 209-220.

[20] Becker T., Schröter H. Die Ausbreitung des Borkenkäferbefalls im Bereich von Sturmwurf- Sukzessionsflächen. Berichte Freiburger Forstliche Forschung 2001; 26 79 S.

[21] Annila E. Influence of temperature upon the development and voltinism of Ips typographus L. (Coleoptera, Scolytidae). Ann. Zool. Fenn. 1969; 6 161-208.

[22] Coeln M. Grundlage für ein thermoenergetisches Modell zur Fernüberwachung der Borkenkäferentwicklung. Dissertation, Institut für Forstentomologie, Forstpathologie und Forstschutz der Universität für Bodenkultur. Wien; 1997.

[23] Baier P.; Pennerstorfer J.; Schopf A. PHENIPS - A comprehensive phenology model of Ips typographus (L.) (Col., Scolytinae) as a tool for hazard rating of bark beetle infestation. Forest Ecology and Management 2007; 249 171-186.

[24] Anandhi A., A. Frei, D. C. Pierson, E. M. Schneiderman, M. S. Zion, D. Lounsbury, and A. H. Matonse. Examination of change factor methodologies for climate change impact assessment, Water Resour. Res. 2011; 47, W03501, DOI 10.1029/2010WR009104.

[25] Wermelinger B., Seifert M. Analysis of the temperature dependent development of the spruce bark beetle Ips typographus (L.) (Col., Scolytidae). Journal of Applied Entomology. 1998; 122(4) 185-191.

[26] R Development Core Team. R - A Language and Environment for Statistical Comput-
      ing. http://www.R-project.org (accessed 15 September 2010).

[27] System for Automated Geoscientific Analyses. http://sourceforge.net/projects/saga-
      gis/files/SAGA%20-%202.0/SAGA%202.0.4/ (Accessed 15 August 2010).

[28] Lobinger G. Die Lufttemperatur als limitierender Faktor für die Schwärmaktivität
      zweier rindenbrütender Fichtenborkenkäferarten, Ips typographus L. und Pityo-
      genes chalcographus L. (Col. Scolytidae). Anzeiger für Schädlingskunde, Pflanzen-
      schutz, Umweltschutz 1994; 67 14-17.

[29] Doležal P, Sehnal F. Effects of photoperiod and temperature on the development and
      diapause of the bark beetle Ips typographus. Journal of Applied Entomology 2007;
      131 165-173.

[30] Schröter H., Delb H., Metzler B. Die Waldschutzsituation 2003/2004 in Baden-Würt-
      temberg. AFZ/Der Wald 2004; 59 343-345.

[31] Führer E., Nopp U. Ursachen, Vorbeugung und Sanierung von Waldschäden. Wien:
      Facultas-Universitätsverlag; 2001.

[32] Hanewinkel M., Hummel S., Cullmann D. A. Modeling and economic evaluation of
      forest biome shifts under climate change in Southwest Germany. Forest Ecology and
      Management 2010; 259 710-719.

Case Study: Valle Camonica and the Adamello Park

Giacomo Gerosa, Angelo Finco, Stefano Oliveri,
Riccardo Marzuoli, Alessandro Ducoli,
Giambattista Sangalli, Bruna Comini,
Paolo Nastasio, Giampaolo Cocca and
Elena Gagliazzi

Additional information is available at the end of the chapter

1. Introduction

1.1. Present conditions in Valle Camonica and the Adamello Park

1.1.1. Site description and geography

Valle Camonica is a N-S oriented valley located in the Rhaetian Alps, and it is characterized by greatly heterogeneous ecosystems in a territory with elevations ranging from the 390 m a.s.l. of the valley bottom to the 3539 m of the Monte Adamello peak (Figure 1). The Adamello massif covers a large area in the north-eastern side of the valley, and it is covered by one of the largest glacier of the southern Alps (Pian di Neve).

The Adamello Park, founded in 1983, encompasses a surface of 51.000 ha and 19 municipalities in the district of Brescia within the Camonica Valley. It is located in the north-eastern part of Valle Camonica and it covers more than 60% of the valley area

The Park shares its borders with the Adamello-Brenta Natural Park and the Stelvio Natural Park, thus determining one of the widest protected area in Europe (even larger if we consider also the near Engadina National Park).

The land cover includes forests for the 37% of the area and pastures for the 14%. Thirteen percent of surfaces are not cultivated while an large amount of the territory, almost 36%, is unusable because covered by rocks or glaciers.

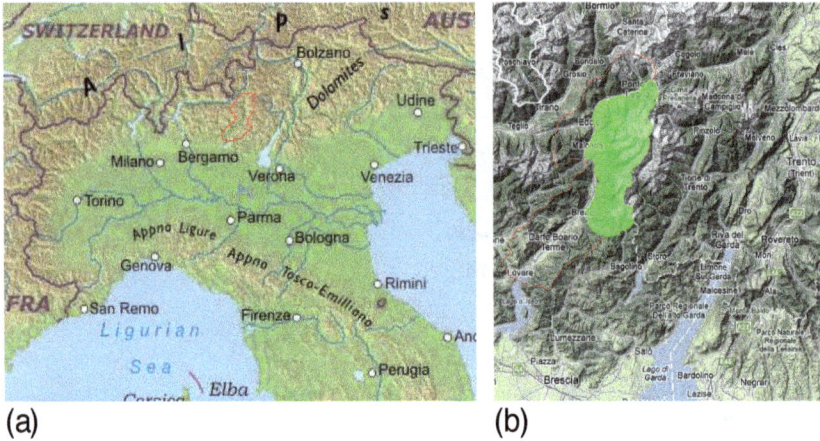

Figure 1. a) Localization of the Valle Camonica in the Lombardy region, Northern Italy. b) Localization of the Adamello park (green area) inValle Camonica (red borders)

1.2. Climate

The climate of the region corresponds to a tempered sub-oceanic type. The rainfalls ranges between a yearly minimum of 900 mm in the valley floor of the southern part of the domain and a maximum of 2.200 mm in the northern part of the Valley. The pluviometric regime show two rain peaks in spring and autumn, and two minima in summer and winter, with an absolute minimum in winter (mainly snow above 1.100 m a.s.l.).

The thermal regime is characterized by an average yearly temperature of 12.5 °C in the valley bottom. The average minimum temperature is registered in January about +2 °C), while the average maximum temperature is recorded in July (around 20 °C). The Iseo lake provides remarkable thermal influence over the southern part of the valley.

Winds have no peculiar characteristics, following the typical breeze regime of a normal alpine valley.

1.3. Vegetation and natural reserves

Following the common italian typological classification, up to 74 forest ecosystem types are represented in the area, classified into 21 wood categories: *Quercus spp.* (oak woods), *Fraxinus ornus-Ostrya carpinifolia, Castanea sativa* (chestnut woods), *Acer psseudoplatanus and Fraxinus excelsior, Tilia cordata, Corylus avellana, Betula pendula, Larix decidua, Pinus cembra, Pinus sylvestris, Pinus mugo, Picea excelsa* and *Fagus sylvatica, Abies alba, Fagus sylvatica, Picea abies, Alnus viridis, Alnus glutinosa e Alnus incana, Robinia pseudoacacia, Salix spp.* and *Populus spp., Laburnum alpinum, Populus tremula, Sorbus aucuparia e Sorbus aria,* plus anthropogenically promoted formations with exotical species and artificial coniferous plantations.

Many Natural Reserves are included in the Park (Figure 2), among which there are fifteen Sites of Community Importance (SCI) plus a national area of special protection for birds (ZPS) which cover approximately 50% of the territory.

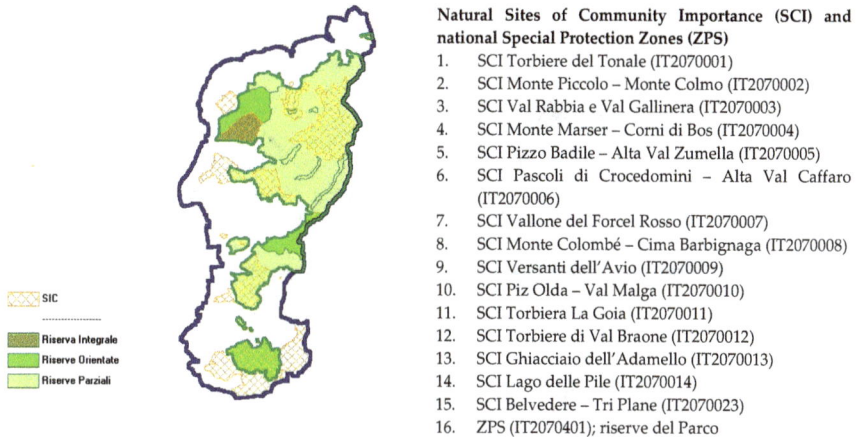

Natural Sites of Community Importance (SCI) and national Special Protection Zones (ZPS)

1. SCI Torbiere del Tonale (IT2070001)
2. SCI Monte Piccolo – Monte Colmo (IT2070002)
3. SCI Val Rabbia e Val Gallinera (IT2070003)
4. SCI Monte Marser – Corni di Bos (IT2070004)
5. SCI Pizzo Badile – Alta Val Zumella (IT2070005)
6. SCI Pascoli di Crocedomini – Alta Val Caffaro (IT2070006)
7. SCI Vallone del Forcel Rosso (IT2070007)
8. SCI Monte Colombé – Cima Barbignaga (IT2070008)
9. SCI Versanti dell'Avio (IT2070009)
10. SCI Piz Olda – Val Malga (IT2070010)
11. SCI Torbiera La Goia (IT2070011)
12. SCI Torbiere di Val Braone (IT2070012)
13. SCI Ghiacciaio dell'Adamello (IT2070013)
14. SCI Lago delle Pile (IT2070014)
15. SCI Belvedere – Tri Plane (IT2070023)
16. ZPS (IT2070401); riserve del Parco

SIC

Riserva Integrale
Riserve Orientate
Riserve Parziali

Figure 2. The natural reserves of the Adamello Park (green lines in the map) and list of the SCI and ZPS included in the Park (right).

1.4. Forestry

The forest surface of the Park is mainly covered with coniferous stands or mixed stands with a coniferous prevalence (more than 70% of the forest surface).

Picea excelsa is clearly dominant, with a standing volume of approximately 1.200.000 m³. Then *Larix decidua* represented with 320.000 m³, *Abies alba* with 25.500 m³ and *Pinus sylvestris* with 16.500 m³ (less relevant amounts are assigned to *Pinus cembra*, *Pinus mugo* and *Juniperus communis*).

The volumetric estimate of broad-leaved species (synthetic estimate) gives back a stock value of 350.000 m³ with a prevalence of *Castanea sativa*, *Fraxinus ornus* and *Ostrya carpinifolia*.

These data witness a strong characterization of the local forests to a secondary type. Most of the main broad-leaved species are in fact not significantly present and this is certainly due to an intensive management of the woods carried out in the past.

The average offer of timber available from the Park is about 10.000 m³ per year of industrial wood and nearly 500 tons of broad-leaved timber (mainly for fuelwood or poles). The industrial wood is then mainly destined to local lumber mills, which usually sell it as unique assortment (44,8%), saw lumber (20,3%), packaging wood (32,9%) and pulpwood (2%). Coppice wood is on the other hand mostly sold on the local market.

On the whole, the level of the offer of timber is probably under its potential, thus implying the need for an improvement of the system in the area.

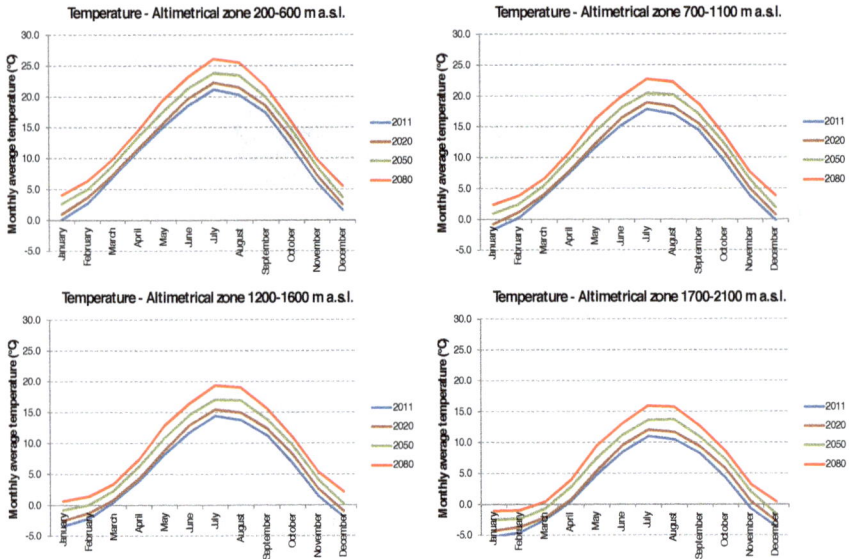

Figure 3. Annual course of monthly averaged temperature for different altimetrical zones. The blue line corresponds to year 2011, the dark red line to year 2020, the green line to year 2050 and the bright red line to year 2080

2. Future scenarios for Valle Camonica and the Adamello Park

2.1. Expected climate

IPCC climate change scenario A1b [1] have been downscaled to the Valle Camonica domain within the framework of the MANFRED project and the results are showed hereafter. All the scenarios foresee a remarkable temperature increase and an overall yearly precipitation decrease in this Alpine area.

The scenario A1b, assumed as a reference since it is an intermediate one and it is the most likely for this area, predicts a temperature increase of 4.0 °C for the lower altimetrical zone (200-600 m. a.s.l.) and an increase of 4.1 °C, 4.2 °C and 4.3 °C for the higher altimetrical zones corresponding respectively to 700-1100 m. a.s.l., 1200-1600 m. a.s.l. and 1700-2100 m. a.s.l. (Figure 3). The highest temperature increase is predicted for August 2080 when the average daily temperatures will likely rise by 5.3 °C over the corresponding values of 2011.

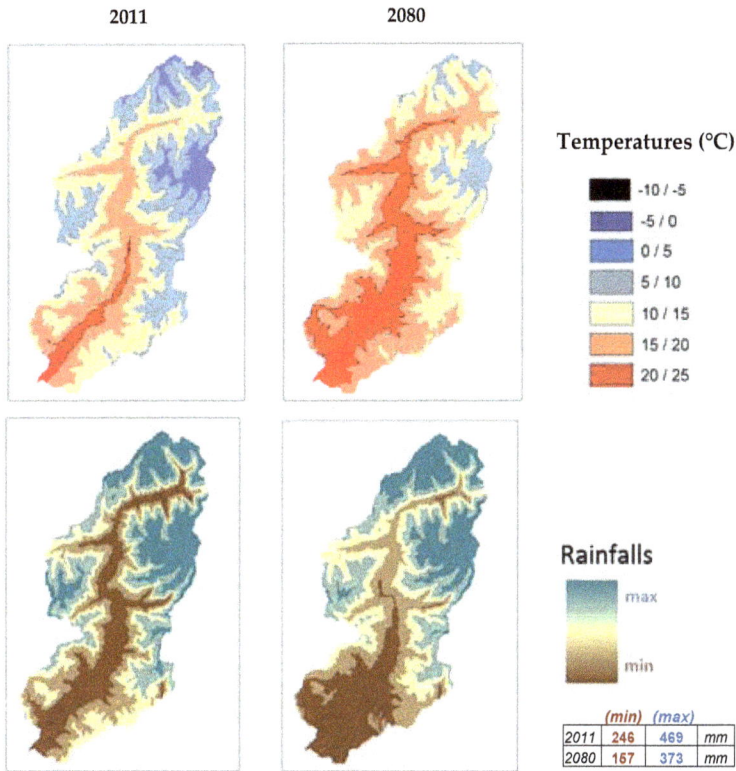

Figure 4. Distribution of the climatic variations in the Summer season in Valle Camonica. Top two maps: summer average temperature for present (year 2011, left) and for future (year 2080, right). Bottom two maps: summer rainfalls distributions for present (year 2011, left) and for future (year 2080, right)

The geographical distribution of these temperature changes can be appreciated in Figure 4 which compares the maps of the the average temperatures in the summer 2011 with their predictions for the summer 2080. While a sensible temperature increase is observed throughout the whole valley, in the year 2080 the highest temperature range (20/25 °C) will be experienced by nearly the half of the southern part of the valley and by the valley floor of the northern part and of the side valleys.

Besides temperature the other parameter which will be significantly affected by climate change in this area is rainfalls. Comparing the future prediction of rainfalls (30 year period between 2071 and 2100) with the present situation, taken as the rainfalls average of the past 30 years (1971-2000), two main facts can be highlighted: a slight increase of rainfalls will characterize the first months of the year (+20%) and the late autumnal period, while a very strong reduction of rainfalls (- 40%) will affects the whole summer period (Figure 5)

Deepening this seasonal trend of rainfalls it can be observed that the increase in the first months is substantially the same for all the altimetrical zones, while the summer decrease will more affect the less elevated areas (Figure 6). Furthermore this decrease will be significantly different throughout the valley. In fact, while on the southern part of the valley less rainfalls will be expected also at high elevation, in the northern part the valley floor will experience an increase in rainfalls (Figure 4).

The decrease in rainfalls will lead to a further consequence, that is the considerably increase of the number of dry days in summer (+ 21 %) and to a lesser extent in autumn (+ 6 %) (Figure 7).

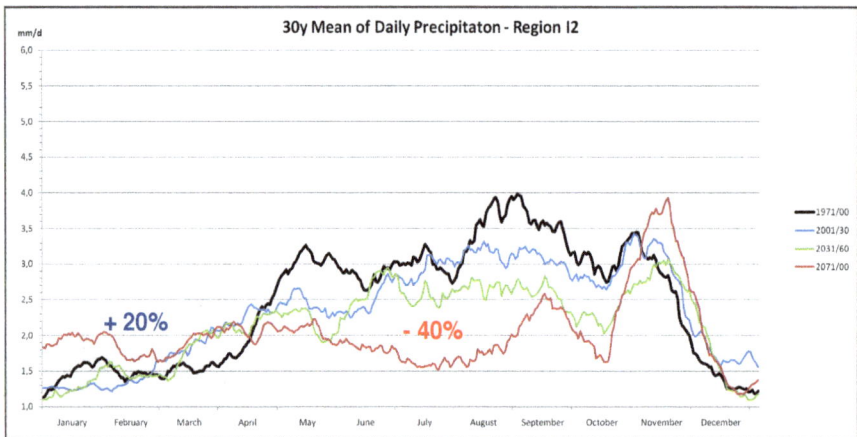

Figure 5. Thirty years mean of daily rainfalls from present to 2100 in Valle Camonica. The black line shows the present situation calculated as the average of the period 1971-2000, the blue line is the 2001-2030 average, the green line is the 2031-2060 average and the red line is the 30 years average between 2071 and 2100. (Elaboration of Zueger and Gebetsroither on downscaled data from Zimmermann)

As a consequence of the increased temperature and of the decreased rainfalls the evapotranspiration of vegetation is expected to decrease. Figure 8 shows the future potential evapotranspiration calculated for the altimetrical zone where higher is the forest coverage (700-1100 m a.s.l.). In that zone, at the present, the vegetation will experience only a very slight water shortage in the half of September. But in the future the water shortage will start earlier, in the mid of June when the plants transpired all the water they received in the previous months (the yellow area in the graph), and it will last until the end of September. After mid June the water supply by rain will not be able to sustain the plants' evapotranspiration demand, and then an increased stomatal control on water losses and a reduced overall actual evapotranspiration could be foreseen.

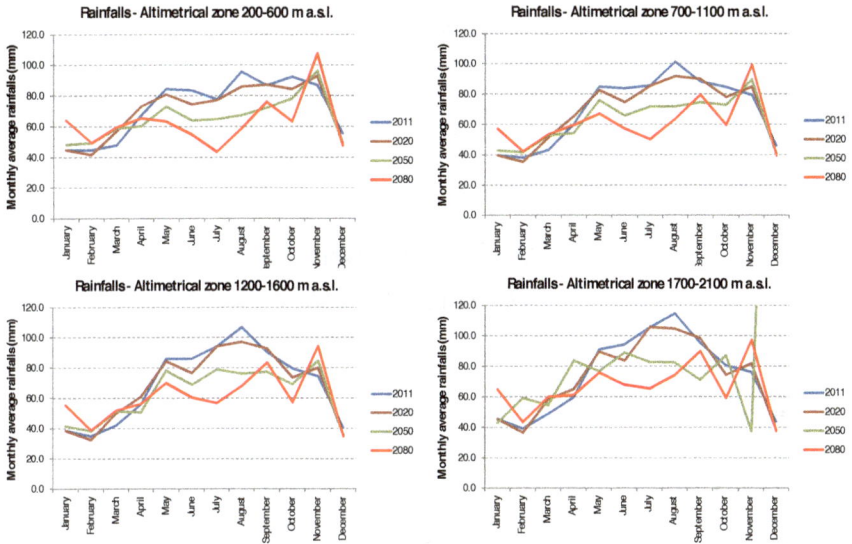

Figure 6. Annual course of monthly averaged rainfalls for different altimetrical zones. The blue line corresponds to year 2011, the dark red line to year 2020, the green line to year 2050 and the bright red line to year 2080

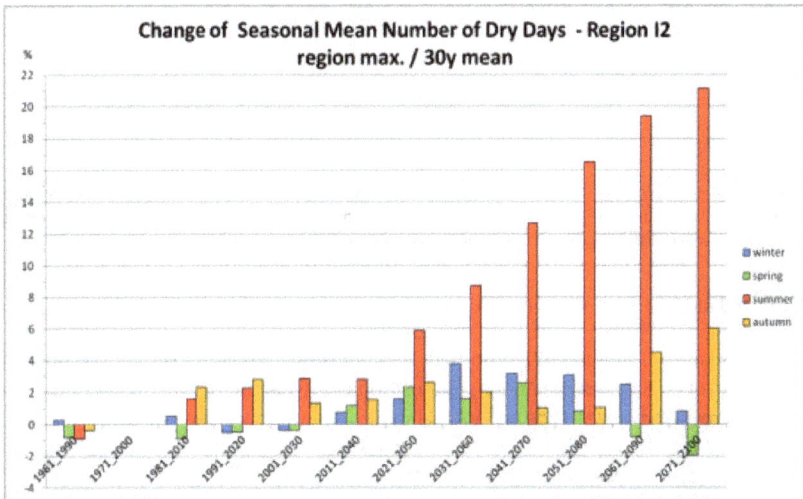

Figure 7. Change of seasonal mean number of dry days. The mean was calculated on a 30 years basis. Blue histograms are winter, green ones are spring, red ones are summer and the yellow ones are autumn

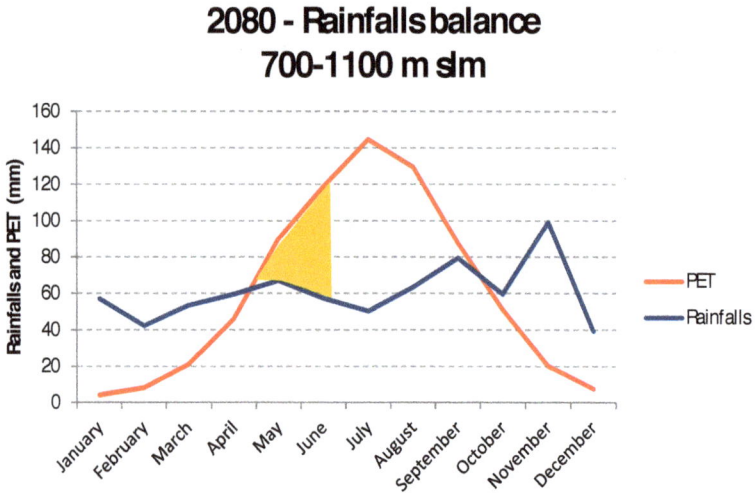

Figure 8. Pluviometric balance for the year 2080 between 700 and 1100 m a.s.l. Rainfalls and potential evapotranspiration (PET) are represented. The yellow area represents the water reservoir which compensates the water deficit when the evapotranspiration demand is higher than the water received with rainfalls.

2.2. Expected shift of potential vegetation

In a recent study G. Pignatti [2] proposed an interesting methodology for analyzing the possible effect of climate change on the composition of the Italian forests. All the Italian forest areas were screened by forest type and local climate. Pignatti found that the distribution of the different forest types can be described by taking into account a simple climatic descriptor: the average temperature of the coldest month of the year.

Based on the average temperature of January, six different climatic zones were defined (A to E: <8°C, -8°/-4°C, -4°/0°C, 0°/+4°C, 4°/8°C and >8°C) to which correspond a different class of forest type (*Alpinetum, Picetum, Fagetum, Castanetum, Cold Lauretum* and *Warm Lauretum*).

The employment of the Pignatti's methodology gives an idea of the potential distribution areal of the different forest types, and by using the temperature forecasts of the climatic models it allows to realize whether new types of vegetation might find a suitable climate for them in a given area. It must be remarked that these results does not necessary mean that these new types of vegetation will onset in that area at the given future year, but that a vegetation migration process will start and that the given specific area will be a future destination of those vegetation types. However the process can be very slow, and it could be further slow or even stopped by the competition with the pre-existing vegetation which – within certain limits - could adapts to the new climatic conditions.

This methodology was applied to the Valle Camonica domain using the temperature distribution predictions of the years 2011, 2020, 2050 and 2080 (Figure 9). It can be observed that in 2080 the climatic area corresponding to the spruce (*Picetum*) should almost disappear, while the beech climatic area (*Fagetum*) will shift to higher elevations. The climatic area of chestnut (*Castanetum*) will significantly expand throughout the whole valley, particularly in the valley floor, and the warmer climate will create a suitable areal for the evergreen oak *Q. ilex* (*Lauretum*) in the valley bottom of the southern part of the Val Camonica domain.

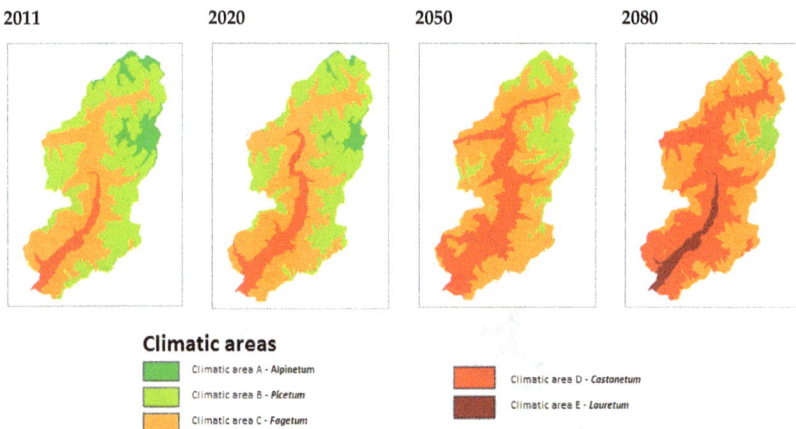

Figure 9. Climatic areas for the reference years 2011, 2020, 2050, 2080 following the methodology proposed by Pignatti

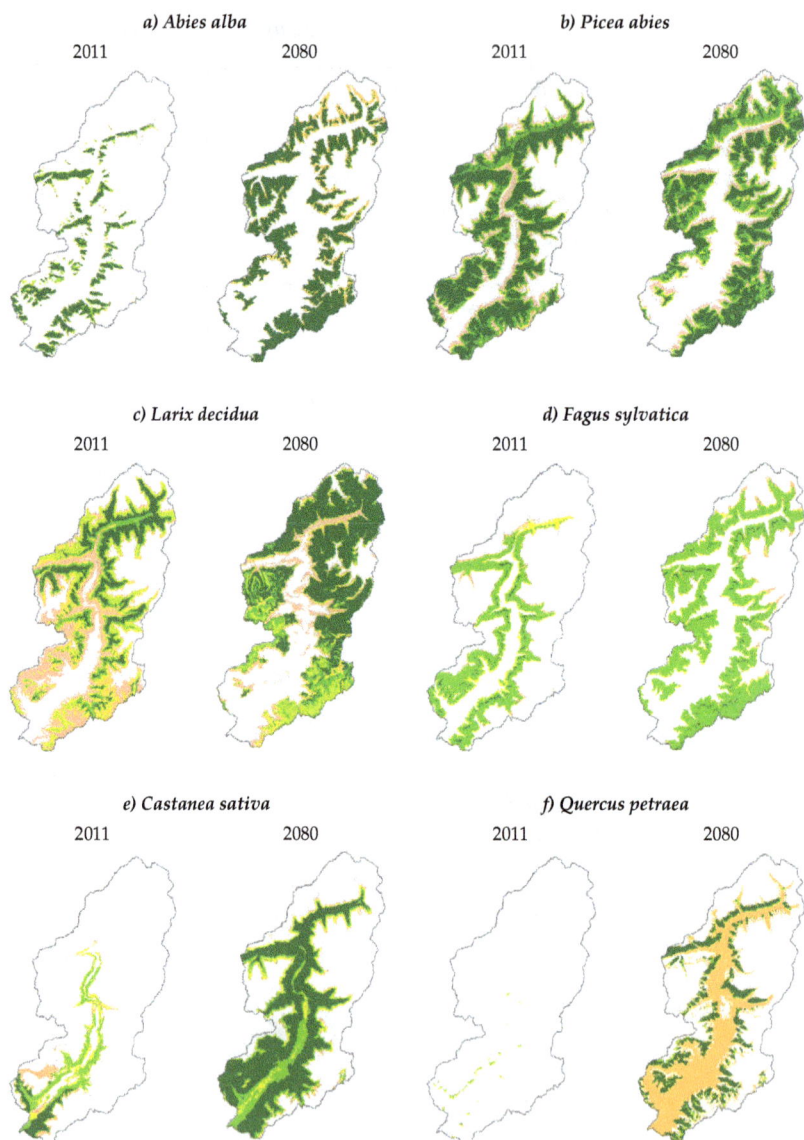

Figure 10. Areals evolution of six different tree species from present (2011) to future (year 2080): a) *Abies alba*, b) *Picea abies*, c) *Larix decidua*, d), *Fagus sylvatica* e) *Castanea sativa*, f) *Quercus petraea*. The greener the color the higher the probability to find the species in each point; lighter areas are less suitable for the species and white areas are not suitable areas.

These results are substantially confirmed by a more detailed climatic envelops forecasts performed by Zimmerman within the MANFRED framework for the 6 main forest tree species of Europe. Using ecological and climate models, Zimmerman downscaled to the valle Camonica domain the future potential areals of *Abies alba, Picea abies, Larix decidua, Fagus sylvatica, Castanea sativa and Quercus petraea* and, which are showed in Figure 10.

It was worth noticing, again, that the areals does not represent the future spatial distribution of the species, but only a region where climatic conditions will be suitable for their growth, regardless of the ecological competition with the pre-existing species. Figure 10 shows a vegetational shift towards higher elevation for *Abies alba* and *Fagus sylvatica*, with a slight expansion of the present coverages. *Castanea sativa* will spread throughout the whole valley floor and in the mountain areas of the southern part of the valley, and *Quercus petraea* might significantly increase its presence in the valley. *Larix decidua* and *Picea abies* will decrease in the southern part of the valley (which will be hotter and with less rainfalls), but these two species will find a better and expanded habitat in the northern part of the valley and at higher elevation.

Where the Zimmerman's and Pignatti's predictions differ are on *Picea Abies*. Even if both predicts an altitudinal shift and a restriction of the spruce areal, Zimmerman does not foresees a disappearance of this species form the Val Camonica domain, but only a slight reduction of it.

3. Climate change impacts: fire and ozone hazards

The predicted pattern of future CC (increasing in summer temperature and winter precipitation) is accounted to produce a shift of conditions (both in time and space) leading to a more severe exposure to natural hazards than present. In this study, current (2011) and future scenario (2080), have been compared analyzing the influence of CC on fire ignition risk and ozone exposition in Valle Camonica.

3.1. Forest fires — 10-years' trend in Valle Camonica (2002-2011)

In the Valle Camonica during the last decade (2002-2011) a total of 246 fires caused a forests loss of about 160 ha/yr[-1]. Number of fires and burned area were particularly high in hot and dry years (e.g. 2002 dry winter and 2003 hot summer) resulting also in much larger fires. Winter and spring are the season mainly affected by the phenomenon both in terms of number and burned hectars (Figure 12). The majority of fires are caused by human action, mostly related to crime acts and negligence whilst natural causes represent only a small percentage of the overall amount. Since the beginning of 2000 thanks also to the National Law on forest fires (L. 353/2000), protective and preventive measures have significantly reduced the number and the average size of fires in respect to the past.

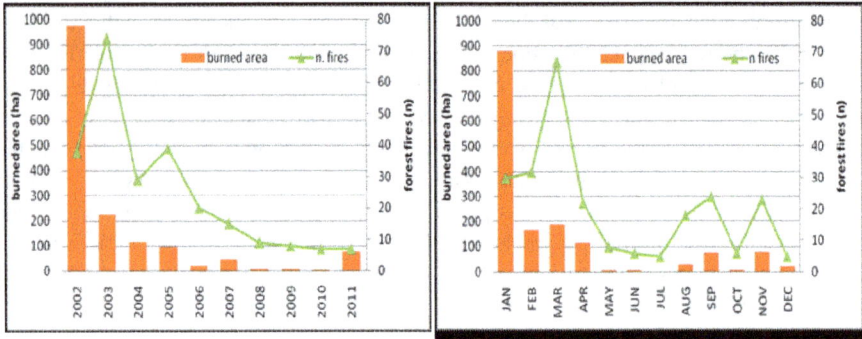

Figure 11. Frequency of fire events and burned area per year (right) and month (left) over the period 2002-2011

3.2. Forest fire risk in Valle Camonica

In this study fire risk has been considered as composed by two parts: hazard and vulnerability. The relationship between the two components is defined as follow:

Risk= Hazard * Vulnerability

According to the fire regime present in Valle Camonica, and in particular the monthly pattern (Figure 11), in our analysis, we evaluated the evolution of fire risk for the so called "fire season", from January to April, where the most part of fires occurred and assumed to occur also in the future. Fire hazard and fire vulnerability have been evaluated by the means of a tool (4.FI.R.E. - FORest FIre Risk Evaluator) developed in the framework of the MANFRED project by ERSAF (*cfr* Chapter 6). The tool, composed by two main modules (4.FI.R.E.- Hazard and 4.FI.R.E.- Vulnerability), allowed us to generate hazard and vulnerability maps for current and future scenario, then combined to achieve the final risk map.

3.2.1. Fire hazard

Current and future scenario were produced by using the module 4.FI.R.E. - Hazard. The methodology for hazard calculation provided in the tool is based on [3]. Conceptually, the production of the hazard map is made of three simple steps (Figure 12): i) Layers pre-processing; ii) Monte Carlo simulations to produce Monte Carlo scores maps and lookup tables; iii) Layer overlay.

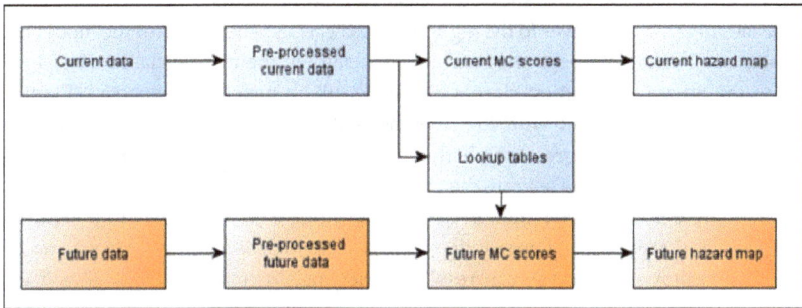

Figure 12. R.E. - Hazard. Workflow

4FI.R.E.–Hazard has been designed to allow the creation of hazard maps both for current and future scenarios. In the first case, the Monte Carlo scores maps are calculated straight from the input layers. In the second case, the Monte Carlo scores maps are calculated using lookup tables derived by comparing the current input layers and their Monte Carlo score maps. Input data consist of a broad set of variables accounted for influencing the proneness of a specific area to forest fire events (i.e. morphological features, climatic conditions, anthropic elements, Table 1).

List of input variables	Variable used in the model
Map of ignition sources	Yes
Digital Elevation Model	Yes
Land use map	Yes
Roads map	No
Forest map	Yes
Precipitation (interpolated map)	Yes
Temperature (interpolated map)	Yes

Table 1. R.E. – Hazard. List of input variables.

Road map was not included in the final model due to the lack of information for future trends in space and time. Topographic attributes (i.e. elevation, slope and aspect) were derived from a Digital Elevation Model provided by the cartographic service of Lombardy region. In this work we assumed that morphological features will not be subjected to significant changes for the future. Therefore, the same raster datasets have been included in both the model runs (2011 and 2080). Site-specific climatic data (temperature and precipitation) and land use map at 2011 and 2080's scenario provided by WSL were included in the final model. In order to characterize the territory with meteorological variables representative of the fire season, average temperature of March and total precipitation within the period January-April were considered as input data.

Results for the current scenario showed a clear decreasing trend, where the area tends to decrease moving from the lower to the highest hazard classes. The results of the model for the future scenario indicated a potential decrease especially in the middle classes (2-4, 4-6, 6-8) whilst for the lowest and highest classes (0-2 and 8-10) an increase is expected (Figure 13). These findings are consistent with the hypothesis that future climate changes will drive changes in the magnitude of forest fires. In the analysis we carried out, meteorological variables are the most important factors leading the change in fire hazard. It is likely that fire regime of Valle Camonica will face a new situation with an increased likelihood to have a potential shift in the seasonal distribution of forest fire occurrences. Increased temperatures and a net decrease of precipitation, in fact, can result in drier fuels increasing the likelihood of fire ignitions. According the future trend of temperature and precipitation, we can speculate that hotter seasons (e.g. spring and summer) will face in future an higher probability of fire events.

Figure 14 illustrates the hazard maps for 2011 and 2080. Considering the spatial distribution of the phenomenon, the hazard increases is more evident at the lower altitudes where the higher class (8-10) tends to replace the lowest ones.

This result is not surprising, since it is straightly leaded by land use and climate change (in particular temperature) which will cause the treeline shift toward higher altitudes. This process is well documented and many studies [4, 5] reported as alpine ecotones are assumed to be particularly sensitive to altered temperature regimes. and climate warming is expected to drive treelines upslope. As a consequence species more sensitive and prone to forest fires (e.g. chestnut and oak) are expected to extend their spatial range at higher elevations increasing the likelihood of having forest fire occurrences.

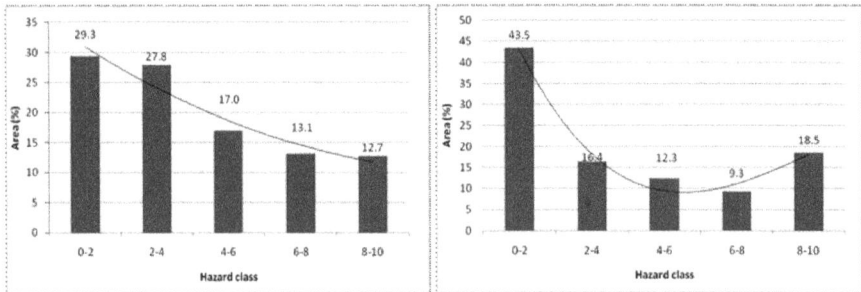

Figure 13. Territorial distribution (%) for each hazard class at 2011 (left) and 2080 (right)

3.2.2. Fire vulnerability

Fire vulnerability for 2011 and 2080 were calculated by using the second of the modules of the tool (4.FI.R.E. - Vulnerability). The process of vulnerability modelling builds on a 4-step design method: i) input selection; ii) layer pre-processing; iii) layer overlay; iv) output creation.

The final output consists of a map resulted from the simple sum (i.e. linear model) of the input layers rescaled to defined vulnerability classes. In the final vulnerability maps (2011 and 2080) percentiles were used to define the reference intervals. As for the hazard model, some variables were not accounted in the final model due to the lack of information for future trends in space and time. Input variables considered in the tool and those included in the model are reported in Table 2.

Protective, productive and naturalistic functions were implemented by using proxy variables (respectively slope, slope and accessibility, presence of protected areas).

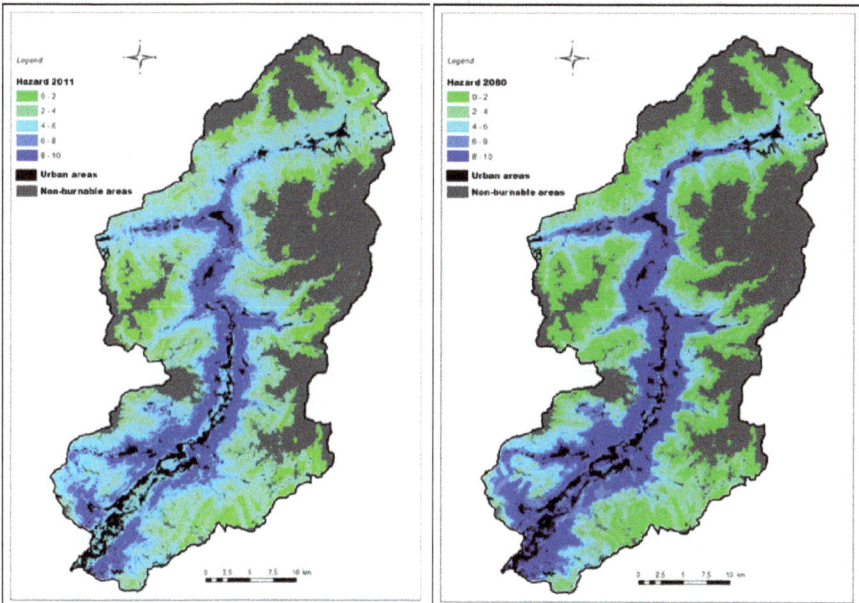

Figure 14. Hazard maps for current (left) and future (right) scenario

The final maps for Valle Camonica (Figure 15) shows the significant vulnerability along the valley bottom given to the presence of areas more accessible and characterized by anthropic presence (urban areas). The scenario for 2080 depicts small changes for vulnerability of urban elements. It is important to notice that the results of the analysis are affected by the input data. In this sense, it is expected that urban areas (the most vulnerable landscape elements) will not be subjected to significant changes during the analyzed time span. The map for 2080 reflects this small change in the highest class. It is also possible to appreciate, for the future scenario, the pattern of vulnerability increase for protected areas. These areas are expected to face an increase of vulnerability probably due to the expansion of forest within those regions currently occupied by non-forest habitats (e.g. meadows and pasture). The natural afforestation and/or re-afforestation of areas currently not covered by forest will increase, in the future scenario, the vulnerability of the territory.

3.2.3. Fire risk

The fire risk was calculated by combining the two maps resulted from the analysis of hazard and vulnerability for current and future scenario. As for vulnerability, percentiles were used to define the risk classes for 2011 and 2080.

List of input variables	Variable used in the model
Forest types resistance	No
Forest types resilience	No
Protective function (proxy slope)	Yes
Productive function (proxy slope and accessibility maps)	Yes
Naturalistic function (proxy protected areas)	Yes
Touristic function	No
Carbon stock function	No
Presence of vulnerable elements (urban areas)	Yes
Exposition (potential damage level of due to the distance from the forest areas)	Yes
Resident population map	No

Table 2. FI.R.E. - Vulnerability. List of input variables

The final risk maps of Valle Camonica (Figure 16) follow the results of hazard and vulnerability. Risk is higher for those areas located at the bottom of the valley and along the mountainsides where the forest are likely to increase their presence in the future.

The major consequences at territorial level are related to the increased safety risk for people, structure and infrastructure. Forest fires in Valle Camonica, as in the most part of the Alps, have a limited direct impact in respect to others natural hazards (e.g. debris flows). On the other hand, forest fires could lead or favor important secondary consequences. In mountainous areas severe forest fires, damaging the forest canopy as well as the soil, could result in increased runoff or sudden snowmelt, which can put homes and other infrastructures below a burned area at risk of safety. Many authors [6, 7] reported how debris flows can be one of the most hazardous consequences of rainfall on burned slopes.

Therefore, the future increase of fire risk should not be considered only for its direct impact but to the future impacts on the overall territorial system of Valle Camonica. Appropriate management strategies for forests and preventive measures for forest fire fighting must be addressed to reduce the risk of fires but bearing in mind also other fire-related or post-fire effects.

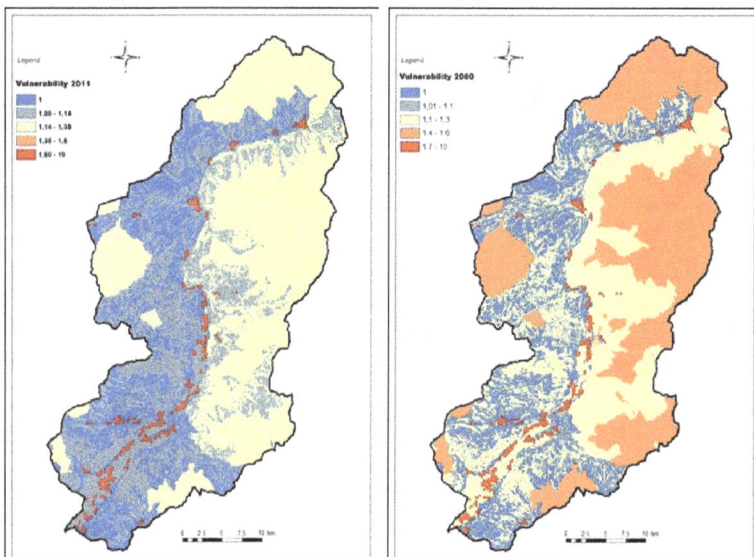

Figure 15. Vulnerability maps for current (left) and future (right) scenario

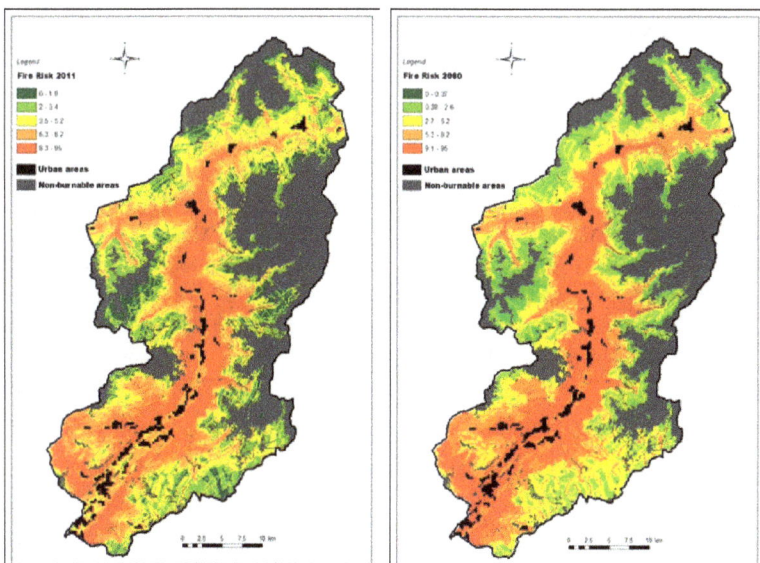

Figure 16. Risk maps for current (left) and future (right) scenario

3.3. Ozone

Ozone is a secondary pollutant produced by photochemical reactions between nitrogen oxides (NO$_x$) and volatile organic compounds (VOC). No direct ozone sources, both anthropogenic and natural, exist. The photochemical reactions that produce ozone are burst by UV radiation, leading, as a consequence that ozone concentrations are typically higher in summer and in mountain regions, where the breeze cycle can transport the ozone precursors from much polluted areas, like it usually happens in the Southern side of the Alps.

Ozone phytotoxicity for vegetation is a well-known problem [8]. The penetration of ozone through stomata is the most harmful way to which the vegetation is exposed. In fact, once ozone enters the substomatal cavity, it quickly reacts and oxidative radicals like hydrogen peroxide, superoxide and similar compounds are produced. These compounds produce many injuries like the membranes disruption through the peroxidation of the membrane lipids and the oxidation of the reduced groups of biomolecules [9, 10].

All these microscopic effects can produce, in sensitive species, visible injuries at leaf level, such as bronzing, chlorosis and necrosis, increased transparency at crown level, a reduced root growth and an increased susceptivity to both biotical and abiotical stresses. Both ozone sensitive and non sensitive species to ozone can experience a productivity reduction as an ozone effect.

At ecosystem level ozone can cause a change in species composition being disadvantageous to the more sensitive ones; furthermore functional changes like a reduced carbon storage capability, an increase in water loss and a reduction of the forest capability of stabilizing mountain slopes can be observed.

3.3.1. The ozone risk assessment procedures: Level I and Level II assessment

The ozone hazard for vegetation had been widely studied since several decades. The United Nations Economic Commission for Europe (UN/ECE) gave birth in 1979 to the Convention on Long-range Transboundary Air Pollution (LRTAP). In the LRTAP framework the Gothenburg protocol (1999) was approved, fixing the critical levels of ozone for vegetation with target values for the following years. A following EU directive (2002/03) acknowledged the Gothenburg protocol, introducing the AOT40 (Accumulated Ozone Exposure over a threshold of 40 ppb) index. The AOT40 is defined as:

$$AOT40 := \sum_{\substack{[O_3]_i > 40\,ppb \\ RadGlob > 50W/m^2}} ([O_3]_i - 40) \cdot \Delta t \tag{1}$$

that is the sum of the ozone concentrations (when they are above 40 ppb) minus the threshold of 40 ppb multiplied by the concentrations evaluation time, which is usually one hour [11]. This sum is calculated only during daylight hours in order simulate the stomatal behavior of vegetation.

AOT40 is employed to map ozone hazard. Then the hazard map is overlapped with the vegetation coverage map (i.e. a vulnerability map) to obtain the so called Level I risk assessment for ozone.

The AOT40 critical level was set by EU for vegetation is 18000 μg m^{-3} h over a period of six months, from 1st April to 30th September. This critical level can be expressed also in terms of ppb h, and its value is 9000 ppb h.

The EU directive gives recommendations to the member state about how to measure ozone and how to calculate AOT40 but, since the monitoring networks are mostly dedicated to estimate the risk for the human health rather than vegetation, ozone data in mountain regions are not so often available, leading to a not proper evaluation of the ozone hazard for forests. Three main measuring options are available to fill this knowledge gap on ozone hazard: mobile laboratories (Figure 17a), passive samplers (Figure 17b) and modeling predictions. All these approaches have their pros and cons: mobile labs can collect hourly data but they require high electric power so they cannot be easily employed in remote areas; passive samplers can be easily employed in remote areas but their time resolution is too coarse, usually one week; modeling predictions are necessary to spatialize the data but they require several measurements to validate the model outputs.

From a regulative point of view AOT40 is, at the moment, the only adopted index to evaluate the ozone hazard, but its scientific soundness is under criticisms. Being a mere exposure index, AOT40 does not take into account the physiology of the vegetation which is exposed to ozone. The ozone damages to vegetation are produced by the molecules entering the plants through the stomata, but plants can regulate stomatal opening as a response to different environmental conditions. The magnitude of the negative effects on vegetation is related to the real amount of this pollutant taken up through stomata, i.e. the dose or stomatal flux, and not simply to the ozone molecules which are surrounding the plants. So high ozone concentrations in the air do not necessarily lead to high ozone doses, thus representing only a potential risk for plants (more correctly an hazard).

The UN/ECE scientific community is hence moving toward a Level II approach based on stomatal fluxes. The Level II approach overpasses the pure risk concept leading to the direct estimation of the negative effects on plants and allowing an evaluation of the related economical losses.

The level II risk assessment requires both models and (better) measurements. Direct measurements of ozone fluxes can be performed by means of the eddy covariance technique, which at the moment is the best available one. Nevertheless these measurements allow to estimate only local ozone risk, but they are of capital importance to parameterize and validate the models, which, once all the input data are available, allow to estimate ozone risk at regional, national or continental level.

The eddy covariance technique is based on the atmospheric turbulence and it requires specific instrumentation, which must be able to measure at least ten times per second the three wind components, the air temperature and the ozone and the other gases concentrations. This instrumentation is mounted above the studied ecosystem.

The ozone fluxes are calculated from the rapid measurement by the following Eq. 2:

$$F_{O_3} = \overline{w'O_3'}$$
(2)

where w is the vertical component of the wind intensity and O_3 is ozone concentration; primed variables mean the fluctuations around their 30 minutes averages which is represented as overscript bar. Additional meteorological measurements are useful for a better comprehension of the stomatal ozone uptake process.

In order to estimate how much ozone enters through plant stomata (the most harmful pathway) water fluxes are used as a tracer, assuming that ozone can enter through plant stomata only when vegetation is transpiring. So the stomatal flux $F_{O3,stom}$ is derived from F_{O3} by using the evapotranspiration data and the Penman-Monteith equation [12] following the procedure described by Gerosa et al. [13].

The calculation of the stomatal flux allows then to estimate the ozone dose (which is 'simply' the sum of the stomatal ozone fluxes $F_{O3,stom}$) and the Phytotoxical Ozone Dose (POD$_1$), which is the cumulated dose over an instantaneous flux threshold of 1 nmol O_3 m^{-2} s^{-1}:

$$POD_1 = \sum_{\forall F_{O_3,stom} > 1} (F_{O_3,stom} - 1) \cdot \Delta t$$
(3)

The POD$_1$ have been introduced in the UN/ECE scientific community because it takes into account the internal capability of the vegetation to detoxify part of the ozone entering through the stomata. Furthermore, many experiments have showed that the POD$_1$ is better correlated with the biomass reduction than the simple ozone dose, allowing thus to estimate the harmful effects of ozone on vegetation.

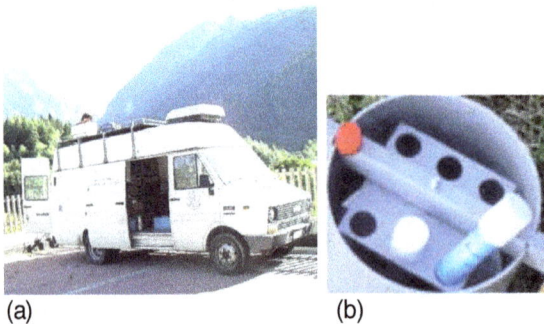

(a)   (b)

Figure 17. The Mobile Laboratory (a), and Passive Samplers for ozone (b) employed in the MANFRED project

3.3.2. Ozone risk assessment in Valle Camonica

Even though ozone concentrations have been estimated to be quite high in the whole mountain areas of the Lombardy region, as reported for instance by Gerosa et al. [14], only since 2007 an ozone monitoring station had been running in Valle Camonica, in Darfo. Furthermore, this station is located at a position which is not very suitable for the ozone hazard assessment for vegetation (low elevation and nearby a crowded main road). The lack of information on ozone hazard for forest in Valle Camonica, one of the widest forested area of the Region, lead to the choice of this valley as a focus for this topic.

In this project both Level I and II approaches were performed in Valle Camonica, and several measurements had been running in 2010 and in 2011.

The results from an extensive ozone monitoring campaign throughout the valley and the following mapping exercise and some results from the micrometeorological tower installed in Paspardo over a *Larix Decidua* forest are presented in this section.

Level I risk assessment had been realized by running a six month field campaign with a mobile lab and passive samplers (Figure 17) located in 11 sites throughout all the valley (Table 3): 10 of them were placed in remote forest areas while the remaining one was placed in Darfo, near the only automatic monitoring station available in Valle Camonica (Regional environmental agency, ARPA). The sites elevation ranged between 300 m and 1800 m a.s.l. spatially covering all the forest areas of the valley. Table 3 shows the coordinates of the sites and other additional data. Two passive samplers were exposed for one week in each site, inside a protective shelter, the first exposure began at 6th April 2010 and the last one (the 26th one) ended at 6th October 2010. The exposed samplers were analyzed and weekly averages of ozone concentration were obtained.

In order to calculate the AOT40 it was necessary to estimate hourly ozone concentrations starting from the weekly averages following the methodology proposed in [15] and reported in chapter 9.

Then, weekly values of AOT40 were mapped on the Valle Camonica domain using a geostatistical technique known as ordinary kriging, which requires a model of the spatial data variability estimated from the semivariogram plots. Details on all the applied methodology, as well as the resulted AOT40 map for the whole summer semester, can be found in chapter 9 (*cfr* Figure 10).

The map of the ozone risk for forests in Valle Camonica (Figure 18a) has been obtained by overlapping the AOT40 map (Figure 10 in Chapter 9) to the forest covered areas in the valley.

It can be observed that all the valley forests resulted under ozone risk because the AOT40 was almost everywhere well above the EU critical level for forest protection (9000 ppb h). The lowest ozone exposure was experienced by the forest located in the central part of the valley and in the southern part of the valley floor. In the southern mountain areas the critical level was exceeded from two up to five times in the conifer forests (Figure 18c) and up to four times in the broadleaves ones (Figure 18a), with the highest exceedances in the most elevated areas.

ID	Site name	Latitudine	Longitudine	Municipality	Elevation [m asl]	Relative Elevation within 5 km
1	Borno	45°56'36.79"	10°10'59.86"	Borno	1000	750
2	Crocedomini	45°56'30.34"	10°19'29.59"	Breno	800	550
3	Mortirolo	46°13'08.14"	10°19'36.67"	Monno	1300	650
4	Gavia	45°17'15.64"	10°31'08.16"	Ponte di Legno	1780	600
5	Darfo	45°52'29.64"	10°10'41.58"	Darfo Boario Terme	320	20
6	Malonno	46°05'30.97"	10°18'41.49"	Malonno	800	400
7	Montecampione	45°49'24.61"	10°09'52.19"	Piancamuno	980	680
8	Temù	46°14'07.94"	10°28'25.20"	Temù	1380	240
9	Pescarzo	46°02'02.08"	10°19'37.19"	Capo di Ponte	640	300
10	Mù	46°10'57.92"	10°20'19.75"	Edolo	800	340
11	Niardo	45°58'1.69"	10°19'31.11"	Niardo	600	280

Table 3. Passive sampler location in the valle Camonica campaign.

The northern part of the valley experienced the highest ozone exposure which ranged between four to eightfold the critical level.

Figure 19 summarizes the ozone risk for the valley forests by reporting the distribution of the broadleaves and conifers areas as a function of the AOT40 values. It can be realized that 98% of the forest areas experienced an ozone exposure between one and five times the critical level set by EU.

Higher exposures are experienced almost only by conifers since so high AOT40 values are typical of the high elevations sites where the high UV intensity strongly influences the ozone formation. Two third of the conifers stands (67%) suffer an ozone exposure above two times the critical level (CL), and the more represented exposure classes for conifers are those between 2 an 3 time the CL. Compared to the conifers, only one third of the broadleaves stands (exactly 40%) falls in ozone exposure classes above two times the CL, and the more represented exposure class for broadleaves is exactly the AOT40 class equal to 18'000 ppb h.

The magnitude of the exceedances from the critical level could be taken as an indicator of the level of ozone risk for the vegetation growing in a given area. From this point of view it could be concluded that the forest vegetation of Valle Camonica is subjected to a significant ozone risk, particularly the conifers. However, again, it must be remembered that AOT40 represents only a potential damage and a more appropriate ozone risk assessment should take into account ozone fluxes [11], as highlighted by UN/ECE.

Ozone risk was thus evaluated also with a Level II risk approach, by running a micrometeorological tower in Paspardo (Figure 20) in the summer periods of 2010 and 2011. This allowed, for the first time in the Alps, to study the ozone uptake of a larch forest using the eddy covariance technique.

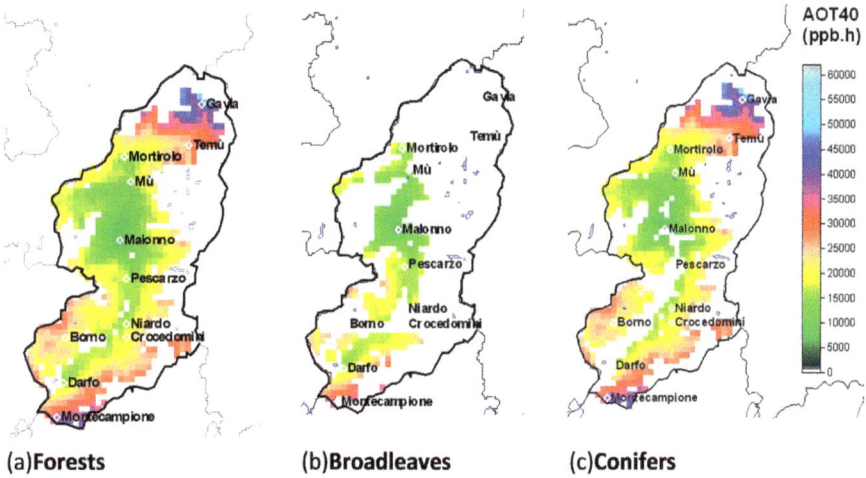

Figure 18. Ozone risk maps for forest vegetation in Valle Camonica. The AOT40 of daylight hours measured in the summer semester 2010 has been reported for each valley area of 1x1 km² where a forest greater than 1 hectare were present. The sites where the passive samplers were located are also indicated. a) All forest areas; b) Broadleaves forests; c) Conifers forests

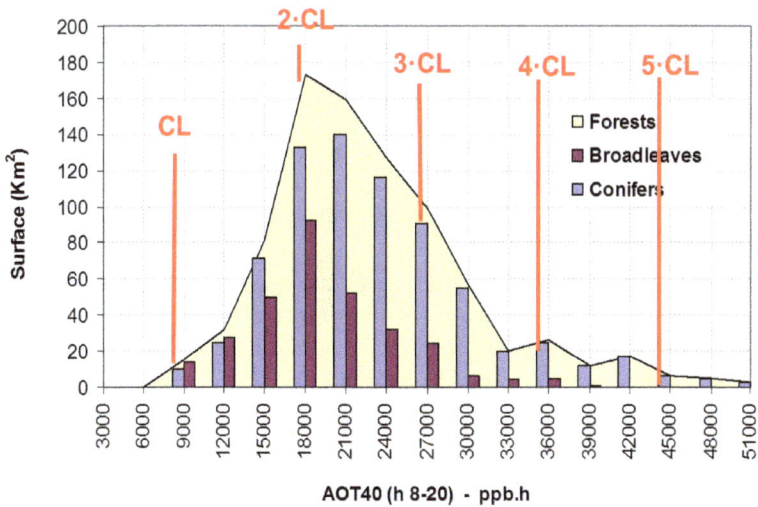

Figure 19. Distribution of forest, broadleaves and conifers coverage as a function AOT40 values. The red lines indicates the times of the exceedances of the critical level

The data processing methodology followed to get to the stomatal flux of ozone is almost complex and has been described in Chapter (8). It included a careful flux partition among the different ozone sink processes, either stomatal uptake and deposition on non-transpiring surfaces or removal by gas-phase reaction in the trunk space, and a separation of the larch uptake from the ozone uptake by the understorey grass.

The result is the cumulated ozone stomatal dose, calculated as POD_1 according to the recent UN/ECE recommendations, reported in Figure 21. The graph compares the accumulation dynamics of POD_1 and AOT40 along the 2010 summer season. The different behavior of these two indexes highlight the criticisms moved to the AOT40. While the POD_1 grew almost linearly for all the measuring period, following the physiological activity of larch, the AOT40 grew more irregularly following the alternation of the photochemical episodes. For example AOT40 had a almost negligible increase in July when ozone was low, but greatly increased in August in concomitance with some photochemical episodes, and then had a very slight increase in the last weeks of the measuring period.

The final POD_1 reached 17.9 mmol m^{-2} while the AOT40 increase stopped at 5150 ppb h. Since the final AOT40 did not overpass the critical level for plants, no risks for plants could be concluded adopting a Level I approach for the studied forest. However, these conclusion differ from how could be argued looking at the stomatal dose of the Level II approach.

Hypothesizing that the dose-effect relationship for larch - which does not yet exist - was the same of the POD_1-effect relationship described for Norway spruce, a loss of biomass increase of -4.5% could be estimated for our larch ecosystem in the measuring period. This relationship published in the UN/ECE mapping manual, in-fact, reports a decrease of 2% of biomass growth every 8 mmol m^{-2} of POD_1 received by trees, and this compared to the growth exhibited by trees which did not receive ozone at all. Further details on these measurement can be found in Chapter 8.

This means that our larch ecosystem had grown a 4.5 % less because of the ozone uptake in the measuring period. This could appear a small ozone effect, but considering that larch has low growing rates the effects could be significant on long term. Moreover the less energy available for growth imply a possible enhanced vulnerability of larches to other stress factor, both biotic or abiotic.

Coming to the future, any performed attempt of estimating future ozone fluxes was almost impossible. The time scale of the output of climatic and chemical scenarios (one day for meteorological data and one month for ozone concentrations) is too coarse to allow any significant estimate and only some general remarks can be given.

The two scenarios used to estimate the future AOT40 levels, substantially foresee no significant increase in future ozone concentrations. But the climatic changes is expected to affect the ozone-plants interactions, and thus the ozone uptake. The strong decrease of summer rainfalls, for example, will cause a reduction of the stomatal openings due to the water shortage, thus preventing the vegetation from ozone damages.

On the contrary the increase in winter and spring rainfalls may cause an increase of the ozone uptake in evergreen species.

Figure 20. The micrometeorological tower running in Paspardo in 2010 and 2011 (a), and a view of the larch forest around the tower (b)

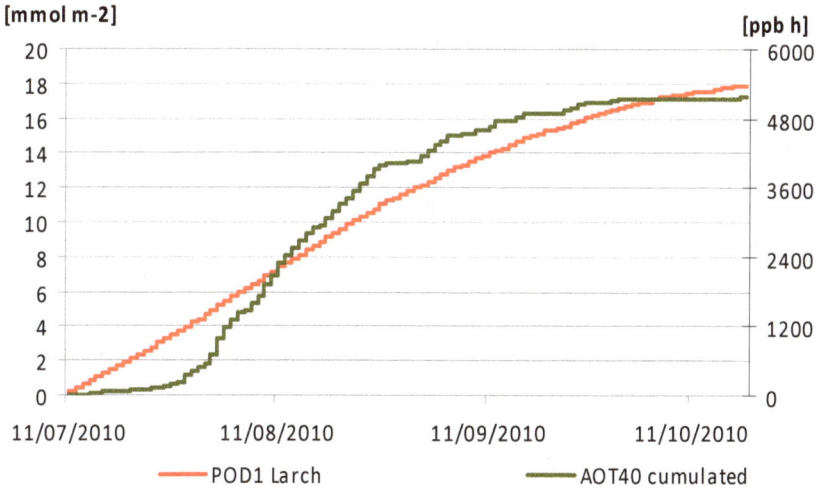

Figure 21. Cumulated POD₁ (red line, left y-axis) and AOT40 (green line, right y-axis) in the measuring period (three months) of the summer 2010.

4. Forestry towards the climate change scenario: dealing with uncertainties

The uncertainty that characterizes the approach to forest management for the next century requires a profound revision of the proven strategies of European forest policy. The trends in climate change are among the main factors of this uncertainty, but not the only ones.

At least two more factors should be considered:

* the enhancing in the magnitude of extreme events: both for the presence of more favorable conditions for pest outbreaks and for augmented vulnerability and stress conditions for the trees;

* the evolution of timber and forest products market: it is hard to foresee the future behavior of BRIC countries about this specific market and what the global response will be.

Some elements that could support the development of a new forest policy course, able to deal with the uncertainty, are here indicated.

It is at present a widespread opinion that forest policy should proceed towards open man-agement systems and provide methods and programs that shall be flexible and easily adapt-able to rapid changes in context.

On the other hand temporal coherence and transparency in the goals are strongly required. The decisions taken today have to be supported by continuity. Once established, the objectives of the national forest planning should be absolutely clear and rapidly adaptable to the current focus. Obviously this possibility is ensured by a strongly efficient monitoring system.

A full transparency in the goals should also be in the future a requirement for every subsidiarity intervention, configuring every investment into a rational and coherent scheme.

Moreover we need the full coordination of all matters dealing with territory (agriculture, construction, facilities, tourism, industry, etc.). Forest governance forms that have not been adequately harmonized and integrated with other soil-consuming activities might in fact prove themselves meaningless and totally ineffective.

Finally a further admonition has to be done, concerning illegal forms of forest exploitation (currently available data indicate that at least 50% of withdrawals in the tropical area are illegal).

Initiatives such as codes of regulation and certificated partnerships are just a few examples of good and effective practices developed to cope with illegality through responsibility.

It is also desirable to amend some existing criticality that could exacerbate as the climate scenario gets more and more constraining.

First of all, a situation of high uncertainty urge for a more accurate monitoring system and conditions which are easier to keep under a continuous control.

For this reason it should be promoted the bundling of the properties against the widespread fragmentation of forest parcels, both in the case of private plots and in public ones and to limit forest utilizations carried on without technical address. In this context, the Park can represent the meeting point between the different realities that operate in forest management.

Moreover all the situations that imply a non-ordinary management should be accurately monitored. Where programmatic choices provide free natural evolution (integral reservoirs, peculiar formations, inaccessible stands, etc.), an adequate system of phytosanitary and hydrogeological monitoring must be implemented.

5. Present and future trends in the technical management of Adamello Park and Valle Camonica Area

The management lines currently implemented in Valle Camonica can already help in facing a situation full of uncertainty as that prospected by the predicted climate change scenarios in the valley, since they are oriented to strengthen the forest formations and make them more stable in terms of specific composition through adequate treatment forms and the corrections of some critical situations.

5.1. Specific management indications

The specific management indications can be summarized in three main issues, as follows.

a.   To address forest policy giving priority to naturalistic and landscaping functions to mountainside forest management, promoting at the same time timber production encouraging conversion towards high-stand forests.

While preserving a production management for forests on a large surface of the Valle Camonica, a policy of re-naturalization of the areas, with a multifunctional approach, has already been carried on by the Park administration in the last decade and should be further improved.

It is believed moreover that the same policy should be adopted at the EU level, though it is clear that such a shift of orientation would require a significant change in the administrative and managerial approach to forestry; nevertheless, although not backed by direct economical evaluations, it would be the most reasonable way to be pursued.

b.   To evaluate the situation of those formations and species that are the most exposed to climate change.

The forecasts suggest a very significant extension of warm phytoclimatic ranges; therefore the species and formations more susceptible to water stress (for instance, gorge and crags formations) will be vulnerable and will trend to thin out, not necessarily replaced by thermophile species.

Going along with the climate change scenarios, which show a contraction and an shift towards higher altitudes of Norway Spruce (*Picea abies*) and Chestnut (*Castanea sativa*) range, it is desirable:

- to keep on removing coniferous from lower-medium mountainsides and to renaturate secondary woods of Norway spruce and chestnut;

- to improve the treatments in chestnut abandoned formations. In the last years these kinds of woods lacked of the traditional active management: as a result of the consequent chaotic growth, these aged chestnut coppices are characterized by a high risk factor, mainly for what concerns fire risk. It is advisable therefore to facilitate the conversion towards high forests and the changeover towards oak and maple.

Also for what concerns Scots pine, it is reasonable to expect a significant shrinking of the species range, especially in gorge and crags situations. Also in this case, the most recommendable substitute is the oak.

c.   Reduce the compositional and structural trivialization of mountainsides promoting and increasing biodiversity. Extreme events increase, induced by biotic and abiotic factors, implies the addressing of proper treatments for mono-cultural and mono-structural forests

5.2. Hints on vegetational composition

From the point of view of the biodiversity and of the composition in species, forestry measures should be addressed in order:

- to promote the conservation species with auxiliary or faunal roles: *Prunus avium, Crataegus monogina, Acer campestre, Betula pendula, Juniperus communis, Laburnum anagyroides, Cornus sanguinea, Cornus mas, Morus alba, Populis tremula, Quercus pubescens, Salix caprea, Taxus baccata, Ulmus glabra, Sambucus nigra, Sambucus racemosa, Sorbus aucuparia e Sorbus aria, Mespilus germanica, Malus sylvestris, Prunus sp., Pyrus pyraster, Quercus pubescens*

- to promote the conservation and diffusion of target species, in order to limit the extreme semplification of the crops and especially to preserve the least competitive species. Specific conservation measures are to be adopted, eventually considering also the total prohibition in use. For what concerns the Adamello Park, target species are: *Quercus sp., Carpinus betulus, Acer pseudoplatanus, Tylia cordata, Fagus sylvatica, Abies alba e Pinus cembra*. As for target forest tipologies, in addition to those representative the single target species (Oak woods, Beech woods, Oak-Hornbeam woods, Sycamore-Lime woods, Fir and Cembran pine woods), it is also recommendable to promote the protection and safeguarding of minors and/or relict species.

For what concerns beech, Oaks and other noble species, it should be noticed that especially Beech range might find benefits from the expected scenario. Since in the territory of Valle Camonica, there is a general lack of seed-bearing trees, the implementation of protection policies to preserve these subjects is essential. *Tilia cordata* and *Acer pseudoplatanus* show greater capacity of seed dispersal. Specific management actions must be addressed to overcame the difficulties of oak regeneration (seed-bearing trees protection, conversion of coppice to high forest for *Quercus robur* x *pubescens* hybrid).

Common Hornbeam *(Carpinus betulus)* and Hop Hornbeam *(Ostrya carpinifolia)* might prove particularly important in limiting the further spread of *Ailantus altissima* and *Robinia pseudoacacia*, possibly fostered by the progressive withering of the valley bottom.

5.3. Hints on technical management

- **High stands management.** It is recommendable to promote the form of high stand management and at the same time the conversion to high stands of coppices no longer actively run or older than 40 years.

- **Coppice management.** Coppice form is advisable only in those cases in which the technical and coltural care are guarantee. A minimum number of sapling has to be assured, but the a priori defined threshold for each species has to be critically revised. In fact, the application of this minimum quantity not always resulted in the best technical option.

- **Natural evolution.** Natural evolution has to be encouraged in those forest types showing structural attributes suggesting the need for treatments despite of hydro-geological and/or topographic concerns (protection forests). The natural evolution is also suggested for those formations having greater bio-ecological value (new formations, riparian formations, primitive formations, rupiculous and ravine formations, scree slope formations).

- **Site-adapted silviculture.** Promote the adoption of extensive silvicultural practices oriented to the raising of forest variability in terms of composition, structure and cover (selective cutting, seeding cutting patch cutting); intensive practices could be implemented in those situations showing perturbed forest ecosystem (e.g. phytosanitary issues), anthropic composition (artificially afforested plots), hydrogeological instability.

- **Seasonality of forest cutting (treatments).** Forest structural management needs to be tailored on specific objectives accounting for the presence of endangered species (e.g. IUCN red list); preventive measures could be addressed to exclude potential and actual nest sites for birds or core areas for vertebrates (e.g. borrows, breeding areas, lek sites, wetlands, ecc.)

- **Forest fire fighting.** Promote actions and ad hoc forest management practices to reduce forest susceptibility to fire. Silvicultural practices must be addressed to avoid situation of monospecificity, supporting, at the same time, conversion of coppice to high stand forests (highly recommended in chestnuts forests) and presence and maintenance of less represented forest types. The most common silvicultural practices addressed to forest fire management and risk reduction are:

a. thinning, cutting and removal of small trees in presence of forests characterized by high tree densities or physiological stress (may include the removal of dead trees and shrubs);

b. selection cutting performed in adult and monospecific forests to improve structural complexity and to increase the presence of deciduous trees. Increasing the proportion of deciduous trees decrease the likelihood that a ground fire evolves into a crown one;

c. cultural practices adopted in intensively managed coppice, to decrease dead fuel availability;

d. conversions applied in degraded coppice located in areas with high danger of forest fires;

e. reafforestation programs to enhance the restoration of degraded forest habitats in presence of monospecific structure or over-managed areas;

f. environmental cleanup: post-fire treatments to remove dead vegetation; Post-fire restoration generally refers to long-term efforts required to restore habitat quality, resilience, and productivity. According to the main forest function (timber production, disaster protection, recreation, environmental...) measures consist of: cutting of burned trees: cutting of burned trees and release of subjects with the highest survival probability; direct seeding: seeding of herbaceous and woody species to prevent surface run-off; reafforestation in order to mitigate potential increases in runoff and erosion which can occur immediately after a wildfire and promote wide range of forest associations less prone to forest fires, more resilient, productive and with higher biodiversity values

g. construction of fire breaks (vegetated fire breaks, protective strips and fuel breaks).

Acknowledgements

The authors are grateful with all the Adamello Park and the Valle Camonica Mountain Community staffs for their valuable support and their contribution to the fruitful discussions.

We also want to thanks Jean-Paul Rukalski for his work, as well as all the municipality of the Valley for their support during the ozone monitoring campaign. We are, in particular, grateful with the municipality of Paspardo which hosted the flux tower and with its major for her support.

This manuscript has been partially funded by the Catholic University's program for promotion and divulgation of the scientific research.

Author details

Giacomo Gerosa[1], Angelo Finco[1,2*], Stefano Oliveri[1,2], Riccardo Marzuoli[1], Alessandro Ducoli[3], Giambattista Sangalli[3], Bruna Comini[4], Paolo Nastasio[4], Giampaolo Cocca[4] and Elena Gagliazzi[4]

*Address all correspondence to: angelo.finco@unicatt.it

1 Dept. of Mathematics and Physics, Catholic University of the Sacred Hearth, Brescia, Italy

2 Ecometrics s.r.l., Environmental Monitoring & Assessment, Brescia, Italy

3 Comunità Montana Valle Camonica – Parco dell'Adamello, Breno, Italy

4 Regional Agency of Services to Agriculture and Forests (ERSAF), Unit for the Valorisation of Biodiversity and Services to the Agro-Forest ecosystems, Gargnano, Italy

References

[1] IPCC(2007). Climate Change 2007: The Physical Science Basis. Contribution of Working Group I to the Fourth Assessment Report of the Intergovernmental Panel on Climate Change [Solomon, S., D. Qin, M. Manning, Z. Chen, M. Marquis, K.B. Averyt, M.Tignor and H.L. Miller (eds.)]. Cambridge University Press, Cambridge, United Kingdom and New York, NY, USA.

[2] Pignatti, G. (2011). La vegetazioneforestale di fronteadalcuniscenari di cambiamentoclimatico in Italia. Forest@. doi:efor, 0650-008.

[3] Conedera, M, Torriani, D, Neff, C, Ricotta, C, Bajocco, S, & Pezzatti, G. B. (2011). Using Monte Carlo simulations to estimate relative fire ignition danger in a low-to-medium fire-prone region. Forest Ecology and Management , 261, 2179-2187.

[4] Dullinger, S, Dirnbock, T, & Grabherr, G. (2004). Modelling climate change-driven treeline shifts: relative effects of temperature increase, dispersal and invisibility. Journal of Ecology, , 92, 241-252.

[5] Gehrig-fasel, J, Guisan, A, & Zimmermann, N. E. (2007). Tree line shifts in the Swiss Alps: Climate change or land abandonment? Journal of Vegetation Science, , 18, 571-582.

[6] Cannon, S. H, Gartner, J. E, Wilson, R. C, Bowers, J. C, & Laber, J. L. (2008). Storm rainfall conditions for floods and debris flows from recently burned areas in southwestern Colorado and southern California, Geomorphology , 96, 250-269.

[7] Santi, P, & Morandi, L. (2012). Comparison of debris-flow volumes from burned and unburned areas. Landslides. DOIs10346-012-0354-4.

[8] Fuhrer, J, Skarby, L, & Ashmore, M. R. (1997). Critical levels for ozone effects on vegetation in Europe. Environmental Pollution 97 (1-2), 91-106.

[9] Fredericksen, T. S, Joyce, B. J, Skelly, J. M, Steiner, K. C, Kolb, T. E, Kouterick, K. B, Savage, J. E, & Snyder, K. R. (1995). Physiology, morphology, and ozone uptake of leaves of black cherry seedlings, saplings, and canopy trees. Environmental Pollution; , 89, 273-283.

[10] Lee, J. C, Skelly, J. M, Steiner, K. C, Zhang, J. W, & Savage, J. E. (1999). Foliar response of black cherry (Prunusserotina) clones to ambient ozone exposure in central Pennsylvania. Environmental Pollution; , 105, 325-331.

[11] Ece, U. N. Mapping Manual Revision, (2004). UNECE convention on long-range transboundary air pollution. Manual on the Methodologies and Criteria for Modelling and Mapping Critical Loads and Levels and Air Pollution Effects, Risks and Trends. <www.icpmapping.org>.

[12] Monteith, J. L. (1981). Evaporation and surface temperature. Quarterly Journal of the Royal Meteorological Society; , 107, 1-27.

[13] Gerosa, G, Vitale, M, Finco, A, Manes, F, Ballarin-denti, A, & Cieslik, S. (2005). Ozone uptake by an evergreen Mediterranean forest (Quercus ilex) in Italy. Part I: Micrometeorological flux measurements and flux partitioning. Atmospheric Environment; , 39, 3255-3266.

[14] Gerosa, G. BallarinDenti, A., (2003). Regional scale risk assessment of ozone and forests. In: Karnosky D.F., Percy K.E;, Chappelka A.H., Simpson C., Pikkarainen J. (Eds). "Air Pollution, Global Change and Forests in the New Millennium", Elsevier Ltd., , 119-139.

[15] Gerosa, G, Ferretti, M, Bussotti, F, & Rocchini, D. (2007). Estimates of ozone AOT40 from passive sampling in forest sites in South-Western Europe. Environmental Pollution: 145(3), 629-635.

Chapter 19

Case Study Carinthia / Slovenia – Productive Forests Affected by Climate Change

Robert Jandl, Andrej Breznikar, Marko Lekše,
Christian Tomiczek, Silvio Schüler,
Klaus Dolschak and Hans Zöscher

Additional information is available at the end of the chapter

1. Introduction

1.1. The present condition

The test area Ossiach Tauern is located in the province of Carinthia (Figure 1). The landscape of the test site is dominated by gently sloped hills. The Ossiacher Tauern is a W-E mountain ridge at the southern side of Lake Ossiach. It ranges from 500 to 1000 m a.s.l. The steep and moderately steep slopes are covered with forests and are north-facing. The upper part of the Ossiacher Tauern forms a plateau. The bedrock is formed of mica schists which are overlain by glacial moraines [1]. The soils are moderately acidic Cambisols (Brown Earths). The nutrient storage and the water-holding capacity of the soils are high.

The Slovenian test area Solčava - Luče lies in the mountain area of the Kamniško Savinjske Alpe on the border with Austria (Figure 1). High slopes and deep, narrow valeys are predominant types of landscape. The lowest point of the area lies at an altitude of 455 m a.s.l., the highest peaks rise over 2000 m a.s.l.. From a geological point of view the area is characterized by very diverse bedrocks. The Western part is dominated by a variety of carbonate rocks (limestones, dolomites, marbly limestones). The Eastern part consists of silicate bedrock.

The region of the Ossiacher Tauern is by far dominated by forest land. Figure 2 shows the highly developed area close to the southeastern side of the lake with numerous hotels and guest houses. Grassland and cropland is dominant at the northeastern side of the lake and also interspersed in the main tourism area. Forests are located on the slopes of the Ossiacher Tauern. In lower elevation, mixed deciduous forests are dominant, in higher elevation coniferous

Figure 1. The test area Ossiach is located on the southern side of Lake Ossiach; the test area Solčava - Luče is located about 50 km to the SE just across the Austrian/Slovenian border.

forests are taking over. The lower part of Figure 2 shows grassland embedded in the forest in southeast of the test region. This area is located on the plateau of the Ossiacher Tauern and has been traditionally used as pasture land. The Ossiacher Tauern is basically public forest that is managed by the Österreichische Bundesforste AG (http://www.bundesforste.at). The local forest interventions are implemented by the unit in Millstadt/Carinthia.

The region Solčava/Luče makes considerable efforts as a hiking tourist resort. The landscape is indeed spectacular and includes U-shaped glacier valleys. (Logarska dolina, Robanov Kot), high waterfalls (Rinka at the end of Logarska valley) and several forest reserves (Figure 3). The total area of Solcava/Luce comprises 21 367 ha. Forests cover 76% (16 191 ha) of that area. Agricultural land is confined to the immediate surroundings of farms and to the bottom of the valleys. The settlements are dispersed. The local community centres are the villages Solčava and Luče.

The present climate conditions for the village of Ossiach are shown in Figure 4. The mean annual air temperature is 8.4 °C and the annual precipitation exceeds 1000 mm. The growing season is presently humid. Typical for the area is a secondary autumnal precipitation peak. The forested test area ranges to an altitude of more than 1000 m and due to the elevational lapse the temperatures are lower. The Ossiacher Tauern is only in exceptional years affected by storms.

The climate in the Solčava/Luče area is characterized by a combination of the mountain climate, the continental influence from the flatlands towards the east, and even the Mediterranean influence from the south. At the main mountain range of the test area the annual precipitation exceeds 2,000 mm. In the village Solčava the annual rainfall is approximately 1600 mm.

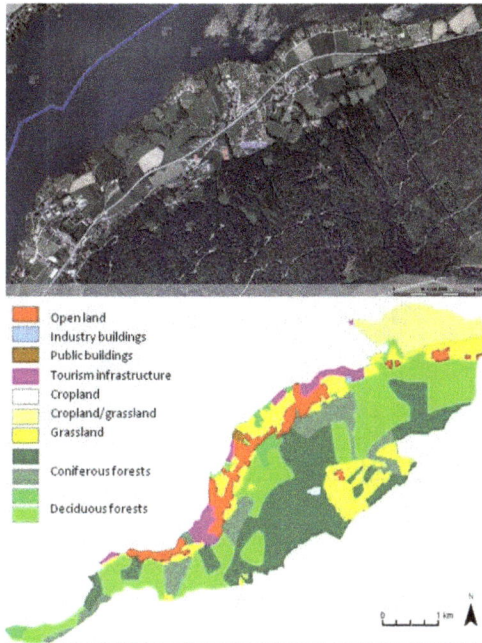

Figure 2. Orthophoto of the village of Ossiach (upper part) and land use categories on the southern side of Lake Ossiach (lower part).

Figure 3. Map of the Solčava/Luče region.

Comparably to the climate in Ossiach, the main precipitation peaks are in June, July and in November. Local observers report recent stronger rainfall peaks in autumn.

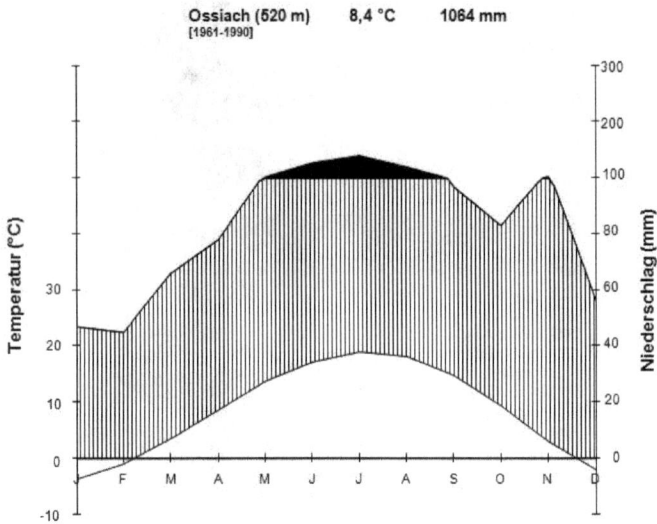

Figure 4. Walter Lieth diagramme for Ossiach based on climate data from 1961-1990.

The average annual temperature in Solčava is 6.4 ° C and already 8.2 ° C in the nearby town of Luče. In the valleys there are very frequent thermal inversions. Heavy snow and sleet usually occur in the altitudinal range of 700 to 900 m, and in mild winters also up to 1200 m. These events cause breakages in the forest canopy. A typical phenomenon of the region are the storms in the Savinja valley and in higher altitudes.

2. Forestry in the area

The potential natural forests of the Ossiacher Tauern would be dominated by beech [2]. This situation is evident in a small natural reserve where no active management of forests in historic times is on the records. In the larger part of the Ossiacher Tauern beech was actively suppressed from the 1950s onward, in order to make room for Norway spruce (Picea abies) and fir (Abies alba) that were demanded by the market. Presently, the forest stands are dominated by spruce with substantial proportions of beech and fir. On moist sites in low elevation and on the plateau alder is playing a local role. Further tree species such as oak, maple and larch are rare. Presently, no major obstacles are presented to timber production. The soils are fertile, the water supply is rich, and the pressure from biotic threats is well under control. The area is not particularly exposed to storms.

Presently, the main forest function is timber production. The Ossiacher Tauern has few designated protection forest and water conservation stands. The region of the Ossiacher

Tauern is an important recreation resort due to its location at Lake Ossiach and the proximity of the provincial capital Villach. In order to honor the expectations of tourists towards the scenery and in order to keep the region suitable for hiking activities a close-to-nature form of management is practiced. Thereby, large openings in the canopy are avoided and natural regeneration is the preferred strategy for the establishment of new forest stands.

The natural regeneration includes spruce, beech and fir with little differentiation in the growth dynamics. Larch only establishes on bare soils and is not abundant. Practical forestry is challenged by the vigor of beech. In natural stand dynamics especially the western part of the region would be dominated by beech at the expense of spruce. The species regenerates vividly and would out-compete spruce and fir. In order to achieve mixed-species stands with a dominance of coniferous trees, an early reduction of beech by cleaning (Läuterung) is advised [3]. -- In the upper parts of the Ossiacher Tauern natural regeneration is working very well. In the lower ranges the vital expansion of the herbaceous vegetation calls for small-scale clear-cuts that are later afforested. The plantlets are produced in a central forest garden and the use of autochthonous planting material is ensured. The plantation needs frequent weeding in the first years. - Selective cuttings and group selection are made on a small-scale, preferably where natural regeneration already had established. The low inclination of the slopes allows the use of tractors and does not require setting up cable systems.

The potential threats to the forests are quite limited. Root rot infestation of Norway spruce caused by *Heterobasidium annosum* is not a problem. Beech can be affected by the formation of red heartwood that economically devalues the stems. The remedy is the early harvesting of beech. A common biotic threat is deer browsing. The population density is rather high. The dense herbaceous vegetation and the abundance of seedlings offer plentiful fodder and allow maintaining a high deer density. However, the damages due to deer browsing are actively monitored and the population density is quickly adjusted when needed. - The pressure from bark beetle is generally low because the trees are vital and as such rather resistant to insect attacks. The forest practitioners have observed an increase in the population density of bark beetle in recent years. Additional counter-measures such as the exposure of catch trees and the installation of pheromone traps are already implemented.

The forest management conditions in the Solčava/Luče area are similar. The major part of the area is located in the montane and subalpline zone, and a substantial part of the area extends above the treeline. The forest stands are mostly coniferous (76%). As a consequence of the thermal inversions conifers also dominate in the valleys. Thermophilic tree species such as oak are rare and are confined to only a few places. The most common forest stands are fir-beech forests (Abieti-Fagetum praealpinum and Luzulo-Abieti-Fagetum). Spruce naturally occurs in the montane and subalpine zone forests and has been actively introduced to the entire region.

The average standing biomass stock of the forests in the Solčava – Luče case study area is 329 m³/ha. Spruce contributes 59%, an additional 10% are contributed by larch (Larix decidua), 5% by fir (Abies alba), and 2% by pine (Pinus silvestris). Among the deciduous tree species beech has the most important role and contributes 19% to the standing stock. Noble hardwoods such as maple, mountain ash, mountain elm, walnut, cherry and lime together represent 3% of stock. Larch is present in mixed forests together with spruce and beech and in pure stands at higher elevations. Fir used to be an important tree in the past, but has strongly declined [4]. Pine occurs mainly in the warmer locations.

The forest surface in the area has been continuously increased in past decades due to reforestation and encroachment of abandoned agricultural land. The consequence was an increase in the standing biomass stock. In addition, the growth rates of the forests increased in the wake of the general trend in Europe [5]. The forests in the Solčava Luče area are mostly private owned (97%). Most common are farm forest estates. These are forests that are an element of an agricultural enterprise.

The main functions of the forests are timber production and protection against natural hazards. Timber is sold mostly as a roundwood, and some is locally further processed to furniture and used as construction wood. In a marketing effort the brand of 'Solčava Larch' has been developed during the last decade in order to raise the commercial value of the timber in recognition of its very high quality. On 12.341 ha or 76% of the area the forests fulfill the purpose of protection. Over 5000 ha are designated protection forests. Their status is declared by a legal act and these forests need to be managed according to special guidelines (Figure 5).

Figure 5. Protection forests in the Solčava Luče area.

The forests are the habitat of a large population of wild-living animals such as deer, roe deer, chamois, and wild boar. Hunting tourism is an important part of a local economy. The high animal density causes frequent damages due to browsing. The efforts of promoting and utilizing natural ecosystem dynamics such as the regeneration of a variety of tree species are therefore compromised.

Practical silviculture engages in a group selection system (Femelhieb) and in the case of mixed stand of beech and silver fir sites also in a selection system. Almost all forests are regenerated naturally. Only in exceptional cases autochthonous seedlings are planted in order to establish high-quality forest stands with a predefined tree-species mixture. Enrichment planting is used to modify the spontaneously developing tree species composition towards an economically desirable tree species mixture.

Tending measures in forest stands are carried out in all age classes. Most common are tendings in young stands. Presently, only 50% of the planned tendings are implemented. It is well understood that this shortcoming has an adverse effect on the quality and stability of future forests.

3. The expected climate

According to the used climate scenarios of the IPCC A1B and B1 the annual precipitation is going to remain unchanged in the next 50 years and is going to decline thereafter. The unambiguous effect provided by climate scenarios is the increase in temperature and the elongation of summer droughts [6,7]. Even the optimistic IPCC scenario B1 shows an increase by more than 2°C in the next 100 years. A consequence is the increase in drought periods during the growing season, both with respect to frequency and duration (Figure 6). Particularly the change in the precipitation regime is controversially discussed among climatologists because the southern part of the Alps poses a considerable challenge to climate modelers [8]. The models used in the described research project predict less precipitation during the summer months especially in the second half of the century. The increase in the air temperature may be around 4°C in the next century. Besides the mere warming effect such an increase in air temperature may lead to more extreme events [9]. A form of extreme events are heat-waves. From presently approx 30 heat days per year an increase to annually more than 70 heat days is predicted. According to the used models the length of heat waves is expected to increase steadily. The heat waves will not be critical in the high mountain regions, but will have adverse effects on the forests in the foothill areas and in the valleys. For the forests in the region of Carinthia and Slovenia it is expected that droughts are going to be more frequent during the growing season, thereby affecting the vigor and the productivity of the forests. The number of dry days during the productive period is expected to increase by 15 % (Figures 7, 8) [7].

4. Climate change impacts

The expected climate change in the transnational case study area will in the long run affect the dominance of tree species. A valuable tool for foresters is given by the climate envelopes where presence/absence data on tree species can be translated into geo-referenced information [10]. In Figure 8 the climate envelopes for three highly relevant tree species within the experimental area are shown. The left yellow arrow indicates the experimental site in Carinthia, the right yellow arrow points to the Solčava Luče area. Green shades indicate an agreement of the used models that the respective species is present, red shades indicate the absence of a tree species according to the models.

Figure 9 suggests that the Ossiacher Tauern will lose the site conditions that are ideally suited for Norway spruce from 2050 onwards. Even beech that is in its optimum range in the region

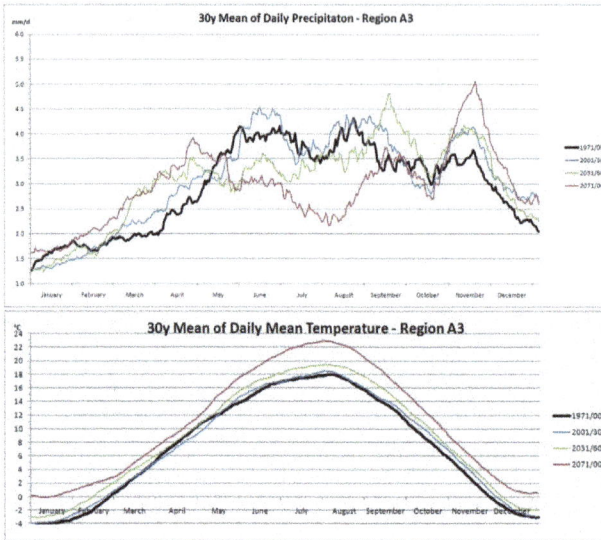

Figure 6. Scenarios for the future precipitation, air temperature, and duration of heat waves for the transnational case study area Carinthia / Slovenia. Upper panel: precipitation, central panel: air temperature; lower panel: duration of heatwaves.

will lose ground and will be only marginally useful for forestry in 100 years. The climate envelope model suggests that oak species will increase their competitivity. It is quite important to interpret these simulations with hindsight to the capabilities and limitations of climate envelope models [10]. These models may not fully capture the local conditions and may not satisfyingly account for the frost situation. Especially late autumn frost events are problematic for oak forests. Valleys with a frequent temperature conversion may prove to be only marginally suitable for oak forests.

Also in the Solčava/Luče area the main climate change impact will also be a change in the tree species composition. Norway spruce is a natural part of forest communities, but in some cases its share is higher than in potential natural tree communities. Most models predict that the share of spruce and beech on the standing stock will decrease. Thermophilic tree such as oak species that are presently still rare will gain ground in the future.

An immediate foreseeable pressure on the forests is exerted by the expected summer droughts. Prolonged dry periods may severely limit the growth rates of trees and the different species cope quite differently with drought. Oaks and pines tend to recover well, whereas beech and spruce carry a legacy of reduced growth rates into the year after a severe summer drought [11].

The predicition of the future pressure from pests and pathogens is subject to very high uncertainties [12]. Quarantaine pests will become more important as consequence of the global exchange of trading goods. Newly arriving pests and pathogens are potentially encountering

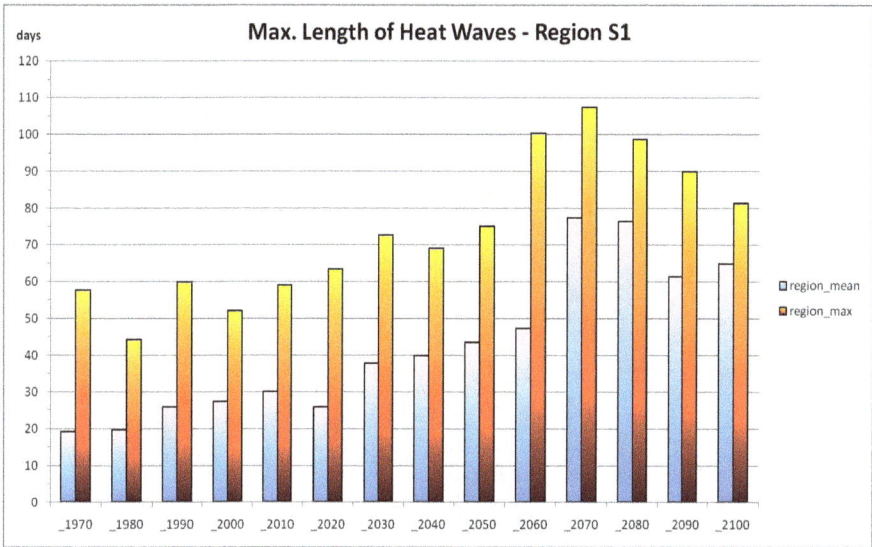

Figure 7. The length of heatwaves is expected to increase.

habitats that would be inaccessible to them. Climate change is opening windows of opportunities for pests and pathogens in areas that may presently be unsuitable for them. Despite these uncertainties foresters need to make evaluations on the risk of biotic forest damages. An attempt to evaluate the biotic risks has been made for the political districts of Austria [13]. Figure 10 shows that the southern part of the country carries, according to experts judgment, a low to moderate risk. Such a mapping needs to integrate numerous factors. Climate change is an important factor, but other influencing factors are of great importance. It has been shown that the increase in forest damages in Europe is by far not only triggered by climate change but that forest management has a strong impact as well [14]. The exposure of forests to biotic risks consists of several climatic factors but also on the tree species composition and its deviation from the potential natural vegetation, the forest road density (accessibility), the workforce available for forestry, and the protective functions of the forests. Assessments like the one presented in Figure 10 also reflect the presently known or anticipated pests and pathogens. The invasion of new species may alter the risk.

Storm is presently on the Ossiach Tauern not a particular problem and there is no indication that it develop into a more pronounced risk in the future. The geographic location of the area does not imply a major storm risk. In the Slovenian part of the area the risk is higher. The area has already experienced several extreme events in the recent past. Examples are a severe flood in Luče in 1990 with an estimated time of recurrence of more than 100 years, a large-scale windthrow on the Črnivec pass in 2008 and frequent snowbreaks in the canopies in the middle

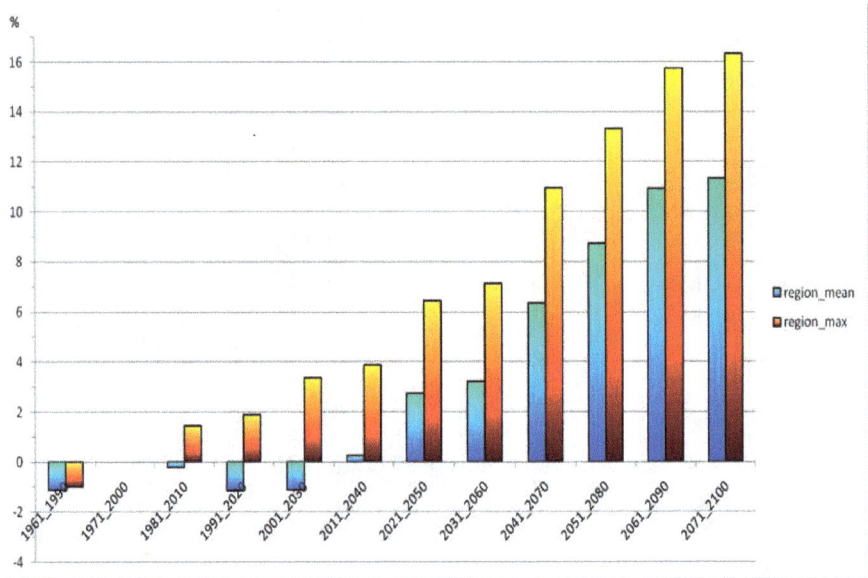

Figure 8. Increase in droughts, expressed as the percentage of the increase in dry days during the growing season. The reference period is 1971/2000.

altitudinal range of the area. In the storm event around 790 hectares of forests were damaged, and 390 hectares were completely destroyed.

By means of a forest productivity model the future development of forests in the Carinthian part of the case study area was assessed. The used models Caldis and Yasso07 are described in [15]. The modeling exercise included the growth assessment of different tree species under the two climate scenarios A1B and B1 as provided by the IPCC [6]. The data for reference stands were taken from the local stand inventory of the forestry enterprise. For simplicity several static assumptions were made:

- The simulation starts with the present tree species composition
- The forest management takes out trees of diameters larger than 45 cm in 2030 and 2090.
- The ingrowth of oak is either favored or suppressed.
- The stand grows under the climate scenario A1B or B1.

Figure 11 shows that the forests are generally growing better under the IPCC scenario B1 than under A1B. This indicates that the limiting factor for forest is not necessarily the temperature and climate change will not necessarily alleviate a thermal constraint. The climate effect is much stronger than the chosen forest management effect. The simulation contains several probabilistic elements such as the mortality due to storms and the regeneration of trees.

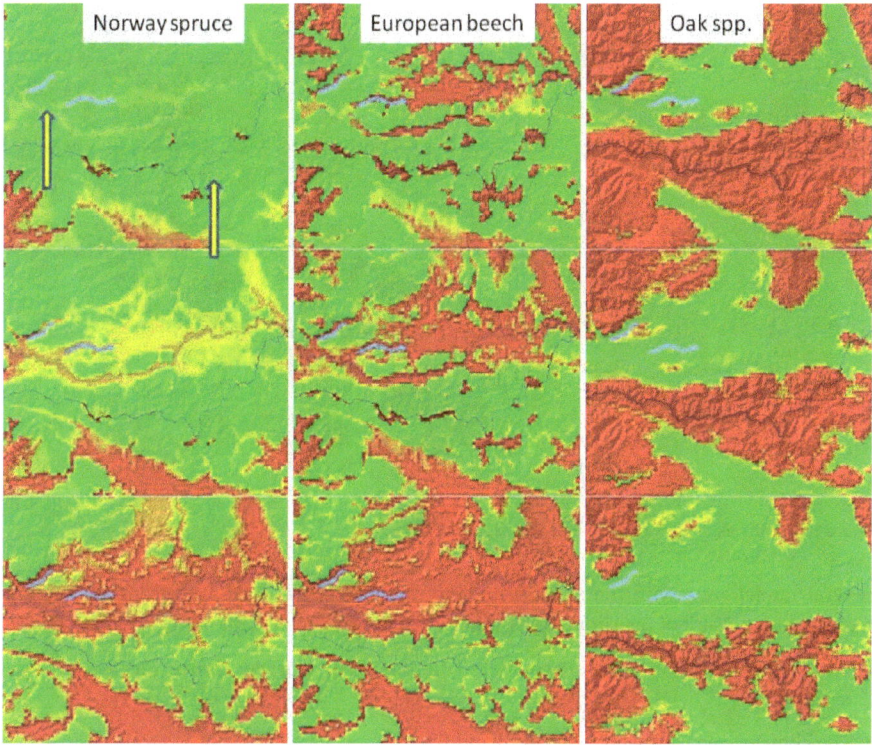

Figure 9. Climate envelopes for spruce (left panel), beech (central panel) and oak (right panel). The upper three maps represent the present conditions, the middle maps represent 2020/50, and the lower maps represent 2051/80. See text for explanation.

Therefore, no clear picture emerges from a single run of the simulation programme. There are several kinks in the temporal trend the stem volume.

Taking the output of the model into a soil carbon simulation model as described in [15] shows again a strong climate effect. The dotted lines representing the B1 scenario with less warming is in several cases the scenario that enables the soils to build up a larger carbon stock. The overarching impact of the forest management strategy on the soil carbon pool is clearly shown. A hypothetical forest without harvesting leads to an accumulation of the soil carbon pool by 10%. However, the effect is only temporary because overly mature stands are experiencing a higher risk of biotic and abiotic damages [16]. In a managed forest a strong peak of carbon is introduced in the assumption that harvesting residues are left behind and the root system of the harvested trees starts to decay. However, due to the reduced aboveground and below-ground litterfall after the harvesting and the effect of the increasingly warm climate the soil carbon pool is gradually declining (Figure 12).

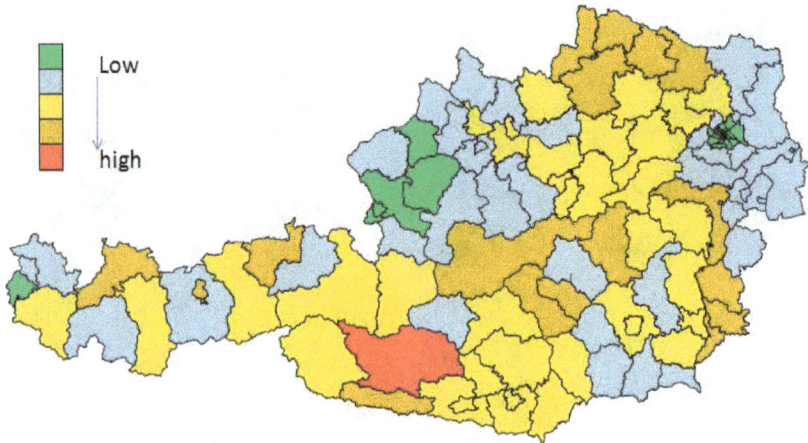

Figure 10. Forest health risk for Austria based on experts opinions.

Several model runs with the same set of input data were averaged in order to obtain a representative result of the simulations. Again, a simplified management regime was chosen:

- Three pure-species stands were compared dominated by Norway spruce, European beech, or oak spp.
- The climate scenarios were IPCC A1B and B1
- All stands were thinned only once and at the same time.

The harvesting was a removal of the thickest trees and no scenarios for the challenges of harvesting and protecting the remaining forest were developed. Furtheron, it was not included whether different timber market situations for spruce, beech, and oak may call for the selection of different diameters of the cut trees and different silvicultural intervention cycles. The effect of the management and the climate is expressed in the deviation from the reference time at the beginning of the simulation.

Figure 13 shows that at the beginning of the simulation period European beech grows better than spruce and oak. The differences are small. The interpretation of the result with respect to beech needs to be made cautiously because the competitivity of the species is extremely high. This property of beech emerges from the underlying data set of the Austrian Forest Inventory and may overestimate the vigor of beech. The graph also shows that in the early phase of the model run the differences between the two climate scenarios are small. After the harvesting intervention beech and oak are recovering and are quickly gaining stem volume again. The growth of spruce is much poorer. Under the warmer climate scenario A1B the growth of spruce starts to decline. The effect is particularly obvious from 2060 on. But even the climate scenario B1 suggests a productivity decline for Norway spruce setting in around the year 2080.

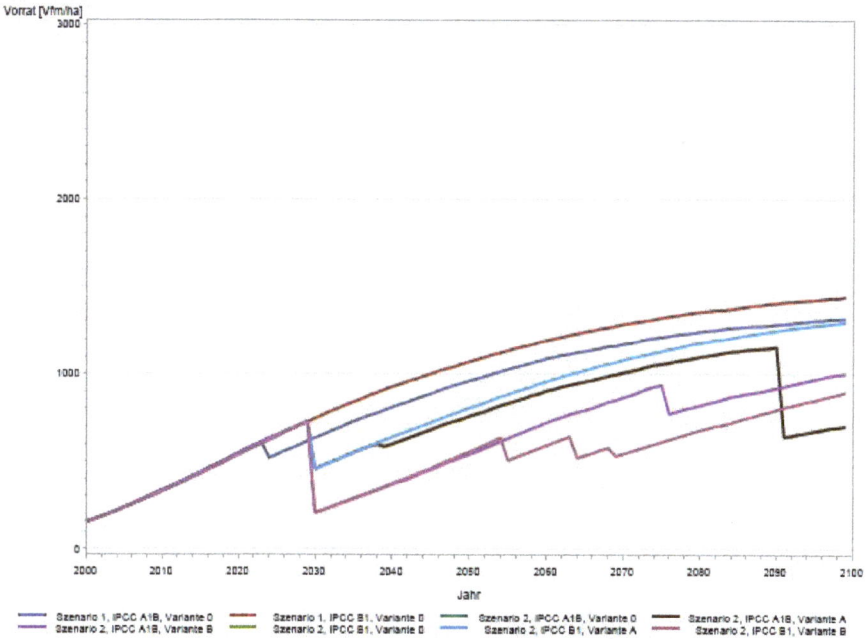

Figure 11. Standing stock of the stem volume of a forest in the Ossiacher Tauern according to different climate and management scenarios.

The simulated development of the forest does not include eventual biotic and abiotic risks arising from pests and pathogens and forest fires. It is assumed that biotic damages are remaining at the present level and that are encoded in the simulation model because data from the Austrian Forest Inventory of the last 30 years were used. An increasing risk of biotic damages is likely [12]. The higher risk needs to be absorbed by an increase in the management intensity, comprising the monitoring of pests and pathogens and the often required small-scale harvesting intervention in order to minimize the propagation of a biotic problem from single trees or small groups of trees to the entire forest stand. The management of forests may become more expensive. A further increase of timber prices and additional markets for the wood sector can alleviate the economic pressure.

Climate change also increases the risk of forest fires. In the entire case study area forest fires are presently rare events and an active strategy of fire suppression pursued. The increase in heat waves and drought periods (Figures 7, 8) can lead to situations that are also not reflected in the parameter selection of the forest growth simulation programme. For practical foresters an alerting system for fire danger may be useful in the future.

The soil carbon pool also responds to climate change and to the forest management decisions as shown in Figure 13. Figure 14 shows that in the first 30 years of the simulation neither the

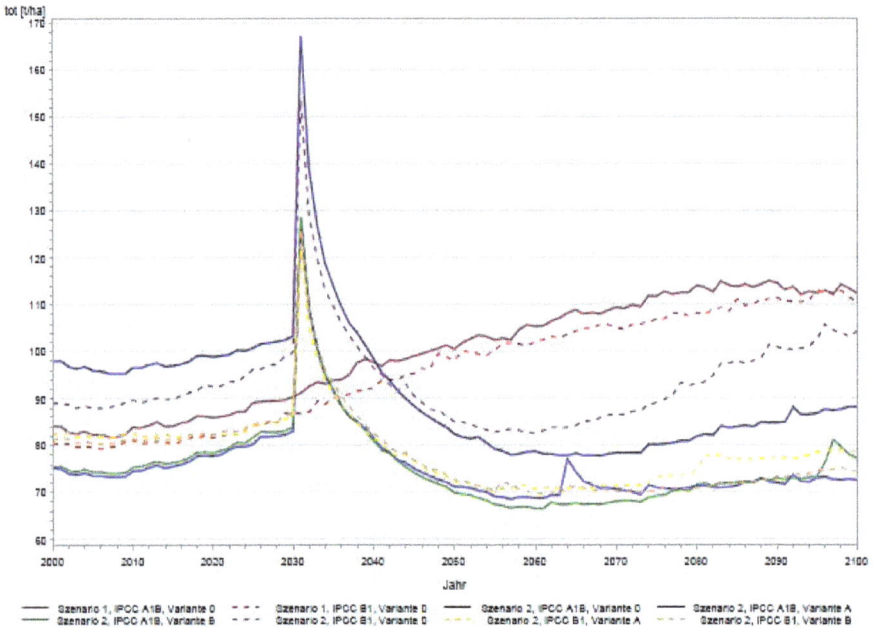

Figure 12. Soil carbon stock of the forest under different climate scenarios and management regimes.

climate scenario nor the selection of tree species leads to a pronounced change in the soil carbon pool. The harvesting operation in the year 2030 leads to different enrichments of the soils with organic matter because the amount of residues differs. Under the assumption of the warmer A1B scenario all soils release carbon into the atmosphere, because the temperature effect stimulates the mineralization of soil organic matter more than the productivity of the forests. In the less warmer B1 scenario beech and oak forests are maintaining their soil carbon pool. The Norway spruce forests loses slightly because the stand is more affected by disturbances thereby reducing the litter input to the soil. Over the course of the simulation period the soils are losing less than 20 tons carbon per hectare. This is a minor carbon loss that may not even be detectable in terrestrial soil surveys.

5. Forestry towards the climate change scenario

Secondary spruce forests on silicatic bedrock represent a known hotspot of biotic risks and have been severely affected by storms and bark beetle attacks in the recent past [17,18]. Rather fast climatic changes will influence the stability of these forests and can compromise their protective functions. In the Solčava Luče area forests are an important part of the economy. On average, forest properties have an area of 26 ha and are parts of agricultural enterprises.

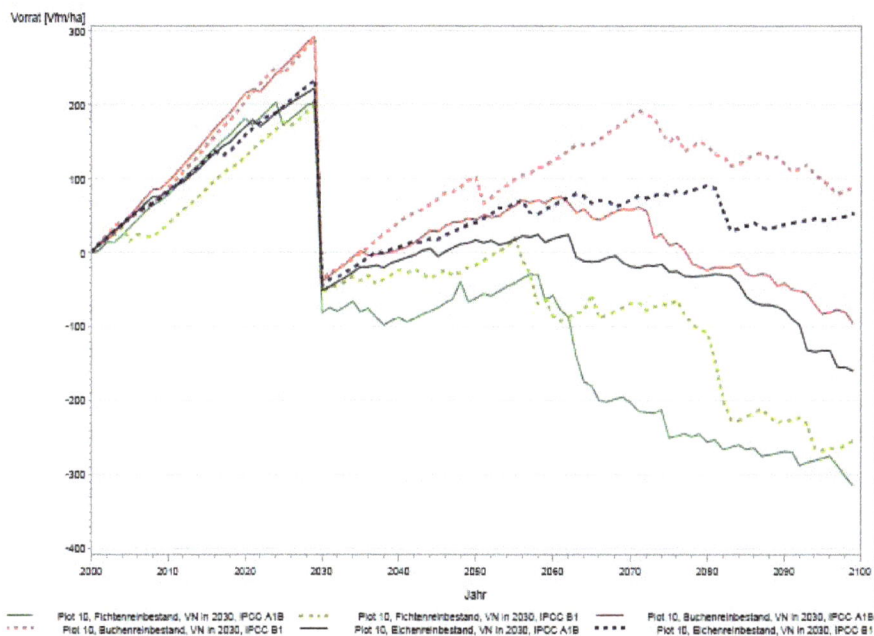

Figure 13. Comparison of the growth performance of Norway spruce (green), European beech (red), and oak (blue) at the Ossiacher Tauern for two climate scenarios (full line: IPCC A1B, dotted line: IPCC B1).

The economic success of these farms relies on the income from forest products. Forestry and wood processing are main economic activities in the area. Practical foresters and forest owners are not yet provided with stringent guidelines as an aid how to deal with climate change. However, comprehensive forest management plans are including advice on forest management based on site and silvicultural classes.

The economic setting in Ossiach is somewhat different. In Figure 15 the distribution of the workforce in the entire province Carinthia and in the community of Ossiach are shown. Although forestry is a highly important part of the regional economy the sector employs not many persons as their primary profession. In the community of Ossiach most people are working in the tertiary sector of the economy.

The main approach in practical forest management in the Solčava/Luče area is the stimulation of a change in the tree species composition. This is achieved by enrichment planting in cases where the natural regeneration of the desired tree species is not happening. The strategy is the introduction of tree species with a wide ecological amplitude. Emphasis is placed on using the most appropriate provenances within a tree species assuming that they can adapt to the future conditions. It also is widely accepted that spruce monocultures will not have the expected flexibility for coping with climate change. A promising method is underplanting pure spruce

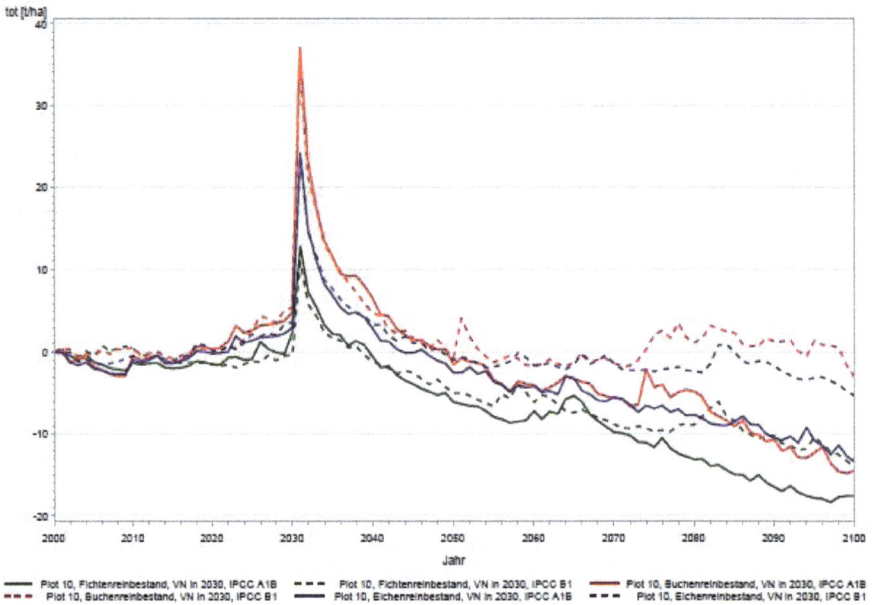

Figure 14. Temporal trend of the soil carbon pool for two climate scenarios (full line: IPCC A1B, dotted line: IPCC B1) for stands of Norway spruce (green), European beech (red), and oak (blue) at the Ossiacher Tauern. The reference soil carbon pool is the status at the beginning of the simulation period.

forests with beech and other broadleaved trees. Even a low proportion of deciduous tree can significantly increase the biological and mechanical stability of those stands [2,3]. The aim of the underplanting is the establishment of a network of tree clusters or stands with a rich tree species composition, based on site conditions and in line with climate change predictions. These clusters are the backbone of forests with a high stability and a low vulnerability towards negative effects of climate change.

In the Ossiach region a change of the tree species composition is achieved by stimulating the natural regeneration of existing species. Given that beech is very competitive, the natural dynamics are in favour of mixed-species stands. Despite the indication of the increasing pressure on Norway spruce as shown in Figure 9 it is assumed that Norway spruce will play an important role in the forestry of the future. The need for an increase in the silvicultural activities is already understood. However, the call for an increased use of deciduous forests by forest ecologists is not yet reconciled with the expected future demand on the timber market. So far, no convincing economical concept for a forest industry based on timber from deciduous trees has been presented. In recognition that climate change is a slow process forest owners are hesitant to adopt a silvicultural strategy that is yet in its infancy.

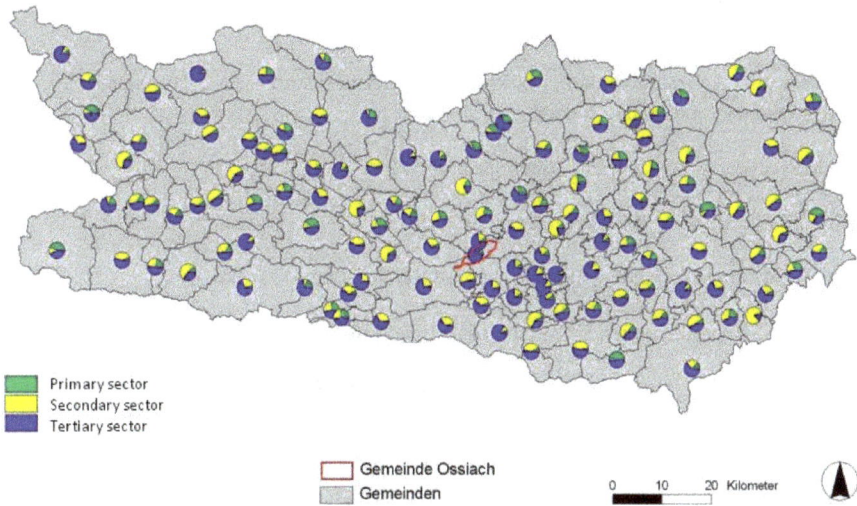

Figure 15. Distribution of work places according to economic sectors. The community Ossiach is indicated by red borders.

The technologies of timber harvesting have been developing at a rapid pace in the past decades. However, dealing with large scale forest damages is still a challenge. Forest access roads are an important element of silviculture. The long-term planning in the Solčava/Luče area includes the increase of the density of forest roads and the introduction of cable line technologies in difficult terrain. Effective risk prevention is based on protocols for pests monitoring, and precise action plans for recommended activities of intervention. There are still many opportunities both in Carinthia and in the Solčava/Luče area to establish protocols for the identification of risk hotspots for presently encountered and emerging problems with pests and pathogens. Promising attempts are databases that are populated with information by an international network of scientists and that are made available as web applications [12,19]. Concepts for the management and silvicultural treatment of protection forests are of crucial importance for the Solčava/Luče area. The specific education of practical foresters has already been recognized and capacity building is among the main tasks of the forestry service.

Acknowledgements

This chapter is an outcome of the Interreg project Manfred, conducted within the Alpine Space programme.

Author details

Robert Jandl[1*], Andrej Breznikar[2], Marko Lekše[2], Christian Tomiczek[1], Silvio Schüler[1], Klaus Dolschak[3] and Hans Zöscher[4]

*Address all correspondence to: robert.jandl@bfw.gv.at

1 Forest Research Center (BFW), Vienna, Austria

2 Zavod za Gozdove Slovenije/Slovenia Forest Service, Ljubljana, Slovenia

3 Institute of Forest Ecology, University of Applied Life Sciences (BOKU), Vienna, Austria

4 Forest Training Center - Forstliche Ausbildungsstätte (FAST) Ossiach, Ossiach, Austria

References

[1] Hauser, Ch. Geologische Karte der Republik Österreich. Erläuterungen zu Blatt Villach- Assling. Eigenverlag der Geologischen Bundesanstalt, Wien, 201-210.

[2] Mayer, H. Wälder des Ostalpenraumes-- Standort, Aufbau und waldbauliche Bedeutung der wichtigsten Waldgesellschaften in den Ostalpen samt Vorland. G. Fischer Verlag, Stuttgart, (1974).

[3] Mayer, H. Waldbau auf soziologisch-ökologischer Grundlage. G. Fischer Verlag, Stuttgart, (1984).

[4] Jurij Diaci (2011). Silver Fir Decline in Mixed Old-Growth Forests in Slovenia: an Interaction of Air Pollution, Changing Forest Matrix and Climate, Air Pollution- New Developments, Anca Maria Moldoveanu (Ed.), 978-9-53307-527-3InTech, Available from: http://www.intechopen.com/books/air-pollution-new-developments/silver-fir-decline-in-mixed-old-growth-forests-in-slovenia-an-interaction-of-air-pollution-changing-

[5] Spiecker, H, Mielikäinen, K, Köhl, M, & Skovsgaard, J. Growth Trends in Europe-Studies from 12 countries. EFI Reports Springer Verlag New York, (1996). , vol. 5

[6] Zimmermann, N. E, Gebetsroither, E, Züger, J, Schmatz, D, & Psomas, A. Future Climate of the European Alps. Chapter 3, InTech, (2012).

[7] Gebetsroither, E, & Züger, J. Drought hazard estimations according to climate change in the Alpine area. Chapter 11. Intech, (2013).

[8] Auer, I, Prettenthaler, F, Böhm, R, & Peske, H. Eds.) Zwei Alpentäler im Klimawandel. Innsbruck University Press, (2010). alpine space- man & environment , 11

[9] IPCCField, C.; Barros, V.; Stocker, T.; Qin, D.; Dokken, D.; Ebi, K.; Mastrandrea, M.; Mach, K.; Plattner, G.-K.; Allen, S.; Tignor, M. & Midgley, P. (Eds.). Managing the Risks of Extreme Events and Disasters to Advance Climate Change Adaptation. A Special Report of Working Groups I and II of the Intergovernmental Panel on Climate Change. Cambridge University Press, (2012). pp.

[10] Zimmermann, N. E, Jandl, R, Hanewinkel, M, Kunstler, G, Kölling, C, Gasparini, P, Breznikar, A, Meier, E. S, Normand, S, Ulmer, U, Gschwandtner, T, Veit, H, Naumann, M, Falk, W, Mellert, K, Rizzo, M, Skudnik, M, & Psomas, A. Potential future ranges of tree species in the Alps. Chapter 4. InTech, (2013).

[11] Beck, W. Auswirkungen von Trockenheit und Hitze auf den Waldzustand in Deutschland-- waldwachstumskundliche Ergebnisse der Studie im Auftrag des BMEL. DVFFA-- Sektion Ertragskunde, Jahrestagung 2010, (2010).

[12] Griess, H.; Veit, H. & Petercord, R. Cerbu, G.; Hanewinkel, M.; Gerosa, G. & Jandl, R. (Eds.) Risk assessment for biotic pests under prospective climate conditions 5 InTech, 2013

[13] Tomiczek, C, & Schweiger, C. Assessment of the regional forest protection risk in Austria on forest district level. Forstschutz aktuell, (2012)., 2012(54), 2-4.

[14] Seidl, R, Schelhaas, M, & Lexer, J. M. J. Unraveling the drivers of intensifying forest disturbance regimes in Europe. Global Change Biology, (2011)., 2011(17), 2842-2852.

[15] Dolschak, K, Jandl, R, & Ledermann, T. Coupling a forest growth model with a soil carbon simulator. Chapter 13. InTech, (2013).

[16] Taverna, R, Hofer, P, Werner, F, Kaufmann, E, & Thürig, E. COEffekte der Schweizer Wald- und Holzwirtschaft. Szenarien zukünftiger Beiträge zum Klimaschutz. Bundesamt für Umwelt, (2007)., 2.

[17] Gschwantner, T, & Prskawetz, M. Sekundäre Nadelwälder in Österreich. BFW-Praxisinformation, (2005)., 2005(6), 11-13.

[18] Gardiner, B, Blennow, K, Carnus, J, Fleischer, M, Ingemarson, P, Landmann, F, Lindner, G, Marzano, M, Nicoll, M, Orazio, B, Peyron, C, Reviron, J. -L, Schelhaas, M. -P, Schuck, M. -J, Spielmann, A, & Usbeck, M. T. Destructive Storms in European Forests: Past and Forthcoming Impacts. EFI Reports, (2010).

[19] Oliveri, S, Pregnolato, M, & Gerosa, G. A new webGIS platform dedicated to forest extreme events in the Alps: aims and functionalities. Chapter 10, InTech, (2013).

Chapter 20

Managing Alpine Forests in a Changing Climate

Peter Brang, Andrej Breznikar, Marc Hanewinkel,
Robert Jandl and Bernhard Maier

Additional information is available at the end of the chapter

1. Introduction

There is mounting evidence that Alpine forest ecosystems will not be able to fully absorb the changes in site factors associated with climate change, such as higher temperatures, more intensive drought stress and associated biotic impacts since these changes exceed the adaptive capacity of the trees. The projected changes in temperature by 2.2 to 5.1 K from 1980 to 1999 to 2080 to 2099, for the A1B scenario in southern Europe [1], correspond to an altitudinal shift of 300 to 700 m in a mountain landscape, if a lapse rate of 0.6 to 0.8 K per 100 m is assumed. Such altitudinal shifts are very often associated with a profound change in tree species composition. This climatic change is projected to occur within about 100 years and thus faster than the average lifetime of a tree in Alpine regions. Widespread tree mortality such as in many forests worldwide [2] is not an unrealistic scenario, and first signs of such phenomena were found in dry inner Alpine valleys [3].

Climatic change is also likely to profoundly alter biotic interactions between trees and pathogens, mostly in favor of the latter [4,5]. An example for negative impacts of climate change on forests is the devastative outbreak of mountain pine beetles in parts of Canada [4]. Large-scale tree mortality has affected the regional forest sector, but has also, at a global scale, influenced the political position of Canada towards climate change treaties. The forest sector, which for long had been a substantial C sink, turned into a large C source. The implications for the national greenhouse gas balance of Canada were such that Canada has withdrawn from the Kyoto Accord. In contrast to such negative effects of climate change on the forests' role as a C sink, there is increasing evidence for positive effects of higher temperatures on tree growth in cool environments [6] with abundant water supply and reasonable soil fertility.

Healthy forests are important for the timber industry and provide jobs in rural areas. Moreover, they are delivering numerous ecosystem services such as protection against natural hazards, carbon sequestration, drinking water, and habitat for a rich biodiversity of plants and animals. Managing Alpine space forests for these ecosystem goods and services in the face of a changing climate poses a number of novel challenges, which all increase the uncertainty in management. For instance, there is no recent guiding example for such rapid environmental change in forest ecosystems. The projected future site conditions are often not present in current landscapes, which means that analogies of potential future states are missing. Moreover, elevated rates of nitrogen deposition cause unprecedented trophic conditions at many sites [7]. The consequences of these remarkable changes which occur simultaneously and often interact are difficult to predict. The real adaptive capacity of the forest ecosystems, and in particular of the current generation of adult trees and of their offspring, are unclear. While ongoing research, e.g. in genetics of forest trees [8,9], is likely to reduce these uncertainties, forest practitioners will still need to make management decisions on the basis of incomplete knowledge and limited guidance.

In this uncertain situation, a 'change management' is needed which aims at ensuring that Alpine space forests continue to provide their goods and services, without major interruptions. Identifying and applying silvicultural strategies which can achieve this is the challenge for forest scientists and managers. Since the uncertainties are large, any management scheme should not be deterministic, but prepared for continuous readjustment, flexible and adaptive: designed to be adaptable to changes in the environmental factors, in the forest itself and also in changes in the anticipated further change.

The term 'adaptive management' (AM) [10-12] describes such a flexible management approach. AM should ensure continuous improvements through iterative cycles of planning, managing, monitoring, and revised planning. In contrast to traditional forest management, AM starts from the assumption that current knowledge about ecosystem functioning is limited and the future even more uncertain, and that current management approaches therefore need continuous revision. A critical element of AM is extensive stakeholder involvement. In many regions of the Alpine space this aspect still needs to be improved.

In this chapter, we want first to present current management approaches in Alpine forests. Second, we will outline adaptation measures which have been discussed, and are partly already being implemented. Third, we will examine how current management should be complemented with ideas from AM to enhance the adaptive capacity of Alpine space forests in the face of a changing climate.

2. Silvicultural practices are diverse

The silvicultural practices currently applied in Alpine space forests are highly diverse. The extremes are the intensive management of plantation forests, mostly of Norway spruce (*Picea abies* (L.) Karst.), or of coppice forests with short rotations, and the complete abandonment of any timber cutting on steep-slope forests with limited accessibility. However, management practices with moderate intensity prevail. Regeneration is often natural, stands are selectively thinned, cut sizes in final cuts are small (i.e. < 1 ha), and substantial forest areas such as the whole of Slovenia and some Swiss regions are managed maintaining permanent forest cover in forestry [13,14].

Silvicultural or forest management guidelines exist in several Alpine regions and countries. They have often been developed for use by forest managers of a specific region and are written in local languages (Table 1). Only the most recent guidelines among those listed do address adaptation to climate change. Some of the guidelines are also unpublished. For instance, the Slovenia Forest Service produced forest management guidelines for all forests, and part of regional forest management plans and plans for forest management units. It has also to be noted that there is usually a gap between guidelines and their practical implementation.

3. Adaptation measures for Alpine space forests

In the context of climate change, adaptation measures are measures that either decrease the probability of damage caused or triggered by climate change, exploit opportunities associated with climate change, or increase the adaptive capacity of the forests in the face of a changing climate. With adaptive capacity, we mean the ability of forest ecosystems to change continuously in composition and structure without breakdown (resistance, [26]) or to rebuild themselves, possibly with different composition and structure, after disturbances caused or triggered by climatic influences (resilience).

Region	Domain of application	Scope	Reference
Austria	Austrian national forests	Silvicultural guidelines based on ecological principles	[15]
Mühlviertel and Sauwald (Upper Austria)	All forests of the region	Tree species choice in the context of climate change	[16]
Vorarlberg (Austria)	All forests of Vorarlberg	Description of forest types and silvicultural guidelines	[17]
Baden-Württemberg (Germany)	All forests of Baden-Württemberg	Silvicultural guidelines based on forest development types	[18]
Bavaria (Germany)	Mountain forests of Bavaria	Silvicultural guidelines	[19]
Bavaria (Germany)	Bavarian state forests	Silvicultural guidelines in the face of climate change	[20]
Switzerland	All protection forests of Switzerland	Hazard- and forest type-specific silvicultural guidelines for protection forests	[21]
Canton of St. Gallen (Switzerland)	All forests of the Canton St. Gallen	Silvicultural guidelines in the face of climate change	[22]
Savoie (France)	Forests of the northern French Alps	Silvicultural guidelines for protection forests	[23]
Southern Alps (France)	Forests of the southern French Alps	Silvicultural guidelines for protection forests	[24]
Valle d'Aosta (Italy)	Pine forests in the eastern Alps	Silvicultural guidelines	[25]

Table 1. Selection of silvicultural guidelines for Alpine space forests.

Many adaptation measures have been proposed (e.g. [12,27-31]). On a European level, more than 440 adaptation measures were compiled in a database [32]. No adaptation measures have been elaborated specifically for all Alpine space forests, which is not surprising since these forests are very diverse. Here, we describe four important adaptation measures, their justification and domain of application, and their probable effectiveness.

3.1. Adaptation measure 1: Increasing tree species richness by variable regeneration cuts, enrichment planting, tending, and controlling ungulate browsing

This measure aims at enhancing tree species richness at the stand scale, or for maintaining it if it is already high. The main argument for high species richness is an increased resistance to disturbances such as drought or storm [33-36], and higher resilience once a disturbance has occurred [37]. This argument is sometimes called 'insurance hypothesis' [38]. Intimate species mixtures (Figure 1) can be achieved through diverse regeneration cuts which create diverse niches or seedlings [28]. Enrichment planting in young growth stands can also be used to increase species richness, as well as tending such stands to rescue rare species or those which are able to withstand a warmer and drier climate [28]. In most cases, a complete change of the tree species composition of a forest cannot be justified and must be considered as risky. Usually, this means doing without natural regeneration (and thus requires large investments) and assuming specific future site conditions (which are in practice not exactly known).

Increasing species richness can also be achieved by introducing new tree species. It has been shown that some species initially do very well in a new environment because they escape their enemies when being moved to a new range [39]. The combination of release from enemies and inherent growth ability produces a synergistic effect that explains initial high performance of invaders [39,40]. Due to its high productivity, tolerance to summer drought and long-term successful use in central Europe, the green subspecies of Douglas fir (*Pseudotsuga menziesii* spp. *menziesii*) is increasingly used as an alternative to Norway spruce. Mixtures of Douglas fir with other species may maximize the delivery of ecosystem services such as timber production and carbon sequestration [41]. Other tree species are currently tested in new planting trials [42].

Increasing species richness may also involve correcting mistakes of the past. In the Alpine Space, man-made Norway spruce forests have often replaced former mixed beech (*Fagus sylvatica* L.) and silver fir (*Abies alba* (Mill.) L.) forests. These artificial forests are vulnerable to windthrow, pest outbreaks, drought and soil deterioration, and their conversion to more natural mixed-species stands [43,44], e.g. by underplanting of broadleaved trees [45], helps to make them more resistant to a warmer and drier climate.

Measures to increase species richness are most effective if applied in the regeneration and young growth stages. During the pole and timber tree stages, the potential for adjusting species richness is restricted to maintaining those species that have not been lost in earlier stages of stand development [28].

In the Alpine region, with is large altitudinal gradients and large forest cover, the preconditions for unassisted migration of the existing tree species are generally good. Relatively fast upslope migration, following the rising temperatures, should be possible for many tree species [46]. A

Figure 1. Mixed steep-slope forests with Norway spruce, silver fir, sycamore, ash and beech. Montreux, Switzerland, 1100 m a.s.l.

problem may occur near valley bottoms where species from warmer and drier climates may need to migrate over larger distances and even over mountains.

An important obstacle to high species richness is ungulate browsing [47,48], which also often limits enrichment planting. In many regions the forests are under pressure from browsing ungulates, and hunting objectives are not aligned with forest management. Selective browsing inhibits the development of additional stand-stabilizing tree species, and the entire discussion on the benefits of mixed species forests remains futile.

It must to be noted that the species adapted for future climates are often less competitive in the current climate, e.g. sessile oak (*Quercus petraea* (Mattuschka) Liebl.) in beech thickets, which makes their maintenance by tending costly. Moreover, for vast high-altitude Alpine forests, the number of tree species able to thrive in these cold environments is currently limited to one or two species, e.g. Norway spruce or larch (*Larix decidua* L.) and Swiss stone pine (*Pinus cembra* L.).

3.2. Adaptation measure 2: Increasing genetic diversity by natural regeneration, long regeneration periods and mixing provenances

It is currently difficult to predict what degree current tree populations are able to absorb climate change. Their phenotypic plasticity [8,49] may be large, but insufficient if the environmental change occurs too quickly [50]. Natural regeneration coupled with long regeneration periods (Figure 2)is well suited to regenerate stands with a high seedling density from many mother trees [51,52], and thus to ensure large genetic variation. Enriching existing populations with provenances from other sites, usually drier and warmer sites, may also increase genetic diversity. However, this seems rather a measure for the future since it is not well established for which species and with which provenances this may be most effective. Like the first adaptation measure, increasing genetic diversity focuses on the regeneration and young growth stages.

Figure 2. Long regeneration periods promote genetic diversity of the new generation. Small gap in a silver fir-Norway spruce stand with natural regeneration, Aegeri, Switzerland, 1100 m a.s.l.

3.3. Adaptation measure 3: Increasing structural diversity by single-tree selection, target-diameter or conversion cutting

Stands composed of trees with different age and size are structurally diverse either vertically in canopy layers or horizontally in adjacent patches. High structural diversity reduces the probability of stand-replacing disturbances, which may, in protection forests, entail the complete loss of the protective effect. For instance, storms affect small trees less likely than large ones (e.g. [53], or bark beetles may only attack trees of a certain size range. Moreover, vertically structured stands are more resilient after disturbance since advance regeneration will quickly be released [21]. Silvicultural measures to increase structural diversity are well established [43,44], but may entail increment losses [44] and trigger disturbances since they impair stand resistance, in particular to wind and snow break. The advantages of high structural diversity, in particular vertically, are considerable in forests where permanent cover is needed such as in steep-slope protection forests [54], but may be less important in timber production forests. Measures to increase structural diversity are applicable to all developmental stages apart from young growth (Figure 3). A change of the silvicultural system from even-aged to uneven-aged silviculture is, however, not justified as response to climate change alone, but should rather be based on other lines of argument [28].

Figure 3. Conversion thinning in an homogeneous Norway spruce afforestation. The intervention aims at promoting structural heterogeneity by very small patch cuts. High stumps are left to prevent snow gliding. Tschamutt, Switzerland, 1600 m a.s.l.

3.4. Adaptation measure 4: Reducing rotation length

Old stands are more vulnerable to disturbance, in particular by wind, since susceptibility is correlated with stand height [53,55-56] which in turn is correlated with stand age. Shorter rotation lengths or even premature felling are a preventive measure in stands with high disturbance risks (Figure 4) and can reduce damage [28]. A shorter rotation length may also be a measure to adapt to faster growth associated with higher temperature in cool environments, and with high inputs of atmospheric nitrogen.

Figure 4. An unstable pure Norway spruce afforestation: a candidate stand for reduced rotation length. Gaschurn, Austria, 980 m a.s.l.

Among the four adaptation measures presented here, the first (increasing species richness) is probably most effective since species largely differ in their reaction to many site factors. Increasing species richness is also often mentioned in management guidelines [16,20,22]. The second adaptation measure (increasing genetic diversity) seems premature for widespread application, the third (increasing structural diversity) most effective in steep-slope protection forests, and the forth (reducing rotation length) restricted to stands with high disturbance risks.

4. Towards the implementation of adaptive management

AM as presented in the first section of this chapter, as a process of continuous improvements through iterative cycles of planning, managing, monitoring, and revised planning, is not currently practiced in the Alpine Space in a systematic manner. However, some elements of AM are increasingly adopted since its advantages are obvious in a situation with high uncertainty. One such element of AM are repeated forest inventories, which are established means of monitoring in many regions of the Alpine space. Another element is the increased use of silvicultural training plots [21]. However, practicioners still carry out many silvicultural tests without proper design and documentation, which largely limits their usefulness.

A general barrier to the implementation of AM is a widespread skepticism to scientific evidence. This attitude is understandable since science is, for many practicioners, a blackbox, in particular when it comes to modeling. Moreover, scientific results reflecting general large-scale phenomena and future trends often contradict the local observed development. This skepticism cannot easily be overcome, and is even partly a reasonable attitude since scientists, which are trained to focus on details and often lack a realistic perception of the management context, tend to over-interpret their findings. However, a strong belief in a large adaptive capacity of forests, based on past experience, must also be questioned. For instance, increased fire hazard on the northern slopes of the Alps and invasions of pathogens are quite likely [57, 58], but difficult to believe in the absence of past evidence.

A strategy of maintenance of existing forest structures, as response to a changing climate, seems risky since it is based on three assumptions which may prove wrong: i) small adverse impacts of climate change, ii) high stand resistance to climatic stress and iii) a high likelihood that silvicultural interventions will help to maintain forest structure [12]. The strategy of 'passive adaptation' [12], which means to suspend planned management interventions and to rely on spontaneous adaptation processes, i.e. natural succession and species migration, seems equally risky. These two strategies may be appropriate for forests with low economic importance or with no or only little protection function. However, if we assume more severe climate scenarios such as the A1B or A1F1 scenario [1], and if important production or protection functions are at stake, which is often the case in Alpine space forests, an active adaptation strategy [12] seems more appropriate.

Given the large uncertainties associated with climate change impacts on Alpine forests, we do at present not advocate large investments in adaptation measures with uncertain effectiveness [28]. Yet, total in-action is also inappropriate. The changes imposed by climate change are of a magnitude that calls for open-mindedness and readiness to revise existing management concepts. Conversely, simple silvicultural rules such as 'I rely on natural regeneration only', 'I use only native species' or 'I never use premature felling' seem unwise since they largely restrict the range of management tools which can be used to influence forest development, and to ensure the provision of ecosystem services.

Rather than already prescribing unconsolidated new silvicultural rules, capacity building efforts in the forest community, in interaction with stakeholders, seem important. This is crucial for the implementation of adaptive management which also involves systematic testing of alternative silvicultural treatments and success monitoring, but in less depth than commonly used in research. This creates a new task for researchers as they should support the implementation of AM by managers. The implementation of simple field tests and systematic monitoring of outcomes requires robust protocols and systematic documentation [59]. This task is new to many forest managers, and needs support from research. This is also new for many scientists and requires a thorough understanding not only of forest ecosystem dynamics, but also of the decision-making process including the information available to the other stakeholders involved and their interests, beliefs and attitudes.

Research can play an important role in AM. Current research in forest management is focusing on predicting impacts of future climates on tree species composition, forest dynamics, and the resulting consequences for forest products (in particular timber) and services. The phenotypic plasticity of trees in a changing climate is studied in genetics. Powerful simulation models enable reviewing our understanding of complex interactions in forest ecosystems, and testing management strategies assuming different climate scenarios. All these efforts are likely to provide important contributions to the revision of existing decision-making tools needed in the context of a changing climate, or the creation of new ones.

5. Conclusions

Climate change poses a novel challenge for forest science and management. AM as defined in this chapter seems particularly suitable to deal with the large uncertainty involved. Forest managers should increasingly see themselves as part of a learning community and perceive the establishment, maintenance and documentation of silvicultural field tests as part of their core business. Capacity building will also help to overcome simplistic management rules, and the resulting awareness for climate change issues is a precondition for acceptance of increased monitoring efforts, which will be needed for early detection of climate change impacts, including new pests and diseases. In particular regarding these transnational issues, cross-border networking is necessary.

In any adaptation strategy, it is also important to correctly weigh different problems. For instance, ungulate browsing [47,48] may completely prevent putting increased tree species richness into practice.

The adaptation measures recommended in this chapter must be regarded as preliminary. The uncertainty about climate development and associated reactions of the forest is large, in particular with regard to disturbances and biological invasions. This calls for robust strategies which are likely to ensure the delivery of ecosystem products and services for different outcomes, preparedness for surprise, and a flexible and adjustable management approach.

Author details

Peter Brang[1*], Andrej Breznikar[2], Marc Hanewinkel[1], Robert Jandl[3] and Bernhard Maier[4]

*Address all correspondence to: brang@wsl.ch

1 WSL Swiss Federal Institute of Forest, Snow and Landscape Research, Birmensdorf, Switzerland

2 Slovenia Forest Service, Ljubljana, Slovenia

3 Institute of Forest Ecology, Austrian Forest Research Center (BFW), Vienna, Austria

4 Stand Montafon, Schruns, Austria

References

[1] Solomon, S, Qin, D, Manning, M, Chen, Z, Marquis, M, Averyt, K. B, Tignor, M, & Miller, H. L. editors. Contribution of Working Group I to the Fourth Assessment Report of the Intergovernmental Panel on Climate Change. Cambridge, United Kingdom and New York, NY, USA: Cambridge University Press; (2007).

[2] Allen, C. D, Macalady, A. K, Chenchouni, H, Bachelet, D, Mcdowell, N, Vennetier, M, Kitzberger, T, Rigling, A, Breshears, D. D, Hogg, E. H, Gonzalez, P, Fensham, R, Zhang, Z, Castro, J, Demidova, N, Lim, J. H, Allard, G, Running, S. W, Semerci, A, & Cobb, N. A global overview of drought and heat-induced tree mortality reveals emerging climate change risks for forests. Forest Ecology and Management (2010). , 259(4), 660-684.

[3] Dobbertin, M, Wermelinger, B, Bigler, C, Bürgi, M, Carron, M, Forster, B, Gimmi, U, & Rigling, A. Linking increasing drought stress to Scots pine mortality and bark beetle infestations. The scientific world journal(2007). , 7, 231-239.

[4] Kurz, W. A, Dymond, C. C, Stinson, G, Rampley, G. J, Neilson, E. T, Carroll, A. L, Ebata, T, & Safranyik, L. Mountain pine beetle and forest carbon feedback to climate change. Nature (2008). , 452, 987-990.

[5] Engesser, R, Forster, B, Meier, F, & Wermelinger, B. Die Bedeutung von forstlichen Schadorganismen im Zeichen des Klimawandels. Schweizerische Zeitschrift für Forstwesen (2008). , 159(10), 344-351.

[6] Leal, S, Melvin, T. M, Grabner, M, Wimmer, R, & Briffa, K. R. Tree-ring growth variability in the Austrian Alps: The influence of site, altitude, tree species and climate. Boreas (2007). , 36(4), 426-440.

[7] Butterbach-Bahl, K, Nemitz, E, & Zaehle, S. Nitrogen as a threat to the European greenhouse balance. In: Sutton MA, Howard CM, Erisman JW, Billen G, Bleeker A, Grennfelt P, van Grinsven H, Grizzetti B, editors. The European Nitrogen Assessment. Cambridge: Cambridge University Press; (2011). , 434-462.

[8] Kapeller, S, Schüler, S, Kraigher, H, Huber, G, Karopka, M, Wohlgemuth, T, & Colin, E. Provenance trials in Alpine range- review and perspectives for applications in climate change. In: Cerbu G, editor. Management strategies to adapt Alpine space forests to climate change risks. Rijeka: InTech; (2012).

[9] Wang, T, Hamann, A, Yanchuk, Y, Neill, O, & Aiken, G. S. Use of response functions in selecting lodgepole pine populations for future climates. Global Change Biology (2006). , 12, 2404-2416.

[10] Holling, C. S. Adaptive Environmental Assessment and Management. Chichester: John Wiley & Sons; (1978).

[11] Spittlehouse, D. L, & Stewart, R. B. Adaptation to climate change in forest management. BC Journal of Ecosystems and Management (2003). 4(1) 1-11

[12] Bolte, A, Ammer, C, Lof, M, Madsen, P, Nabuurs, G. J, Schall, P, Spathelf, P, & Rock, J. Adaptive forest management in central Europe: Climate change impacts, strategies and integrative concept. Scandinavian Journal of Forest Research (2009). , 24, 473-482.

[13] Schütz, J. P. Geschichtlicher Hergang und aktuelle Bedeutung der Plenterung in Europa. Allgemeine Forst- und Jagdzeitung (1994). ; 165(5-6) 106-114

[14] Pommerening, A, & Murphy, S. A review of the history, definitions and methods of continuous cover forestry with special attention to afforestation and restocking. Forestry (2004). , 77, 27-44.

[15] Weinfurter, P. Waldbauhandbuch. Eine Orientierungshilfe für die Praxis. Österreichische Bundesforste AG; (2004).

[16] Jasser, C, & Diwold, G. Baumartenwahl im Mühlviertel- Empfehlungen für das Wuchsgebiet Mühlviertel und Sauwald. Oberösterreichische Landesregierung; (2011).

[17] Amann, G, Schennach, R, Kessler, J, Maier, B, & Terzer, S. Handbuch der Vorarlberger Waldgesellschaften. Gesellschaftsbeschreibungen und waldbaulicher Leitfaden. Amt der Vorarlberger Landesregierung. Abteilung Forstwesen; (2010).

[18] Ministerium Ländlicher Raum Baden-Württembergeditor. Richtlinie landesweiter Waldentwicklungstypen. Stuttgart; (1999).

[19] Bayerisches Staatsministerium für ErnährungLandwirtschaft und Forsten, editor. Grundsätze für die Waldbehandlung im bayerischen Hochgebirge; (1982).

[20] Bayerische Staatsforsten, editor. Waldbauhandbuch Bayerische Staatsforsten. Bewirtschaftung von Fichten- und Fichtenmischbeständen im Bayerischen Staatswald (Stabilität- Strukturreichtum- Klimaanpassung). Regensburg: Bayerische Staatsforsten; (2008).

[21] Frehner, M, Wasser, B, & Schwitter, R. Sustainability and success monitoring in protection forests. Guidelines for managing forests with protective functions. Partial translation by Brang P, Matter C. Environmental Studies (2007).

[22] Kantonsforstamt St. Gallen, editor. Waldpflege und Waldverjüngung unter dem Aspekt der Klimaveränderung. Strategiepapier. Empfehlungen des Forstdienstes des Kantons St.Gallen; (2008).

[23] Gauquelin, X, & Courbaud, B. Guide des sylvicultures de montagne- Alpes du nord françaises. Grenoble: Cemagref, ONF; (2006). 289 p.

[24] Ladier, J, Rey, F, Calès, G, Simon-teissier, S, & Quesney, T. Guide des Sylvicultures de Montagne pour les Alpes du Sud françaises. 1. Gestion des forêts à rôle de protection contre les aléas naturels. Paris: ONF; (2012). 135 p.

[25] Vacchiano, G. editor. Il deperimento del Pino silvestre nelle Alpi occidentali: natura ed indirizzi di gestione. Arezzo: Regione Piemonte, Regione Autonoma Valle d'Aosta, Compagnia delle Foreste; (2008). 128 p.

[26] Grimm, V, & Wissel, C. Babel, or the ecological stability discussions: an inventory and analysis of terminology and a guide for avoiding confusion. Oecologia (1997). , 109(3), 323-334.

[27] Millar, C. I, Stephenson, N. L, & Stephens, S. L. Climate change and forests of the future: managing in the face of uncertainty. Ecological Applications (2007). , 17(8), 2145-2151.

[28] Brang, P, Bugmann, H, Bürgi, A, Mühlethaler, U, Rigling, A, & Schwitter, R. Klimawandel als waldbauliche Herausforderung. Schweizerische Zeitschrift für Forstwesen (2008). , 159, 362-373.

[29] Lindner, M, Maroschek, M, Netherer, S, Kremer, A, Barbati, A, Garcia-gonzalo, J, Seidl, R, Delzon, S, Corona, P, Kolstrom, M, Lexer, M. J, & Marchetti, M. Climate change impacts, adaptive capacity, and vulnerability of European forest ecosystems. Forest Ecology and Management (2010). , 259(4), 698-709.

[30] FAO, editor. Forests and climate change. Rome: Food and Agriculture Organization; (2012).

[31] Lexer, M. J. Waldwirtschaft im Klimawandel- ein Hintergrundbericht der CIPRA. Schaan: CIPRA; (2012).

[32] Kolström, M, Lindner, M, Vilén, T, Maroschek, M, Seidl, R, Lexer, M. J, Netherer, S, Kremer, A, Delzon, S, Barbati, A, Marchetti, M, & Corona, P. Reviewing the science and implementation of climate change adaptation measures in European forestry. Forests (2011). , 2(4), 961-982.

[33] Von Lüpke, B, & Spellmann, H. Aspects of stability, growth and natural regeneration in mixed Norway spruce-beech stands as a basis of silvicultural decisions. In: Olsthoorn AFM, Bartelink HH, Gardiner JJ, Pretzsch H, Hekhuis HJ, Franc A, Wall S. (eds) Management of mixed-species forest: silviculture and economics. Wageningen: DLO Institute for Forestry and Nature Research (IBN-DLO); (1999). , 245-267.

[34] Schütz, J. P, Götz, M, Schmid, W, & Mandallaz, D. Vulnerability of spruce (Picea abies) and beech (Fagus sylvatica) forest stands to storms and consequences for silviculture. European Journal of Forest Research (2006). , 125, 291-302.

[35] Knoke, T, Ammer, C, Stimm, B, & Mosandl, R. Admixing broadleaved to coniferous tree species: a review on yield, ecological stability and economics. European Journal of Forest Research (2008). , 127, 89-101.

[36] Puettmann, K. J, Coates, K. D, & Messier, C. A critique of silviculture- managing for complexity. Washington, Covelo, London: Island Press; (2009).

[37] Brang, P. Resistance and elasticity: promising concepts for the management of protection forests in the European Alps. Forest Ecology and Management (2001). 145(1-2) 107-119

[38] Pretzsch, H. Diversity and productivity in forests: evidence from long-term experimental plots. In: Scherer-Lorenzen M, Körner C, Schulze E-D, (eds.) Forest diversity and function. Temperate and boreal systems. Ecological Studies 176. Springer, Berlin, Heidelberg, New York: Springer; (2005). , 41-64.

[39] Blumenthal, D, Mitchell, C. E, Pyšek, P, & Jarošík, V. Synergy between pathogen release and resource availability in plant invasion. Proceedings of the National Academy of Sciences (2009). DOIpnas.0812607106, 106(19), 7899-7904.

[40] Seastedt, T. Traits of plant invaders. Nature (2009). , 459, 783-784.

[41] Prietzel, J, & Bachmann, S. Changes in soil organic C and N stocks after forest transformation from Norway spruce and Scots pine into Douglas fir, Douglas fir-spruce, or European beech stands at different sites in Southern Germany. Forest Ecology and Management (2012). , 269, 134-148.

[42] Schmiedinger, A, Bachmann, M, Kölling, C, & Schirmer, R. Verfahren zur Auswahl von Baumarten für Anbauversuche vor dem Hintergrund des Klimawandels. Forstarchiv (2009). , 80, 15-22.

[43] Schütz, J. P. Opportunities and strategies of transforming regular forests to irregular forests. Forest Ecology and Management (2001). 151(1-3) 87-94

[44] Spiecker, H. Norway Spruce conversion- options and consequences. European Forest Institute Research Report (2004). , 18-269.

[45] Oleskog, G, & Löf, M. The ecological and silvicultural bases for underplanting beech (Fagus sylvatica L.) below Norway spruce shelterwoods (Picea abies (L.) Karst.). Schriften aus der Forstlichen Fakultät der Universität Göttingen und der Niedersächsischen Forstlichen Versuchsanstalt (2005). J. D. Sauerländer's Verlag.

[46] Vitasse, Y, Hoch, G, Randin, C. F, Lenz, A, Kollas, C, & Körner, C. Tree recruitment of European tree species at their current upper elevational limits in the Swiss Alps. Journal of Biogeography (2012). , 39(8), 1439-1449.

[47] Gill RMAA review of damage by mammals in north temperate forests. 1. Deer. Forestry (1992). , 65, 145-169.

[48] Gill RMAA review of damage by mammals in north temperate forests. 3. Impact on trees and forests. Forestry (1992). , 65, 363-388.

[49] Mátyás, C, & Nagy, L. Ujvári Jármay É. Genetic background of response of trees to aridification at the xeric forest limit and consequences for bioclimatic modelling. Forstarchiv (2010). , 81, 130-141.

[50] Hoffmann, A. A, & Sgrò, C. M. Climate change and evolutionary adaptation. Nature (2010). , 470, 479-486.

[51] Finkeldey, R, & Ziehe, M. Genetic implications of silvicultural regimes. Forest Ecology and Management (2004). 97(1-3) 231-244

[52] Finkeldey, R. Genetik, Ökologie, Forstwirtschaft: Zusammenhänge und Perspektiven. Schweizerische Zeitschrift für Forstwesen (2010). , 161(6), 198-206.

[53] Mayer, P, Brang, P, Dobbertin, M, Hallenbarter, D, Renaud, J-P, Walthert, L, & Zimmermann, S. Forest storm damage is more frequent on acidic soils. Annals of Forest Science (2005). , 62, 303-311.

[54] Schönenberger, W, & Brang, P. Silviculture in mountain forests. In: Burley J., Evans J (eds.). Encyclopedia of forest sciences. Amsterdam: Elsevier. (2004). , 1085-1094.

[55] Hanewinkel, M, Breidenbach, J, Neeff, T, & Kublin, E. Seventy-seven years of natural disturbances in a mountain forest area- the influence of storm, snow and insect damage analysed with a long-term time series. Canadian Journal of Forest Research (2008). , 38(8), 2249-2261.

[56] Schmidt, M, Hanewinkel, M, Kändler, G, Kublin, E, & Kohnle, U. An inventory-based approach for modeling single tree storm damage- experiences with the winter storm 1999 in southwestern Germany. Canadian Journal of Forest Research (2010). , 40(8), 1636-1652.

[57] Schumacher, S, Reineking, B, Sibold, J, & Bugmann, H. Modeling the impact of climate and vegetation on fire regimes in mountain landscapes. Landscape Ecology(2006). , 21, 539-554.

[58] Seidl, R, Schelhaas, M-J, & Lexer, M. J. Unraveling the drivers of intensifying forest disturbance regimes in Europe. Global Change Biology (2011). , 17-2842.

[59] Rosa, J, Riou-Nivert, P, Paillassa, E. Guide de l'expérimentation forestière. Principes de base. Prise en compte du changement climatique. Paris: CNPF/IDF. (2011).

Permissions

The contributors of this book come from diverse backgrounds, making this book a truly international effort. This book will bring forth new frontiers with its revolutionizing research information and detailed analysis of the nascent developments around the world.

We would like to thank Gillian Ann Cerbu, Marc Hanewinkel, Giacomo Gerosa and Robert Jandl, for lending their expertise to make the book truly unique. They have played a crucial role in the development of this book. Without their invaluable contribution this book wouldn't have been possible. They have made vital efforts to compile up to date information on the varied aspects of this subject to make this book a valuable addition to the collection of many professionals and students.

This book was conceptualized with the vision of imparting up-to-date information and advanced data in this field. To ensure the same, a matchless editorial board was set up. Every individual on the board went through rigorous rounds of assessment to prove their worth. After which they invested a large part of their time researching and compiling the most relevant data for our readers. Conferences and sessions were held from time to time between the editorial board and the contributing authors to present the data in the most comprehensible form. The editorial team has worked tirelessly to provide valuable and valid information to help people across the globe.

Every chapter published in this book has been scrutinized by our experts. Their significance has been extensively debated. The topics covered herein carry significant findings which will fuel the growth of the discipline. They may even be implemented as practical applications or may be referred to as a beginning point for another development. Chapters in this book were first published by InTech; hereby published with permission under the Creative Commons Attribution License or equivalent.

The editorial board has been involved in producing this book since its inception. They have spent rigorous hours researching and exploring the diverse topics which have resulted in the successful publishing of this book. They have passed on their knowledge of decades through this book. To expedite this challenging task, the publisher supported the team at every step. A small team of assistant editors was also appointed to further simplify the editing procedure and attain best results for the readers.

Our editorial team has been hand-picked from every corner of the world. Their multi-ethnicity adds dynamic inputs to the discussions which result in innovative outcomes. These outcomes are then further discussed with the researchers and contributors who give their valuable feedback and opinion regarding the same. The feedback is then collaborated with the researches and they are edited in a comprehensive manner to aid the understanding of the subject.

Apart from the editorial board, the designing team has also invested a significant amount of their time in understanding the subject and creating the most relevant covers. They scrutinized every image to scout for the most suitable representation of the subject and create an appropriate cover for the book.

The publishing team has been involved in this book since its early stages. They were actively engaged in every process, be it collecting the data, connecting with the contributors or procuring relevant information. The team has been an ardent support to the editorial, designing and production team. Their endless efforts to recruit the best for this project, has resulted in the accomplishment of this book. They are a veteran in the field of academics and their pool of knowledge is as vast as their experience in printing. Their expertise and guidance has proved useful at every step. Their uncompromising quality standards have made this book an exceptional effort. Their encouragement from time to time has been an inspiration for everyone.

The publisher and the editorial board hope that this book will prove to be a valuable piece of knowledge for researchers, students, practitioners and scholars across the globe.

List of Contributors

Robert Jandl and Silvio Schüler
Forest Research Center (BFW), Vienna, Austria

Gillian Cerbu
Forstliche Versuchs- und Forschungsanstalt Baden Württemberg (FVA), Freiburg i. Breisgau, Germany

Marc Hanewinkel
Eidgenössische Forschungsanstalt für Wald, Schnee und Landschaft (WSL), Birmensdorf, Switzerland

Fred Berger
National Research Institute of Science and Technology for Environment and Agriculture (IRSTEA), Grenoble, France

Giacomo Gerosa
DMF, Università Cattolica del Sacro Cuore, Brescia, Italy

Niklaus E. Zimmermann, Dirk Schmatz and Achilleas Psomas
Swiss Federal Research Institute WSL, Birmensdorf, Switzerland

Ernst Gebetsroither and Johann Züger
Austrian Institute of Technology, Vienna, Austria

Eliane S. Meier, Signe Normand and Ulrich Ulmer
Swiss Federal Research Institute WSL, Birmensdorf, Switzerland

Robert Jandl and Thomas Gschwandtner
Federal Research and Training Center for Forests, Natural Hazzards and Landscape BLW, Vienna, Austria

Georges Kunstler
National Research Institute of Science and Technology for Environment and Agriculture IRSTEA, Grenoble, France

Christian Kölling, Maria Naumann, Wolfgang Falk and Karl Mellert
Bayerische Landesanstalt für Wald und Forstwirtschaft LWF, Freising, Germany

Patrizia Gasparini and Maria Rizzo
Consiglio per la Ricerca e la sperimentazione in Agricoltura - Unità di Ricerca per il Monitoraggio e la Pianificazione Forestale, CRA-MPF, Trento, Italy

Andrej Breznikar
Slovenia Forest Service, Lubljana, Slovenia

Holger Veit
Forstliche Versuchs- und Forschungsanstalt Baden-Württemberg FVA, Freiburg, Germany

Mitja Skudnik
Slovenian Forestry Institute, Lubljana, Slovenia

Luca Cetara
Accademia Europea di Bolzano, Bolzano, Italy

Federico Mannoni
Istituto per le Piante da Legno e l'Ambiente – IPLA, Torino, Italy

Holger Griess and Ralf Petercord
Landesanstalt für Wald und Forstwirtschaft Bayern (LWF), Freising, Germany

Holger Veit
Forstliche Versuchs- und Forschungsanstalt Baden-Württemberg (FVA), Department of Biometrics, Freiburg im Breisgau, Germany

Bin You
Forest Research Institute of Baden-Württemberg, FVA, Freiburg, Germany

Mitja Skudnik
Slovenian Forestry Institute, Slovenia

Bruna Comini, Giampaolo Cocca, Elena Gagliazzi, Paolo Nastasio and Enrico Calvo
Regional Agency for Development in Agriculture and Forestry (ERSAF) of Lombardia Region, Italy

Roberto Colombo and Lorenzo Busetto
Remote Sensing of Environmental Dynamics Lab., University of Milano-Bicocca, Italy

B. Di Mauro, Mitja Skudnik, Tomaz Sturm and Andrej Breznikar
Slovenian Forestry Institute - Department of Forest and Landscape Planning and Monitoring, Slovenia

Giacomo Gerosa, Angelo Finco, Antonio Negri and Riccardo Marzuoli
Mathemathics and physics department, Catholic University of the Sacred Heart, Italy

Gerhard Wieser
Department of Alpine Timberline Ecophysiology, Federal Research and Training Centre for Forests, Natural Hazards and Landscape, Innsbruck, Austria

Angelo Finco, Stefano Oliveri and Giacomo Gerosa
Catholic University of the Sacred Heart, Mathematics and Physics Department, Catholic University of the Sacred Heart, Brescia, Italy

Wilfried Winiwarter, Johann Züger and Ernst Gebetsroither
Austrian Institute of Technology GmbH, Vienna, Austria

Stefano Oliveri
Ecometrics srl., Italy

Marco Pregnolato and Giacomo Gerosa
Catholic University of Brescia, Mathematics and Physics Department, Italy

Frederic Berger, Christophe Bigot, Franck Bourrier , Oliver Jancke and David Toe
Istrea, National Research Institute of Science and Technology, Grenoble, France

Luuk Dorren
FOEN,Federal Office for the Environment , Hazard Prevention Division , Bern, Switzerland

Karl Kleemayr
BFW, Federal Research and Training Centre for Forests, Natural Hazards and Landscape, Department for Natural Hazards and Alpine Timberline, Innsbruck, Austria

Bernhard Maier
Stand Montafon, Forest Division, Schruns, Austria

Spela Planinsek
Slovenian Forest Institute, Ljubljana, Slovenia

Gillian Cerbu
FVA, Forest Research Institute of Baden-Wuerttemburg , Freiberg, German

Ernst Gebetsroither, Johann Züger and Wolfgang Loibl
AIT Austrian Institute of Technology GmbH, Foresight & Policy Development Department, Vienna, Austria

Klaus Dolschak
Dept. of Forest Ecology, University of Applied Life Sciences (BOKU), Vienna, Austria

Robert Jandl and Thomas Ledermann
Forest Research Center (BFW), Vienna, Austria

Robert Jandl
Forest Research Center (BFW), Vienna, Austria

Frédéric Berger
IRSTEA, Grenoble, France

Andrej Breznikar
Zavod za gozdove Slovenije, Slovenia Forest Service, Ljubljana, Slovenia

Giacomo Gerosa
DMF, Università Cattolica del Sacro Cuore, Brescia, Italy

Holger Veit and Gillian Cerbu
Forstliche Versuchs- und Forschungsanstalt Baden-Württemberg (FVA), Department of Biometrics, Freiburg im Breisgau, Germany

Marc Hanewinkel
Eidgenössische Forschungsanstalt für Wald, Schnee und Landschaft (WSL), Birmensdorf, Switzerland

Stefan Kapeller andSilvio Schüler
Federal Research and Training Centre for Forests, Natural Hazards and Landscape, Vienna, Austria

Gerhard Huber
Bavarian Office for Forest Seeding and Planting, Teisendorf, Germany

Gregor Božič
Slovenian Forestry Institute, Ljubljana, Slovenia

Tom Wohlgemuth
Swiss Federal Institute for Forest, Snow and Landscape Research WSL, Birmensdorf, Switzerland

Raphael Klumpp
University of Natural Resources and Life Sciences, Vienna, Austria

Laurent Borgniet, David Toe and Frédéric Berger
Institut national de recherche en sciences et technologies pour l'environnement et l'agriculture, Irstea, Grenoble, France

Marta Galvagno and Umberto Morra di Cella
Agenzia Regionale per la Protezione dell'Ambiente della Valle d'Aosta, ARPA, Aosta, Italy

Cinzia Panigada,Roberto Colombo and Simone Gottardelli
Remote Sensing of Environmental Dynamics Laboratory, DISAT, Università degli Studi di Milano Bicocca, Milano, Italy

Ivan Rollet and Flavio Vertui
Corpo Forestale della Valle d'Aosta, Dip. Risorse Naturali e Corpo Forestale, Regione Autonoma Valle d'Aosta, Aosta, Italy

Mario Negro
Forestazione e Sentieristica, Dip. Risorse Naturali e Corpo Forestale, Regione Autonoma Valle d'Aosta, Aosta, Italy

Cédric Fermont
Office National des Forêts, Département de la Drôme, France

Holger Veit and Peter Brang
Department of Biometrics, Forest Research Institute of Baden Württemberg, Freiburg, Germany

Holger Grieb
Department of Forest Management, Bavarian Forest Institute, Freising, Germany

Bernhard Maier
Stand Montafon – Forstfonds, Schruns, Austria

Giacomo Gerosa and Riccardo Marzuoli
Dept. of Mathematics and Physics, Catholic University of the Sacred Hearth, Brescia, Italy

Angelo Finco and Stefano Oliveri
Dept. of Mathematics and Physics, Catholic University of the Sacred Hearth, Brescia, Italy
Ecometrics s.r.l., Environmental Monitoring & Assessment, Brescia, Italy

Alessandro Ducoli and Giambattista Sangalli
Comunità Montana Valle Camonica – Parco dell'Adamello, Breno, Italy

Bruna Comini, Paolo Nastasio, Giampaolo Cocca and Elena Gagliazzi
Regional Agency of Services to Agriculture and Forests (ERSAF), Unit for the Valorisation of Biodiversity and Services to the Agro-Forest ecosystems, Gargnano, Italy

Robert Jandl, Christian Tomiczek and Silvio Schüler
Forest Research Center (BFW), Vienna, Austria

Andrej Breznikar and Marko Lekše
Zavod za Gozdove Slovenije/Slovenia Forest Service, Ljubljana, Slovenia

Klaus Dolschak
Institute of Forest Ecology, University of Applied Life Sciences (BOKU), Vienna, Austria

Hans Zöscher
Forest Training Center - Forstliche Ausbildungsstätte (FAST) Ossiach, Ossiach, Austria

Peter Brang and Marc Hanewinkel
WSL Swiss Federal Institute of Forest, Snow and Landscape Research, Birmensdorf, Switzerland

Andrej Breznikar
Slovenia Forest Service, Ljubljana, Slovenia

Robert Jandl
Institute of Forest Ecology, Austrian Forest Research Center (BFW), Vienna, Austria

Bernhard Maier
Stand Montafon, Schruns, Austria

www.ingramcontent.com/pod-product-compliance
Lightning Source LLC
Chambersburg PA
CBHW072251210326
41458CB00073B/961